D1243514

History of Modern Science and Mathematics

Editorial Board

History of Modern Science and Mathematics

Brian S. Baigrie, Editor

VOLUME IV

New York • Detroit • San Diego • San Francisco • Cleveland • New Haven, Conn. • Waterville, Maine • London • Munich

Charles Scribner's Sons
Publisher: Frank Menchaca
Senior Editor: John Fitzpatrick

Produced by The Moschovitis Group, Inc.
339 Fifth Avenue, New York, New York 10016
www.mosgroup.com

Executive Editor: Valerie Tomaselli
Managing Editor: Sean Pidgeon
Editorial Coordinator: Sonja Matanovic
Design and Layout: Annemarie Redmond
Illustrations: Richard Garratt
Production Assistant: Rashida Allen
Copyediting: Zeiders & Associates
Proofreading: Adams Holman, Paul Scaramazza
Index: Judy Davis

Library of Congress Cataloging-in-Publication Data

History of modern science and mathematics / Brain S. Baigrie, editor.
p. cm.
Includes bibliographical references and index.
ISBN 0-684-80636-3 (set hardcover : alk. paper)
1. Science--History. 2. Mathematics--History. I. Baigrie, Brian S.
(Brian Scott)
Q125 .H6298 2002
509--dc21

2002004042

Table of Contents

Volume III

Volume IV

Number Theory

Israel Kleiner

Pre-1543 Roots

Number theory, the study of properties of the positive integers, which was broadened in the nineteenth century to include other types of numbers, is one of the oldest branches of mathematics. It has fascinated both amateurs and mathematicians throughout the ages. The subject is tangible, the results are usually simple to state and understand, and they are often suggested by numerical examples. Nevertheless, they are frequently very difficult to prove. "It is just this," said Carl Friedrich Gauss (1777–1855), one of the greatest mathematicians of all time, "which gives number theory that magical charm that has made it the favorite science of the greatest mathematicians." To deal with the many difficult number-theoretic problems, mathematicians had to resort to—often to invent—advanced techniques, mainly from algebra, analysis, and geometry. This gave rise in the nineteenth and twentieth centuries to distinct branches of number theory, such as algebraic number theory, analytic number theory, transcendental number theory, geometry of numbers, and arithmetic of algebraic curves. While number theory was considered for over three millennia to be one of the "purest" branches of mathematics, without any applications, it has found important uses in the twentieth century in such areas as cryptography, physics, biology, and graphic design.

The study of Diophantine equations (so named after the Greek mathematician Diophantus, fl. A.D. 250) has been a central theme in number theory. These are equations in two or more variables, with integer or rational coefficients, for which the solutions sought are integers or rational numbers. The earliest such equation, $x^2 + y^2 = z^2$, dates back to Babylonian times, about 1700 B.C. This equation has been important throughout the history of number theory. Its integer solutions are called *Pythagorean triples*. [This is because of the Pythagorean theorem and its converse: Given numbers (or magnitudes) a, b, and c, $a^2 + b^2 = c^2$ if and only if a, b, and c are the sides of a right-angled triangle.]

Records of Babylonian mathematics have been preserved on ancient clay tablets. One of the most renowned tablets is named Plimpton 322. It consists of a table of 15 rows of numbers that (according to most historians of mathematics) is a list of Pythagorean triples. There is no indication of how they have been generated (not by trial), nor why (mathematics for fun?), but the listing suggests (as do other sources) that the Babylonians knew the Pythagorean theorem more than a millennium before the birth of Pythagoras (c. 560 B.C.) and that they studied number theory no later than algebra or geometry, nearly 4000 years ago.

Euclid's (fl. c. 295 B.C.) *Elements* is known mainly for its axiomatic development of geometry. But three of its 13 books (Books VII-IX) are devoted to number theory. Here Euclid introduces several fundamental number-theoretic concepts, such as divisibility, prime and composite integers, and the greatest common divisor (gcd) and least common multiple (lcm) of two integers. A number of basic results are also established. (Given integers a and b, we say that b is divisible by a, or that a divides b, if there exists an integer c such that $b = ac$. A *prime* is an integer greater than 1 that is divisible only by itself. Integers greater than 1 that are not prime are called *composite*.)

The first two propositions of Book VII present the Euclidean algorithm for finding the gcd of two numbers. This is one of the central results in number theory. It is based on the important fact that if a and b are positive integers, there exist integers q and r such that $a = bq + r$, where $0 \leq r < b$. A very significant corollary of the Euclidean algorithm is that if d is the gcd of a and b, then $d = ax + by$ for some integers x and y. Two other basic results in Book VII deal with primes: (1) every integer is divisible by some prime, and (2) if a prime divides the product of two integers, it must divide at least one of the integers.

Book IX resumes the study of primes (among other things). Proposition 20 proves that there are infinitely many primes, a far-from-obvious result. The beautiful, now-classic proof given by Euclid is used in textbooks to this day. One of the most important results in number theory (if not the most important) is undoubtedly the fundamental theorem of arithmetic (FTA; arithmetic and number theory were at one time used interchangeably), which asserts that every positive integer n greater than 1 is a unique product of primes, so that (1) $n = p_1p_2 \ldots p_s$, where the p_i are prime, and (2) if also $n = q_1q_2 \ldots q_t$, with q_j prime, then $s = t$ and (after possibly rearranging the order of the q_j) $p_k = q_k$ for $k = 1, 2, \ldots, s$. There has been some debate among historians as to whether Proposition 14 of Book IX of the *Elements* is equivalent to the FTA. It states that "if a number be the least that is measured [divisible] by prime numbers, it will not be measured by any other prime number except those originally measuring it." In any case, results (1) and (2) above readily yield a proof of the FTA.

Two final noteworthy topics in the *Elements* are perfect numbers and Pythagorean triples. The Pythagoreans of the fifth century B.C. believed that numbers (i.e., positive integers) are at the basis of all things. In particular, they assigned various attributes to specific numbers. For example, 1 was godlike (although not considered a number, it was the generator of all numbers), 2 was feminine, 3 was masculine, and 5 was connected with marriage (since $5 = 2 + 3$). The number 6 was associated with perfection. Since 6 is the sum of all its proper divisors ($6 = 1 + 2 + 3$), any number with this property was called *perfect*. Another perfect number is 28 ($28 = 1 + 2 + 4 + 7 + 14$). Perhaps this number mysticism motivated the Pythagoreans to initiate a serious study of properties of numbers, which later appeared in Books VII-IX of Euclid's *Elements*. In any case, the last result in Book IX, Proposition 36, deals with perfect numbers. Specifically, it shows that if $2^n - 1$ is prime for some integer n, then $2^{n-1}(2^n - 1)$ is perfect. Must all (even) perfect numbers be of this form? And for which n is $2^n - 1$ prime? These questions, suggested by the result above, are discussed in subsequent sections.

Going on to Pythagorean triples, integer solutions of $x^2 + y^2 = z^2$, recall that the Babylonians determined 15 such triples. Euclid gave a formula that generates *all* of them (Book X, Lemma 1 preceding Proposition 29), namely $x = a^2 - b^2$, $y = 2ab$, $z = a^2 + b^2$, where a and b are arbitrary positive integers with $a > b$ (although this is not how Euclid put it, since he had no algebraic notation). This implies that there are infinitely many Pythagorean triples.

The other great Greek work in number theory is Diophantus's *Arithmetica*. It is divided into 13 "books," six of which have survived in Greek; four others were recently found in Arabic. The *Arithmetica* contains numerous problems, each of which gives rise to one or more equations, many of degree two or three. Although many of these sets of equations are indeterminate (i.e., have more than one integer or rational solution), Diophantus found in most cases a single positive rational solution. Here are three of his problems:

1. "To divide a given square into two squares" (Book II, Problem 8). This requires solving $a^2 = x^2 + y^2$ for x and y, given a. Diophantus picked $a = 4$ and gave (16/5, 20/5) as the solution. There are, in fact, infinitely many solutions, which can be inferred from his method of solution; he said so explicitly in the body of Problem

19 of Book III. This problem would later motivate Pierre de Fermat (1601–65), one of the foremost mathematicians of the seventeenth century.

2. "To find two numbers such that their product added to either gives a cube" (Book IV, Problem 26). This requires finding x, y, and z such that $xy + x = z^3$. Diophantus expressed y as a function of x; this led him to a cubic in x and z, which he proceeded to solve. The study of cubic Diophantine equations would be fundamental to mathematicians of the eighteenth and subsequent centuries.
3. "Given two numbers, if, when some square is multiplied into one of the numbers and the other number is subtracted from the product, the result is a square, another square larger than the aforesaid square can always be found which has the same property" (Book VI, Lemma to Problem 15). That is, if for fixed a and b, the equation $ax^2 - b = y^2$ has a solution, say $x = p$ and $y = q$, it has another solution, $x = s$ and $y = t$, with $s > p$ and (necessarily) $t > q$. This problem dealt with the important idea of generating a new solution from a given one.

The solutions of many of the problems in the *Arithmetica* required great ingenuity—clever "tricks," as some would have it. However, Leonhard Euler (1707–83) and others saw deep methods embedded in them. Recent historians have reconstructed Diophantus's work in the light of modern developments, suggesting that it contained the germ of an important geometric idea on the arithmetic of algebraic curves, the tangent and secant method. This entailed solving Diophantine equations by finding the intersection(s) of the curves determined by these equations with certain lines (tangent and secant lines). The *Arithmetica* had great influence on the rise of modern number theory, especially on Fermat's work.

Although there was little if any progress in number theory in Europe during the Middle Ages, the Indian and Chinese civilizations were active in the field. The Indians especially were avid number theorists. For example, Brahmagupta (A.D. 598–665) solved the general linear Diophantine equation $ax + by = c$ (a, b, and c are fixed integers), and for certain values of d, the Pell equation, $x^2 - dy^2 = 1$ (d a nonsquare positive integer). A solution of the general Pell equation was given by Bhaskara (1114–85); this important equation is discussed in detail below. By the fourth century A.D., the Chinese had dealt with specific instances of what is today called the *Chinese remainder theorem*, the simultaneous solution of linear congruences, $x \equiv a_1 \pmod{m_1}, \ldots, x \equiv a_k \pmod{m_k}$, with the m_i relatively prime in pairs, although the Chinese did not use the congruence notation. The general problem was solved by Ch'in Chiu-Shao (c. 1202–c. 61). [Given integers a, b, and m, with $m > 1$, a is said to be congruent to b modulo m, written $a \equiv b \pmod{m}$, if $a - b$ is divisible by m.] The congruence notation was introduced by Gauss in 1801. Two integers are relatively prime if they have no common divisors other than 1, that is, if their gcd is 1.] Both Indian and Chinese number-theoretic works were motivated to a large extent by problems in astronomy.

1547–1700

THE FOUNDING OF MODERN NUMBER THEORY: FERMAT

Pierre de Fermat (1601–65) was arguably the greatest mathematician of the first half of the seventeenth century. He made fundamental contributions to analytic geometry, calculus, probability, and number theory; the last was his mathematical passion. In fact, he founded number theory in its modern form. Fermat's interest in the subject was aroused by Diophantus's *Arithmetica*. The book had come to his attention in the 1630s through an excellent Latin translation, with commentaries, by Claude Bachet (1591–1638), a country gentleman of independent means who became very interested in number theory. Several of Fermat's important discoveries in the subject were given in the margins of this translation as commentaries on, and elaborations of, some of Diophantus's results. Most of his other results became known through his extensive correspondence with leading scientists of the day, principally Marin Mersenne (1588–1648) and Pierre de Carcavi (c. 1600–84), who championed Fermat's work by disseminating it in the wider scientific community.

Fermat produced no formal publications in number theory, nor did he give any proofs (save one) in his letters or in the margins of the *Arithmetica*, although he did provide comments and hints. All his claims but one (see below) were later shown to be correct. One of his major aims was to interest his mathematical colleagues in number theory by proposing challenging problems (for which he had the solutions). As he put it, "Questions of this kind [i.e., number-theoretic] are not inferior to the more celebrated questions in geometry [i.e., other branches of mathematics] in respect of beauty, difficulty, or method of proof." But his protestations were to no avail: Mathematicians showed little serious interest in number theory until Euler came on the scene some 100 years later.

Several of Fermat's major results are given below. They turned out to be exemplars of major concerns of number theory.

Fermat's Little Theorem. This theorem says that for any integer a and prime p, $a^p - a$ is divisible by p [nowadays it is expressed in terms of congruences, as $a^p \equiv a \pmod{p}$]. An equivalent, useful way of putting it is that $a^{p-1} - 1$ is divisible by p, provided that a is not divisible by p. This is one of Fermat's most important results and has recently found significant applications in cryptography. Fermat is thought to have become interested in this problem through Euclid's result on perfect numbers, which raised the question of primes of the form $2^n - 1$ (see above). First, Fermat showed that for $2^n - 1$ to be prime, n must be prime (this is easy), and then studied conditions for $2^n - 1$ to have divisors. This led him to the special case $a = 2$ of Fermat's little theorem (i.e., that $2^{p-1} - 1$ is divisible by p if p is an odd prime), and thence to arbitrary a.

Numbers of the form $2^p - 1$ (p prime) are called *Mersenne numbers* (since they were studied by Mersenne) and denoted by M_p. It is easy to see that M_p is *not* prime for every prime p, for example, $M_{11} = 23 \times 89$. Mersenne claimed that for $p = 2, 3, 5, 7, 13, 17, 19, 31, 67, 127$, and 257, M_p *is* prime (these are called *Mersenne primes*). He was wrong about $p = 67$ and 257 (keep in mind that the corresponding M_p are huge numbers), and he missed $p = 61, 89$, and 107 (among those less than 257). By the end of 2000, 38 Mersenne primes were known; the thirty-eighth, for $p = 6{,}972{,}593$, was found June 1, 1999, by one of the 12,000 participants in the Great Internet Mersenne Prime Search [see Bibliography]; it is the first Mersenne prime to have more than a million decimal digits. Are there infinitely many Mersenne primes? This is still an open question, about 350 years after it was posed, although probabilistic arguments suggest that the answer is yes.

Sums of Two Squares. Diophantus remarked that the product of two integers, each of which is a sum of two squares, is again a sum of two squares [i.e., that $(a^2 + b^2)(c^2 + d^2) = (ac + bd)^2 + (ad - bc)^2$, although he did not state it in this generality]. This appears to have prompted Fermat to ask which integers are sums of two squares. Since every integer is a product of primes, the identity above reduces the problem to asking which primes are sums of two squares. Now all (odd) primes are of the form $4n + 1$ or $4n + 3$, and it is easy to see that no prime (in fact, no integer) of the form $4n + 3$ can be a sum of two squares. Fermat claimed to have shown that every prime of the form $4n + 1$ is a sum of two squares and that it is a unique such sum. It is then not difficult to characterize those integers that are sums of two squares. About 200 years later, Carl Gustav Jacobi (1804–51) gave expressions for the number of ways in which an integer can be written as a sum of two squares (e.g., $65 = 8^2 + 1^1 = 7^2 + 4^2$).

Fermat also proved the following related theorems: Every prime of the form $8n + 1$ or $8n+ 3$ can be written as $x^2 + 2y^2$, and every prime of the form $3n + 1$ can be written as $x^2 + 3y^2$. These were no idle results. For example, the latter was used by Euler in his proof of Fermat's last theorem for $n = 3$ (see below). Moreover, these results raised the question of representation of primes in the form $x^2 + ny^2$ for general n (Fermat already had difficulty with $n = 5$). This was an issue with important ramifications, which evolved in the following two centuries into the question of representation of integers by binary quadratic forms, $ax^2 + bxy + cy^2$, one of the central problems in number theory (see below).

Fermat's Last Theorem. In the margin of Problem 8 of Book II of Diophantus's *Arithmetica*, which asked for the representation of a given square as a sum of two squares (see above), Fermat said that unlike that result, "It is impossible to separate a cube into two cubes or a fourth power into two fourth powers or, in general, any power greater than the second into powers of like degree. I have discovered a truly marvelous demonstration, which this margin is too narrow to contain." Fermat was claiming that the equation $z^n = x^n + y^n$ has no (nonzero) integer solutions if $n > 2$. This has come to be known as *Fermat's last theorem* (FLT). (Mathematicians doubt whether Fermat had a proof of this result. In later correspondence on this problem, he made reference only to proofs of the theorem for $n = 3$ and 4.) This "theorem" was perhaps the most outstanding unsolved problem for 360 years. Princeton mathematician Andrew Wiles (b. 1953) gave a proof in 1994 (discussed in detail below).

As mentioned, Fermat gave only one proof in number theory, and that was the case $n = 4$ of FLT, namely that $x^4 + y^4 = z^4$ has no nonzero integer solutions (it is easier than the case $n = 3$). This he did in the context of a problem on Pythagorean triples. What he showed was that "the area of a right-angled triangle whose sides have rational length cannot be a square of a rational number." Here he was responding to a problem raised by Bachet, based on one in Book VI of Diophantus's *Arithmetica*, that of finding a right-angled triangle whose area equals a given number.

It can be shown that the problem above is equivalent to showing that the area of a right-angled triangle with integer sides cannot be a square (integer). Fermat proceeded by assuming that such a triangle was possible and obtained another of the same type but with a smaller hypotenuse. Continuing this process, he obtained an infinite decreasing sequence of positive integers (the hypotenuses of the corresponding right triangles). But this is clearly impossible, which established the result. It follows as an easy corollary that $x^4 + y^4 = z^4$ has no integer solutions.

More important than the result was the method used to establish it, since known as the *method of infinite descent*. Its essence is this: Assume that a positive integer satisfies a given condition and show (by an iterative process) that a smaller positive integer satisfies the same condition; then no positive integer can satisfy the condition. Logically, this is nothing but a variant of the principle of mathematical induction, but it provided Fermat (and his followers) with a powerful tool for proving many number-theoretic results. As he put it with considerable foresight, "This method will enable extraordinary developments to be made in the theory of numbers."

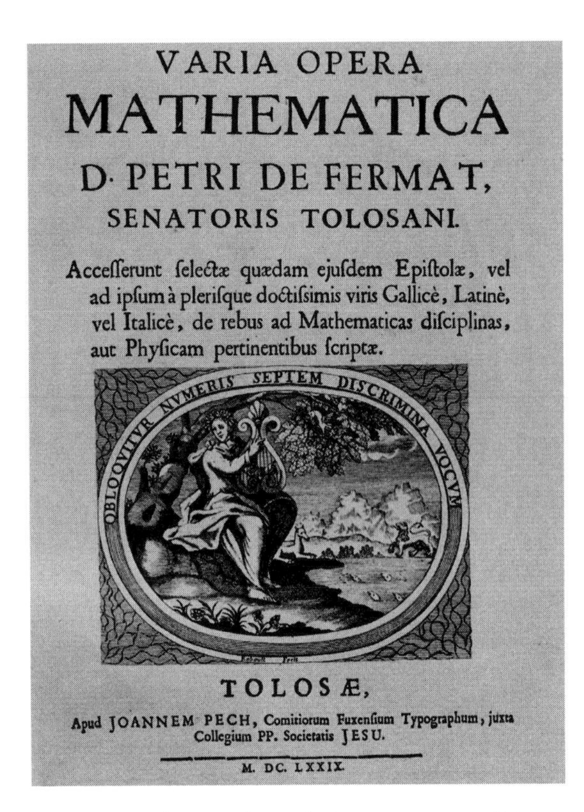

VARIA OPERA
MATHEMATICA
D. PETRI DE FERMAT,
SENATORIS TOLOSANI.

Acceſſerunt ſelectæ quædam ejuſdem Epiſtolæ, vel ad ipſum à plerisque doctiſſimis viris Gallicè, Latinè, vel Italicè, de rebus ad Mathematicas diſciplinas, aut Phyſicam pertinentibus ſcriptæ.

TOLOSÆ,
Apud JOANNEM PECH, Comitiorum Fuxenſium Typographum, juxta Collegium PP. Societatis JESU.
M. DC. LXXIX.

Figure 1. Title page from the Latin version of Pierre de Fermat, Varia opera mathematica *(Toulouse, 1629). © Archivo Iconografico, S.A. CORBIS*

Bachet's Equation. This is the equation $x^2 + k = y^3$ (k an integer), a special case of which was considered by Bachet in his edition of the *Arithmetica*. Fermat found the (positive) solutions for $x^2 + 2 = y^3$ and $x^2 + 4 = y^3$, namely $x = 5$, $y = 3$ for the first equation, and $x = 2$, $y = 2$ and $x = 11$, $y = 5$ for the second. It is easy to verify that these are solutions of the respective equations, but it is rather difficult to show that they are the only (positive) solutions. Bachet's equation plays a central role in number theory to this day (see below).

Pell's Equation. The Pell equation, $x^2 - dy^2 = 1$, was noted in connection with Indian mathematics, although Fermat was probably unaware of that work. "The study of the [quadratic] form $x^2 - 2y^2$ must have convinced Fermat of the paramount importance of the equation $x^2 - Ny^2 = \pm 1$," said the distinguished modern mathematician André Weil (1906–98). Fermat claimed to have shown that the equation has infinitely many integer solutions. This equation, too, has been very important in number theory, even in recent times (see below).

Fermat Numbers. Having investigated when $2^n - 1$ is prime, it was natural for Fermat to consider the same question for numbers of the form $2^n + 1$. It is easy to show that for $2^n + 1$ to be prime, n must be a power of 2. Numbers of the form $2^{2^k} + 1$ are called *Fermat numbers* and are denoted by F_k. Fermat repeatedly asserted in correspondence that the F_k are prime for every k, although he admitted that he could not find a proof. Now, $F_1 = 5$, $F_2 = 17$, $F_3 = 257$, and $F_4 = 65{,}537$ are all prime (they are called *Fermat primes*). However, Euler showed in 1732 that F_5 is not. The proof was not simply a matter of computation (F_5 has 10 digits and one would need a table of primes up to 100,000, unavailable to Euler, to test the primality of F_5 by "brute computational force"). Euler's proof was largely theoretical: He proved that every factor of F_k must be of the form $t \times 2^{k+2} + 1$ (where t is some integer) and hence was able to show that F_5 is divisible by 641; in fact, $F_5 = 641 \times 6{,}700{,}417$. (In general, proving that a given very large number is composite is much easier than factoring it.) It was shown only in the 1870s that F_6 is composite. Various other Fermat numbers have been shown to be composite (this is a difficult problem), but none to be prime. In fact, it is thought that there are *no* Fermat primes other than the four listed above. But it is known (and easy to prove) that any two Fermat numbers are relatively prime. Since every integer greater than 1 is divisible by some prime, this gives a proof (other than Euclid's) that there are infinitely many primes. Fermat primes were shown by Gauss to be closely related to the constructibility of regular polygons (see below).

To conclude our account of Fermat: It is remarkable how well he chose for consideration problems that would become central in number theory. These stimulated the best mathematical minds, including those of Euler and Gauss, for the next two centuries. Without doubt, Fermat is the founder of modern number theory.

Eighteenth Century

EULER

Leonhard Euler (1707–83) was the greatest mathematician of the eighteenth century and one of the most eminent of all time, "the first among mathematicians," according to Joseph Louis Lagrange (1736–1813). He was also the most productive ever. Although "only" four volumes of a projected 80-volume collection of his works are on number theory, they contain priceless treasures, dealing with all existing areas in number theory and giving birth to new methods and results.

Euler was the first to take up the study of number theory in close to 100 years. His love for the subject, like Fermat's, was great. Adrien-Marie Legendre (1752–1833), in the preface to his 1798 book on number theory, put it thus: "It appears . . . that Euler had a special inclination towards . . . [number theoretic] investigations, and that he took them up with a kind of passionate addiction, as happens to nearly all those who concern themselves with them." A considerable part of Euler's number-theoretic work consisted in proving Fermat's results and trying to reconstruct his and Diophantus's methods.

Euler's interest in number theory was apparently stimulated by his friend Christian Goldbach (1690–1764), an amateur mathematician (of Goldbach's conjecture fame; see below) with whom Euler carried on a correspondence over several decades. It began with a letter in 1729, when Euler was 22, in which Goldbach asked Euler's opinion about Fermat's claim that the Fermat numbers are all prime. Euler was skeptical, but it was only two years later that he discovered the counterexample F_5 (see above). It set him on a lifelong study of Fermat's works.

Below are described some of Euler's number-theoretic investigations, focusing on new departures in both methods and results.

Analytic Number Theory. The overriding reason why there was little interest in number theory among mathematicians in the seventeenth and eighteenth centuries was probably the ascendance during this period of calculus as the predominant mathematical field. A major topic was summation of series. Gottfried Wilhelm Leibniz's (1646–1716) result, $1 - 1/3 + 1/5 - 1/7 + \ldots = \pi/4$, fascinated mathematicians. Leibniz and the brothers Jakob Bernoulli (1654–1705) and Johann Bernoulli (1667–1748) attempted to sum the series $\Sigma\, 1/n^2 = 1 + 1/4 + 1/9 + 1/16 + \ldots$, without success. In 1735, Euler prevailed, showing that $1 + 1/4 + 1/9 + 1/16 + \ldots = \pi^2/6$. This was a spectacular achievement for the young Euler, with important consequences for number theory. It helped establish his growing reputation.

Euler next studied the series $\Sigma\, 1/n^{2k}$ for an arbitrary positive integer k and proved the beautiful result that $\Sigma\, 1/n^{2k} = (2^{2k-1}\pi^{2k}|B_{2k}|)/(2k)!$, where the B_i are the Bernoulli numbers, the coefficients in the power series expansion $x/(e^x - 1) = \Sigma\, B_n x^n/n!$ ($|B_{2k}|$ denotes the absolute value of B_{2k}). The Bernoulli numbers are rational (e.g., $B_2 = 1/6$, $B_4 = -1/30$, $B_6 = 1/42$; hence $\Sigma\, 1/n^4 = \pi^4/90$ and $\Sigma\, 1/n^6 = \pi^6/945$). They turned out to be very important in number theory and elsewhere.

The series $\Sigma\, 1/n^{2k+1}$ was a mystery to Euler and remained so to mathematicians of subsequent generations. (Only in 1978 was it shown that $\Sigma\, 1/n^3$ is irrational. In November 2000, it was announced that $\Sigma\, 1/n^{2k+1}$ is irrational for infinitely many k; but it is still not known if $\Sigma\, 1/n^5$ is irrational.) It is probably this lack of knowledge that persuaded Euler to study the function $\zeta(s) = \Sigma\, 1/n^s$ for all real $s > 1$, for which the series converges. This turned out to be a pivotal function in number theory, the zeta function (see below). Euler soon derived what came to be called the *Euler product formula*, $\zeta(s) = \Sigma\, 1/n^s = \prod 1/(1 - p^{-s})$, where the product ranges over all the primes p. This most important identity may be viewed as an analytic counterpart of the fundamental theorem of arithmetic, expressing integers in terms of primes (the FTA is needed in its proof).

An easy consequence of Euler's product formula is yet another proof of the infinitude of primes. [If there were finitely many primes, then letting s approach 1 from the right, $\Sigma\, 1/n^s$ approaches infinity, while $\prod 1/(1 - p^{-s})$ is finite.] Another relatively easy corollary of the product formula is the divergence of $\Sigma\, 1/p$, where the sum is taken over all the primes p. Since $\Sigma\ 1/n^2$ converges, this shows that the primes are "denser" than the squares in the sequence of positive integers (i.e., there are, in some sense, "more" primes than squares).

Euler's introduction of analysis (the study of the continuous) into number theory (the study of the discrete) may at first appear paradoxical. However, it was a crucial development, greatly exploited in the next century. It led to the rise of a new area of study, analytic number theory (see below). As Euler put it: "One may see how closely and wonderfully infinitesimal analysis is related . . . to the theory of numbers, however repugnant the latter may seem to that higher kind of calculus." Another important instance in which Euler began to relate analysis to number theory was his use of "elliptic integrals" (integrals arising when finding the length of an arc of an ellipse) to study Diophantine equations of the form $y^2 = f(x)$, with $f(x)$ of degree three or four (their graphs are called *elliptic curves*; see below). More broadly, building bridges between different, seemingly unrelated areas of mathematics is an important and powerful idea, for it brings to bear the tools of one field in the service of the other. (A very important example of bridge building between algebra and geometry resulted in the seventeenth century in a new field, analytic geometry.) Further examples of the vital interaction between number theory and other areas are discussed below.

Diophantine Equations. This is a vast subfield of number theory, with diverse branches. Following are some examples of Euler's contributions.

The simplest Diophantine equation is $ax + by = c$. In the early 1730s, Euler rediscovered its solution, known to Brahmagupta, Bachet, Fermat, and others. In this context, he proved the important result, from which the solution of the equation follows, that if a and b are relatively prime to c, there is an x relatively prime to c such that $ax \equiv b \pmod{c}$. In connection with this work, Euler proved that for any relatively prime positive integers a and n, $a^{\varphi(n)} \equiv 1 \pmod{n}$, where $\varphi(n)$, the Euler φ-function, denotes the number of positive integers less than n and relatively prime to n. This important result generalized Fermat's little theorem, $a^{p-1} \equiv 1 \pmod{p}$, since for a prime p, $\varphi(k) = p - 1$.

Euler's φ-function is an example of a multiplicative arithmetic function, a function $f: N \to N$ (N the positive integers) such that $f(mn) = f(m)\, f(n)$ for m and n relatively prime. Other important multiplicative arithmetic functions introduced and studied by Euler were $d(n)$, the number of positive divisors of n, and $\sigma(n)$, the sum of the divisors of n. [In this notation, a number n is perfect if $\sigma(n) = 2n$.] Using the multiplicativity of the σ-function, Euler proved (in a posthumous paper) the converse of Euclid's theorem on perfect numbers, namely that if $2^{n-1}(2^n - 1)$ is perfect (necessarily even), then $2^n - 1$ is prime. Thus the theorems of Euclid and Euler (one proved 2000 years after the other) characterize all even perfect numbers, reducing their existence to that of Mersenne primes. It is not known if there are any odd perfect numbers, but if there are, they are huge, larger than 10^{150} (this result dates from the 1980s).

In the latter part of his book *Elements of Algebra* (1770), Euler dealt with various Diophantine equations. An important one was $z^3 = ax^2 + by^2$, for which he developed various techniques. He then specialized it to Bachet's equation, $z^3 = x^2 + 2$, which Fermat claimed to have solved. Euler's solution introduced a new, and most important, technique. He factored the right side of the equation and obtained $z^3 = x^2 + 2 = (x + \sqrt{-2})(x - \sqrt{-2})$. This was now an equation in the domain of *complex integers* of the form $Z(\sqrt{-2}) = \{a + b\sqrt{-2}: a, b \in Z\}$. These possess many of the number-theoretic properties of the ordinary integers Z (see below). Euler exploited this analogy to solve Bachet's equation. Analogy, it should be noted, is a most important mathematical device. Euler was its undisputed master, using it again and again.

With this problem, Euler had taken the audacious step of introducing complex numbers into number theory, the study of the positive integers. "A momentous event had taken place," declared Weil. This foreshadowed the creation of a new field, algebraic number theory. While Euler had earlier wedded number theory to analysis, he now linked number theory with algebra. This bridge building, too, would prove most fruitful in the following century (see below). Following are two other examples of Euler's use of these ideas.

When his attention was drawn in the 1740s to Fermat's claim about $x^n + y^n = z^n$, he called it "a very beautiful theorem." In 1753, he wrote to Goldbach that he had proved it for $n = 3$, but he published a proof only in 1770, in his *Elements of Algebra*. Here he calculated with numbers in the domain $Z(\sqrt{-3}) = \{a + b\sqrt{-3}: a, b \in Z\}$. There was, however, a considerable gap in the proof: The domain $Z(\sqrt{-3})$, unlike $Z(\sqrt{-2})$, does not have the arithmetic properties of Z. Analogy is, indeed, a powerful tool, but it must be used with great caution. Even the likes of Euler can err. (Not much, however, is needed to repair the proof.) Euler also applied his emerging ideas on the use of complex integers for studying number-theoretic problems to quadratic forms of the type $x^2 + cy^2$, by writing them as $(x + y\sqrt{-c})(x - y\sqrt{-c})$.

Partitions. A partition of a positive integer n is a representation of n as a sum of positive integers. For instance, the partitions of 5 are 5, 4 + 1, 3 + 2, 3 + 1 + 1, 2 + 2 + 1, 2 + 1 + 1 + 1, and 1 + 1 + 1 + 1 + 1; the order of the summands is irrelevant. Partition theory is the study of such representations. It is a subfield of additive number theory, which deals with the representation of integers as sums of other integers (e.g., as cubes). Euler initiated the study of partitions in his great book *Introduction to the Analysis of the Infinite* (1748).

Let $p(n)$ denote the number of partitions of n; the object of partition theory is to study the properties of this and related arithmetic functions, for example $p_k(n)$, the number of partitions of n in which each summand is no greater than k. The function $p(n)$ is rather complicated in that it grows enormously; for example, $p(200) = 3{,}972{,}999{,}029{,}388$. Euler's interest in partitions was stimulated by a colleague's letter that asked for the number of ways in which an integer n can be written as a sum of t distinct integers. Euler began to study $p(n)$ by introducing the important notion of its generating function, the formal power series $\sum p(n)x^n$ (no questions of convergence are involved). With its aid, he proved the fundamental result that the number of partitions of an integer in which all summands are odd equals the number of partitions of the integer in which all summands are distinct. Partition theory is now an active and important area of number theory that has recently found applications in physics. A major contributor to the theory in this century was the great Indian mathematician Srinivasa Ramanujan (1887–1920). For example, in 1918 he proved [with Godfrey H. Hardy (1877–1947)] that $\log p(n) \sim c\sqrt{n}$, for some real number c [this means that $\log p(n)/c\sqrt{n} \to 1$ as $n \to \infty$].

Quadratic Reciprocity Law. The quadratic reciprocity law was conjectured, but not proved, by Euler. It came to be one of the central results in number theory. It says (in the language of congruences) that there is a reciprocity relation between the solvability of $x^2 \equiv p \pmod{q}$ and $x^2 \equiv q \pmod{p}$ for any distinct odd primes p and q. Specifically, $x^2 \equiv p \pmod{q}$ is solvable if and only if $x^2 \equiv q \pmod{p}$ is solvable, unless $p \equiv q \equiv 3 \pmod{4}$, in which case $x^2 \equiv p \pmod{q}$ is solvable if and only if x^2

$\equiv q \pmod{p}$ is not. It is the fundamental law when it comes to the solvability of quadratic Diophantine equations (or quadratic congruences).

The quadratic reciprocity law arose from the study of representations of primes by quadratic forms $x^2 + cy^2$, in particular in connection with the representation of p by $x^2 + qy^2$ (where p and q are distinct odd primes). Euler had made some progress on this question in the 1740s but gave a clear formulation of the law of quadratic reciprocity only in 1772. He attached great importance to this conjecture, proved in 1801 by Gauss, and made an important contribution by proving what came to be called the *Euler criterion*: $x^2 \equiv a \pmod{p}$ is solvable (where p is an odd prime not dividing a) if and only if $a^{(p-1)/2} \equiv 1 \pmod{p}$. [If $x^2 \equiv a \pmod{p}$ is solvable, a is said to be a quadratic residue mod p; otherwise, it is a quadratic nonresidue.]

The above results represent only a miniscule part of Euler's contribution to number theory, which themselves are only a miniscule part of his overall contribution to mathematics. They alone, however, would have earned him entry to mathematics' hall of fame.

LAGRANGE

Euler's contemporaries took little interest in number theory, with the sole exception of Joseph Louis Lagrange, much of whose work, especially in number theory, was inspired directly by Euler's. Lagrange became actively interested in number-theoretic problems in the 1760s, although his interest lasted less than 10 years. Among his accomplishments, three stand out: work on Pell's equation, sums of four squares, and binary quadratic forms.

Pell's Equation. Pell's equation, $x^2 - dy^2 = 1$ (d a nonsquare positive integer), is one of the most important Diophantine equations. It is a key to the solution of arbitrary quadratic Diophantine equations; its solutions yield the best approximations (in some sense) to $\sqrt{d}$ [the Pell equation can be written as $(x/y)^2 = d + 1/y^2$, so that for large y, x/y is an approximation to $\sqrt{d}$]; there is a 1–1 correspondence between the solutions of $x^2 - dy^2 = 1$ and the units (invertible elements) of quadratic fields, $Q(\sqrt{d}) = \{s + t\sqrt{d}: s,t \in Q\}$ (see below); and the equation has recently (1970) played a crucial role in the solution of David Hilbert's (1862–1943) tenth problem, the nonexistence of an algorithm for solving arbitrary Diophantine equations. (In 1900, Hilbert presented 23 problems at the International Congress of Mathematicians in Paris. These played a key role in twentieth-century mathematics.)

The Pell equation has been studied often, as seen above. [The name is a misnomer; Euler incorrectly attributed the solution of the equation to the British mathematician John Pell (1611–85).] The first extant record is Archimedes' (287–212 B.C.) cattle problem, which yields (following considerable algebraic simplification) the equation $x^2 - 4{,}729{,}494y^2 = 1$. Among others who dealt with Pell's equation were Bhaskara, William Brouncker (1620–84), Fermat, and Euler. None of these mathematicians, however, showed that their methods of solution would always work.

The definitive treatment was given by Lagrange in the 1760s. This appeared as a supplement to Euler's *Elements of Algebra*. In particular, Lagrange was the first to prove that a solution of Pell's equation always exists. In fact, he gave an explicit procedure for finding all solutions using the continued fraction expansion of $\sqrt{d}$. (It is one thing to prove the existence of solutions, quite another to find them.) Noteworthy was Lagrange's use of irrational numbers to solve equations in integers. He also used complex numbers to solve Diophantine equations, thereby extending some of Euler's ideas. When Euler heard about these approaches, he remarked: "I have greatly admired your method of using irrationals and even imaginary numbers in this kind of analysis which deals with nothing else than rational numbers. Already for several years I have had similar ideas."

Sums of Four Squares. Fermat claimed to have proved that every positive integer is a sum of four squares (of integers, some of which may be 0). Euler was captivated by this result and tried for many years to prove it, without success. Lagrange was (of course) very pleased to have succeeded where Euler had failed. He gave a proof using Euler's identity that a product of a sum of four squares is again a sum of four squares (recall a similar identity about sums of two squares). In 1829, Jacobi used elliptic functions to give an explicit formula for the number of representations of an integer as a sum of four squares.

Binary Quadratic Forms. A binary quadratic form is an expression of the type $f(x,y) = ax^2 + bxy + cy^2$, with a, b, and c integers. The basic question related to such forms is: Given $f(x,y)$, which integers are represented by it; that is, for which integers n are there integers x and y such that $n = ax^2 + bxy + cy^2$? Other interesting questions deal with the number of such solutions for a given n and an algorithm for finding them.

Fermat had considered specific cases of quadratic forms of the type $x^2 + cy^2$, and Euler studied forms of the type $ax^2 + cy^2$, but Lagrange was the first to deal with general quadratic forms. A fundamental observation was that two distinct forms can represent the same set of integers. This is the case, for example, for the forms $x^2 + y^2$ and $2x^2 + 6xy + 5y^2$; to see this, note that $2x^2 + 6xy + 5y^2 = (x + y)^2 + (x + 2y)^2$.

To deal with this phenomenon, Lagrange introduced the fundamental notion of equivalence of forms. Two forms, $f(x,y) = ax^2 + bxy + cy^2$ and $F(X,Y) = AX^2 + BXY + CY^2$, are equivalent if there exists a transformation of the variables, $x = sX + tY$, $y = uX + vY$, such that $sv - tu = \pm 1$ (with s, t, u, v integers). It is easy to see that two equivalent forms represent the same set of integers. An important object is the discriminant $D = b^2 - 4ac$ of a form $f(x,y)$. It is an invariant of the form under equivalence; that is, if f and F are equivalent forms, they have the same discriminant (the converse fails).

Equivalence of forms is an equivalence relation (i.e., it is reflexive, symmetric, and transitive), hence it divides the quadratic forms into equivalence classes. (The terms *equivalent* and *equivalence class* are due to Gauss rather than Lagrange.) Lagrange showed that there are only finitely many inequivalent forms for a given discriminant D; their number is called the *class number*, denoted by $h(D)$. He also described a procedure for finding a "simple" representative for each class, called a *reduced form*.

Lagrange applied his theory to prove results of Fermat and Euler on the representation of primes by quadratic forms, but the theory enabled him to go beyond them. For example, he showed that every prime of the form $20n + 1$ can be represented by the form $x^2 + 5y^2$. This form caused difficulty for both Fermat and Euler. Its discriminant is -20, and $h(-20) = 2$, the other inequivalent form being $2x^2 + 2xy + 3y^2$. Lagrange's comprehensive and beautiful theory of binary quadratic forms was fundamental for subsequent developments in number theory.

LEGENDRE

Adrien-Marie Legendre was one of the most prominent mathematicians of Europe, although not of the stature of Euler or Lagrange. His texts were very influential. In 1798, he published his *Theory of Numbers*, the first book devoted exclusively to number theory. It went into several editions but was soon to be superseded by Gauss's *Disquisitiones arithmeticae* (see below).

Waring's Problem

Given that every positive integer is a sum of (four) squares, a natural question is whether every such integer is a sum of cubes, of fourth powers, and so on. Edward Waring (1736–98), Lucasian professor of mathematics at Cambridge University, conjectured in 1770, on scant numerical evidence, that every positive integer is a sum of nine cubes, 19 fourth powers, and so on. The following came to be known as *Waring's problem*: (i) Given a positive integer k, is every positive integer n a sum of kth powers, $n = x_1^k + x_2^k + \ldots + x_s^k$, where s depends on k but not on n and, if so, (ii) what is the smallest value of s [usually denoted by $g(k)$] for a given k?

An affirmative answer to (i) above was given by Hilbert in 1909. But the determination of $g(k)$ for specific k has proved difficult, although upper bounds for $g(k)$ exist. It has been shown that $g(2) = 4$, $g(3) = 9$, $g(4) = 19$, $g(5) = 37$, and $g(6) = 73$, but the value of $g(7)$ is not known.

Many of Legendre's results were found independently by Gauss, and serious priority disputes arose between them. Legendre's proofs, moreover, left much to be desired, even by mid-eighteenth-century standards. For example, he discovered the law of quadratic reciprocity, unaware of Euler's prior discovery, and gave a proof based on what he viewed as a self-evident fact, namely the existence of infinitely many primes in any arithmetic progression $an + b$ ($n = 1, 2, 3, \ldots$, with a and b relatively prime). This was a very difficult result, proved subsequently by Peter Lejeune Dirichlet (1805–59) using deep methods in analysis. Legendre was chagrined when Gauss (who gave a rigorous proof) claimed the result as his own. In connection with this law, Legendre introduced the useful and celebrated Legendre symbol (a/p) (it does *not* denote division), with p an odd prime and a an integer not divisible by p: $(a/p) = 1$ if $x^2 \equiv a \pmod{p}$ is solvable and $(a/p) = -1$ if it is not. In terms of this symbol, the law of quadratic reciprocity can be stated succinctly as $(p/q)(q/p) = (-1)^{(p-1)(q-1)/4}$, where p and q are distinct odd primes.

An important achievement was Legendre's proof (in his 70s) of Fermat's last theorem for $n = 5$ and his conjecture that $\pi(x) \sim x/(A \log x + B)$, where $\pi(x)$ denotes the number of primes less than or equal to x, and $f(x) \sim g(x)$, read "$f(x)$ is asymptotic [approximately equal] to $g(x)$," means that $\lim(f(x)/g(x)) = 1$ as $x \to \infty$. This conjecture was refined by Gauss and became known as the *prime number theorem* (see below).

Finally, "One of Legendre's main claims to fame [in number theory]" (according to Weil) is the result that the equation $ax^2 + by^2 + cz^2 = 0$ has a solution in integers not all zero if and only if $-bc$, $-ca$, $-ab$ are quadratic residues mod a, b, c, respectively, where a, b, c are integers not of the same sign and abc is square-free.

Nineteenth Century

GAUSS'S *DISQUISITIONES ARITHMETICAE*

Gauss, the greatest mathematician of the nineteenth century, contributed to many areas of mathmetics, but number theory, the queen of mathematics (according to him), was his greatest mathematical love. As he put it in an 1838 letter to Dirichlet: "I place this part of mathematics [number theory] above all others (and have always done so)." His supreme masterpiece was *Disquisitiones arithmeticae* (Arithmetical Investigations), published in 1801 but completed in 1798, when he was 21.

Pre-nineteenth-century number theory consisted of many brilliant results but often lacked thematic unity and general methodology. In the *Disquisitiones*, Gauss supplied both. He systematized the subject, provided it with deep and rigorous methods, solved important problems, and furnished mathematicians with new ideas

Figure 2. Portrait of German mathematician and astronomer Carl Friedrich Gauss. © Bettmann/ CORBIS

to help guide their research for much of the nineteenth century. The following are several of the far-reaching concepts and results in the *Disquisitiones*.

Whereas little attention was given to formal proof during the eighteenth century, in the nineteenth century there was an emergence of a critical spirit in which rigor and abstraction began to play fundamental roles. Gauss was one of its early exponents. The fundamental theorem of arithmetic, a cornerstone of number theory, was undoubtedly known to its pioneers, but Gauss was the first to state the theorem explicitly and to give a rigorous proof. Perhaps Fermat, Euler, and others thought the result too obvious to mention (but the proof is far from trivial).

Gauss was also the first to define the fundamental notion of congruence, introducing the notation "≡" in use today. He chose it deliberately because it is similar to the notation "=" for equality. In fact, congruence has many of the same properties as equality [e.g., if $a \equiv b \pmod{m}$ and $c \equiv d \pmod{m}$, then $a + c \equiv b + d \pmod{m}$]. As mentioned, analogy is a powerful tool, and the simple notation that Gauss introduced for congruence had a great impact on number theory. Gauss himself exploited the analogy in important ways, in particular in his theory of congruences of the second degree, $ax^2 + bx + c \equiv 0 \pmod{m}$, where a, b, c, and m are integers, with $m > 1$. This can be reduced to the congruence $ax^2 + bx + c \equiv 0 \pmod{p}$, for prime divisors p of m, and by completing the square (as for quadratic equations), to the congruence $y^2 \equiv d \pmod{p}$. This was the congruence (as seen above) at the heart of the pivotal quadratic reciprocity law, which Gauss called the *golden theorem*. He was the first to give a rigorous proof. In fact, he gave *six* proofs, hoping that one of them might generalize to "higher" reciprocity laws, dealing with solutions of $y^n \equiv d \pmod{p}$ for $n > 2$ (see below). Gauss considered his work on quadratic reciprocity to be one of his most important contributions to number theory.

By far the largest part of the *Disquisitiones* was the powerful and beautiful but difficult theory of binary quadratic forms. Major strides were made by Lagrange (see above), but Gauss brought the theory to perfection. Here he introduced the important and deep concepts of genus and composition of forms. (Two forms are said to be in the same genus if there is a nonzero integer that is representable by both. If the integers m and n are representable by forms f and g, respectively, then mn is representable by the composition of f and g.) An important criterion for representability of integers by quadratic forms is the following: If n is "properly" representable by $f(x, y) = ax^2 + bxy + cy^2$ (i.e., $n = ax^2 + bxy + cy^2$ for some relatively prime integers x and y), and D is the discriminant of f, then $x^2 \equiv D \pmod{4|n|}$ is solvable; and if $x^2 \equiv D \pmod{4|n|}$ is solvable, n is properly representable by some form with discriminant D. This result was probably a strong motivation for Gauss's interest in quadratic residues.

According to Weil, the theory of quadratic forms "remained a stumbling-block for all readers of the *Disquisitiones* [for more than 60 years]." Dirichlet made it more accessible in his *Lectures on Number Theory*, published by his student Richard Dedekind (1831–1916) in 1863. This motivated a generation of mathematicians to try to come to grips with its ideas. In 1871, Dedekind reinterpreted the theory of binary quadratic forms in terms of his just-created theory of algebraic numbers; in particular, he established a correspondence between quadratic forms of discriminant D and the ideals of the quadratic field $Q(\sqrt{D})$, under which the product of ideals corresponds to the composition of quadratic forms (see below).

The last part of the *Disquisitiones*, a beautiful blend of algebra, geometry, and number theory, dealt with cyclotomy, the division of a circle into n equal parts. An important aspect of this work was a characterization of regular polygons constructible with straightedge (unmarked ruler) and compass: A regular polygon of n sides is so constructible if and only if $n = 2^k p_1 p_2 \ldots p_s$, where the p_i are distinct

Scientific Biography: Carl Friedrich Gauss

Gauss was born in Brunswick, Germany. He was a most precocious child and joked later in life that he could count before he could talk. In 1792, he entered the Collegium Carolinum, studying classical languages and, on his own, the works of Isaac Newton (1642–1727), Euler, and Lagrange. In 1795, when he enrolled at the University of Göttingen, he was still undecided about which of his two loves—philology or mathematics—to pursue as a career. He opted for mathematics the following year, after having shown that the regular polygon of 17 sides is constructible with straightedge and compass. This was the first discovery of a constructible regular polygon in over 2000 years, when the Greeks constructed those of sides 3, 4, 5, and 15. Another early landmark was Gauss's proof in 1799 of the fundamental theorem of algebra, the existence of a complex root for any polynomial with real coefficients. The proof eluded the likes of Jean le Rond d'Alembert (1717–83), Euler, and Lagrange. It earned Gauss a doctorate from the University of Helmstedt.

Gauss broke new ground in all areas of mathematics to which he turned: algebra, analysis (both real and complex), geometry (differential and non-Euclidean), number theory, probability, and statistics. To his contemporaries, he was the prince of mathematicians, and is, by universal acknowledgment, one of the three greatest mathematicians of all time [the other two are Archimedes (c. 287–212 B.C.) and Newton].

Ideas no less profound and far-reaching than those in Gauss's published works were found in his mathematical diary. This is a remarkable 19-page document of 146 very brief (often cryptic) entries with discoveries covering the years 1796 through 1814. The diary became public only in 1898. The publication of almost any one of these entries would have made the author famous. Some of the entries anticipated major creations of nineteenth-century mathematics: complex analysis, elliptic function theory, and non-Euclidean geometry.

Gauss's writings are in the spirit of rigorous modern mathematics. "It is demanded of a proof that all doubt become impossible," he wrote to a friend. And he practiced what he preached. His proofs were elegant and polished, often to the point where all traces of his method of discovery were removed. "He is like a fox who effaces his tracks with his tail," deplored Niels Henrik Abel (1802–29).

The finished product of Gauss's research gives, in particular, no indication of his great skill in, and love of, computation. Some of his deepest theorems in number theory were inspired by calculation, for example, the quadratic reciprocity law and the prime number theorem. It was probably a striking combination of remarkable insight, formidable computing ability, and great logical power that produced a mathematician whose ideas are still bearing rich fruit two centuries after he burst on the mathematical scene.

Fermat primes (see above). The vertices of a regular polygon of n sides inscribed in the unit circle are the roots of the equation $x^n - 1 = (x - 1)(x^{n-1} + x^{n-2} + \ldots + x + 1) = 0$. The polynomial $f(x) = x^{p-1} + x^{p-2} + \ldots + x + 1$, p prime, now called the p-th *cyclotomic polynomial*, was central to Gauss's characterization. It has since been an essential object in number-theoretic studies.

ALGEBRAIC NUMBER THEORY: GAUSS, KUMMER, AND DEDEKIND

Algebraic number theory is the study of number-theoretic problems using the concepts and results of abstract algebra, mainly those of groups, rings, fields, and ideals. In fact, some of these abstract concepts were invented to deal with problems in number theory. The initial inroads into the subject were made in the eighteenth century by Euler and Lagrange, in their use of "foreign" objects such as irrational and complex numbers to help solve problems about integers (see above). But the fundamental breakthroughs were achieved in the nineteenth century. Two basic problems provided the early stimulus for these developments: reciprocity laws and Fermat's last theorem.

The quadratic reciprocity law, the relationship between the solvability of $x^2 \equiv p \pmod{q}$ and $x^2 \equiv q \pmod{p}$, with p and q distinct odd primes, is (as mentioned above) a central result in number theory. One of the major problems of number theory in the nineteenth century was its extension to higher analogs, which would describe the relationship between the solvability of $x^n \equiv p \pmod{q}$ and $x^n \equiv q \pmod{p}$ for $n > 2$. (The cases $n = 3$ and $n = 4$ give rise, respectively, to what are known as *cubic* and *biquadratic reciprocity*.) Gauss opined that such laws cannot even be conjectured within the context of the integers. As he put it: "Such a theory [of higher reciprocity] demands that the domain of higher arithmetic [i.e., the domain of integers] be endlessly enlarged." This was, indeed, a prophetic statement.

Gauss himself began to enlarge the domain of higher arithmetic by introducing (in 1832) what came to be known as the *Gaussian integers*, $Z(i) = \{a + bi: a, b \in Z\}$. He needed these to formulate a biquadratic reciprocity law. The elements of $Z(i)$ do indeed qualify as "integers" in the sense that they obey all the crucial arithmetic properties of the "ordinary" integers Z: They can be added, subtracted, and multiplied, and most important, they obey a fundamental theorem of arithmetic: Every nonzero, noninvertible element of $Z(i)$ is a unique product of primes of $Z(i)$, called *Gaussian primes*. [The Gaussian primes are those elements that cannot be written nontrivially as products of Gaussian integers; for example, $8 - i = (2 + i)(3 - 2i)$, where $2 + i$ and $3 - 2i$ are Gaussian primes.] A domain with a unique factorization property such as the above is called a *unique factorization domain* (UFD). Gauss also formulated a cubic reciprocity law, and to do that he introduced yet another domain of "integers," the cyclotomic integers of order 3, $C_3 = \{a + bw + cw^2: a, b, c \in Z\}$, where $w = (-1 + \sqrt{3}\,i)/2$ is a primitive cube root of 1 ($w^3 = 1, w \neq 1$). This, too, turned out to be a UFD.

Recall that Fermat, in the seventeenth century, proved Fermat's last theorem, the unsolvability in nonzero integers of $x^n + y^n = z^n$, $n > 2$, for $n = 4$. Given this result, one can readily show that it suffices to prove FLT for $n = p$, an odd prime. Over the next two centuries, the theorem was proved for only three more cases: $n = 3$ (Euler, in the eighteenth century), $n = 5$ (Legendre and Dirichlet, independently, in the early nineteenth century), and $n = 7$ [Gabriel Lamé (1795–1870), in 1837]. A general attack on FLT was made in 1847, again by Lamé. His idea was to factor the left side of $x^p + y^p = z^p$ into linear factors (as Euler had already done for $n = 3$) to obtain the equation $(x + y)(x + yw)(x + yw^2) \ldots (x + yw^{p-1}) = z^p$, where w is a primitive pth root of 1 ($w^p = 1, w \neq 1$). This is an equation in the domain of cyclotomic integers of order p, $C_p = \{a_0 + a_1w + a_2w^2 + \ldots + a_{p-1}w^{p-1}: a_i \in Z\}$. Lamé now proceeded as Euler and others had done before him, that is, he used the arithmetic of the domain C_p and thereby "proved" FLT (the approach is analogous to Euler's proof of Bachet's equation, $x^2 + 2 = y^3$).

Well, not quite. The proof hinged on knowing that the arithmetic properties of Z do, indeed, carry over to C_p and that, most importantly, C_p is a UFD. When Lamé presented the proof to the Paris Academy of Sciences, Joseph Liouville (1809–82), who was in the audience, took the floor to point out precisely that. Lamé responded that he would reconsider his proof but was confident that he could repair it.

Alas, this was not to be. Two months after Lamé's presentation, Liouville received a letter from Ernst Kummer (1810–93) informing him that while C_p is, indeed, a UFD for all $p < 23$, C_{23} is not. (It was shown in 1971 that unique factorization fails in C_p for *all* $p > 23$.) But all hope was not lost, continued Kummer in his letter: "It is possibly to rescue it [unique factorization] by introducing new kinds of complex numbers, which I have called *ideal complex numbers*. I considered long ago the application of this theory to the proof of Fermat's [Last] Theorem and I succeeded in deriving the impossibility of the equation $x^n + y^n = z^n$ [for all $n < 100$]."

Kummer "rescued" unique factorization in C_p by adjoining to it "ideal numbers" and thereby established FLT for all $p < 100$. That was quite a feat, considering that during the previous two centuries, FLT had been proved for only three primes. It would take another century and more for further crucial progress on FLT (see below). [The notion of *ideal numbers* is complicated, but the following example may give an idea of what is involved. Let D be the set of even integers. Here 100

= 2 × 50 = 10 × 10, where 2, 10, and 50 are primes in D, that is, cannot be factored in D, so that D does not possess unique factorization. If we adjoin the "ideal" number 5 to D (it does not exist in D), unique factorization will have been restored to the element 100. For then 100 = 2 × 50 = 2 × 2 × 5 × 5 and 100 = 10 × 10 = 2 × 5 × 2 × 5. To restore unique factorization to every element of D, infinitely many ideal numbers will have to be added.] Kummer was also very interested in higher reciprocity laws. These, too, give rise to the cyclotomic integers C_p. His introduction of ideal numbers was motivated at least as much by these considerations as by FLT.

Kummer's work was brilliant, but it left unanswered important questions (as any good work ought to do). In particular, can one simplify his complicated theory, and more importantly, can one extend it to other domains that arise in various number-theoretic contexts, for example, the quadratic domains, important in the study of quadratic forms? These are the domains $Z_d = \{a + b\sqrt{d}: a,b \in Z\}$, if $d \equiv 2$ or 3 (mod 4), or $Z_d = \{a/2 + (b/2)\sqrt{d}$: a and b are both even or both odd$\}$, if $d \equiv 1$ (mod 4). And they are not, as a rule, UFDs. For instance, $Z_{-5} = \{a + b\sqrt{-5} : a, b \in Z\}$ is not, since (e.g.) $6 = 2 \times 3 = (1 + \sqrt{-5})(1 - \sqrt{-5})$, and 2, 3, $1 + \sqrt{-5}$, and $1 - \sqrt{-5}$ are primes in Z_{-5}.

The comments above give rise to two fundamental questions: (1) What are the domains for which a unique factorization theorem (UFT) is to hold? (2) What shape is such a theorem to take? [It clearly cannot state that every element in such a domain as has been determined in (1) is a unique product of primes, since this would disqualify many of the quadratic domains.] It took Dedekind about 20 years to answer these two questions (the first was the more difficult).

Dedekind's work, given in Supplement X to the second edition (1871) of Dirichlet's *Lectures on Number Theory*, was revolutionary in its formulation, its grand conception, its fundamental new ideas, and its modern spirit. As for our concerns here, to answer (1) above, namely, to determine the domains in which a UFT would obtain, Dedekind first had to define "fields of algebraic numbers"; the domains in question would be identified as distinguished subsets of these fields. (A field is a set with two operations, addition and multiplication, both associative and commutative, a distributive law, a zero, an additive inverse for each element, an identity, and a multiplicative inverse for each nonzero element; examples are the rational, real, and complex numbers. A complex number is algebraic if it is a root of a polynomial with integer coefficients; for example, $\sqrt{1/2 - \sqrt{3}}$ is algebraic since it is a root of $f(x) = 4x^4 - 4x^2 - 11$.)

The fields of algebraic numbers needed for Dedekind's theory were sets of the form $Q(a) = \{q_0 + q_1a + q_2a^2 + \ldots + q_na^n\}$, where q_i are rational numbers and a is an algebraic number. The $Q(a)$ are, in fact, fields, and all their elements are algebraic numbers. The domains for which a UFT was sought were defined to be "the integers of $Q(a)$," those elements $I(a)$ of $Q(a)$ that are roots of monic polynomials with integer coefficients. (A polynomial is monic if the coefficient of the highest-degree term is 1.) Such elements are called *algebraic integers*. (For example, $\sqrt{15} + 3$ is an algebraic integer since it is a root of the polynomial $x^2 - 6x - 6$.) They behave like integers in the sense that they can be added, subtracted, and multiplied, but they do not, in general, form UFDs. They do, however, form commutative rings (sets with two operations, addition and multiplication, both associative and commutative, a distributive law, a zero, an additive inverse for each element, and an identity).

The $I(a)$ are vast generalizations of the domains of integers that were considered above, namely the Gaussian integers, the cyclotomic integers, and the quadratic integers (also, of course, the ordinary integers). Having defined the $I(a)$, Dedekind's second major task was to formulate and prove a UFT in $I(a)$. It turned out to be the following: Every nonzero ideal in $I(a)$ is a unique product of prime ideals. [An ideal in $I(a)$ is a subset K closed under addition, subtraction, and multiplication by all elements of $I(a)$; this is a generalization of Kummer's ideal numbers. K is a prime ideal if $ab \in K$ implies that $a \in K$ or $b \in K$; this is a generalization of a prime number. For example, mZ, the set of all integer multiples of m, is an ideal in Z for every integer m; $3Z$ and $5Z$ are prime ideals, $15Z$ is not, but it is a unique product of prime ideals, $15Z = (3Z)(5Z)$.] The domains $I(a)$ are examples of Dedekind domains. These are integral domains (i.e., commutative rings in which the product of two nonzero elements is nonzero) in which every nonzero ideal is a unique product of prime ideals. They play an important role in number theory.

To summarize the events that have been described: After more than 2000 years in which number theory meant the study of properties of the (positive) integers, its scope became enormously enlarged. One could no longer use the term *integer* with impunity; it had to be qualified—a "rational" (ordinary) integer, a Gaussian integer, a cyclotomic integer, a quadratic integer, or any one of an infinite species of other algebraic integers [the various $I(a)$]. Moreover, powerful new algebraic tools were introduced and brought to bear on the study of these integers—fields, commutative rings, ideals, prime ideals, and Dedekind domains. A new subject emerged—algebraic number theory, of vital importance to this day.

ANALYTIC NUMBER THEORY: PRIME NUMBER THEOREM, RIEMANN HYPOTHESIS, AND DIRICHLET'S THEOREM

Algebra was not the only "foreign" subject that invaded number theory in the nineteenth century. Analysis was another. The bridge building between number theory and analysis began with Euler in the eighteenth century and gave rise in the nineteenth to a new field, analytic number theory. The broad context for the introduction of analytic methods into number theory was the problem of the distribution of primes among the integers. Euclid had shown that there are infinitely many primes, but do they follow a discernible pattern? This question baffled mathematicians for centuries.

Numerical evidence showed that the primes are spread out irregularly among the integers, in particular that they get scarcer—but not uniformly—as the integers increase in size. For example, there are eight primes between 9991 and 10,090 and 12 primes between 67,471 and 67,570. Furthermore, arbitrarily large gaps exist between primes. For example, it is easy to produce a sequence of $10^9 - 1$ consecutive composite integers, namely $10^9! + 2, 10^9! + 3, \ldots, 10^9! + 10^9$. On the other hand, considerable evidence suggests that there are infinitely many pairs of primes p, q as close together as can be, namely such that $q - p = 2$ (they are called *twin primes*). This apparent irregularity in the distribution of primes prompted Euler in the eighteenth century to observe that "mathematicians have tried in vain to this day to discover some order in the sequence of prime numbers, and we have reason to believe that it is a mystery which the human mind will never penetrate."

Euler's pessimism was in an important sense unjustified. It is true that there is no regularity in the distribution of primes considered individually, but there is regularity in their distribution when considered collectively. Instead of looking for a rule that will generate successive primes, one asked for a description of the number of primes in a given interval. Specifically, if $\pi(x)$ denotes the number of primes less than or equal to x (where x is any positive real number), the goal was to describe the behavior of the function $\pi(x)$. By inspecting the primes among the first three million integers, Gauss conjectured that $\pi(x)$ is asymptotic to $x/\log x$, $\pi(x) \sim x/\log x$ (the log is to the base e). [The irregularity in the distribution of primes would preclude an exact formula for $\pi(x)$.] If $\pi(x) \sim x/\log x$ is rewritten in the form $\pi(x)/x \sim$

$1/\log x$ this says, roughly speaking, that the probability of picking a prime from the first x integers is approximately $1/\log x$ and that the approximation improves with the size of x.

Gauss's conjecture, made at the start of the nineteenth century and now known as the *prime number theorem* (PNT), was proved at the century's end. It is remarkable that such a complex distribution as exhibited by the primes would be "modeled" by such a simple expression as $x/\log x$. Philip J. Davis (b. 1923) and Reuben Hersh (b. 1927), in *The Mathematical Experience* (1995), refer to the PNT as "one of the finest examples of the extraction of order from chaos in the whole of mathematics."

A major step toward the proof was taken by Georg Bernhard Riemann (1826–66) in the middle of the nineteenth century. His central idea was to extend the zeta function, introduced a century earlier by Euler, to complex variables, so that now $\zeta(s) = \Sigma\, 1/n^s$ was defined for complex numbers s. The series converges for those s for which $\mathrm{Re}(s) > 1$ [$\mathrm{Re}(s)$, the real part of $s = a + bi$, is a], and Euler's product formula $\Sigma\, 1/n^s = \prod 1/(1 - p^{-s})$ continues to hold for these s. Most important, the function $\zeta(s)$ can be extended to all complex numbers (by a method called *analytic continuation*). It has come to be known as the *Riemann zeta function*.

Riemann showed that the study of $\zeta(s)$ leads to information about $\pi(x)$. He conjectured that all the "nontrivial" roots of $\zeta(s)$ (the trivial roots are the negative even integers) are on the line $\mathrm{Re}(s) = 1/2$, i.e., that $\zeta(s) = 0 \Rightarrow s = 1/2 + bi$, b a real number. [The line $\mathrm{Re}(s) = 1/2$ is known as the *critical line*. It has recently been shown that more than a billion roots of $\zeta(s)$ lie on it.] This conjecture, still open 150 years later, is called the *Riemann hypothesis*. Arguably the most celebrated unsolved problem in mathematics, it has numerous implications in all branches of number theory. In particular, Riemann showed that it implies the PNT.

The proof of the PNT was given (independently) in 1896 by Jacques Hadamard (1865–1963) and Charles de la Vallée-Poussin (1866–1962). They proved a much weaker result than the Riemann hypothesis, namely that $\zeta(s) \neq 0$ for $\mathrm{Re}(s) = 1$, and showed that this implies the PNT (in fact, it is equivalent to the PNT). Both relied on Riemann's work, as well as on advanced techniques in the theory of functions of a complex variable developed subsequently. In the 1940s, in a most unexpected development, Paul Erdös (1913–96) and Atle Selberg (b. 1917) gave an "elementary" proof of the PNT (a far from simple proof, but one not using complex analysis).

Euclid proved that there are infinitely many primes. This can be rephrased to say that there are infinitely many primes in the arithmetic sequence $2n + 1$ ($n = 0, 1, 2, 3, \ldots$). In 1837, Dirichlet proved a grand generalization of this result by showing that *any* arithmetic sequence $an + b$ ($n = 0, 1, 2, 3, \ldots$), with a and b relatively prime, contains infinitely many primes. To do this, he introduced far-reaching ideas from analysis—in particular, the very important L-series, $L(s, X) = \Sigma\, X(n)/n^s$ (s is a real number greater than 1, and the Dirichlet character X is a function that associates with each integer a, relatively prime to n, an nth root of 1, and satisfies certain properties). He showed that if $L(1, X) \neq 0$ (where X is not the principal character), $an + b$ has infinitely many primes (compare the proof of the PNT). He applied similar ideas from analysis to prove other results, for example that $am^2 + bmn + cn^2$ contains infinitely many primes ($m, n = 0, 1, 2, 3, \ldots$), where a, b, and c are fixed relatively prime integers. Dirichlet was the first person to introduce deep methods of analysis into number theory, providing new perspectives on the subject, and may therefore be considered the founder of analytic number theory.

Many questions about the distribution of primes remain open. For example, it is not known if $n^2 + 1$ contains infinitely many primes ($n = 0, 1, 2, 3, \ldots$), although it is easy to show that no polynomial $p(n)$ in a single integer variable n contains only primes. But polynomials in two variables seem easier to handle. For example, it was shown in the 1960s that $m^2 + n^2 + 1$ yields infinitely many primes as m and n range over the positive integers. In 1996, J. Friedlander and H. Iwaniec (both dates unknown) showed that $m^2 + n^4$ contains infinitely many primes, and in 1999, D. R. Heath-Brown (dates unknown) proved the same for $m^3 + 2n^3$. The proofs used ideas from analysis and other areas and were considered remarkable achievements. Among other unsolved problems about primes are the following: Is every integer greater than 2 a sum of two primes as the evidence suggests? (This is the celebrated Goldbach conjecture, outstanding for 250 years.) Are there infinitely many primes of the form $2p + 1$, where p is a prime? Is there a prime between n^2 and $(n + 1)^2$? Is $\pi(x + y) \leq \pi(x) + \pi(y)$ for every x and y? Are there infinitely many twin primes? Mersenne primes? Fermat primes? When all is said and done, perhaps Euler's comments about the mysterious character of the primes are warranted.

Twentieth Century

FERMAT'S LAST THEOREM

The twentieth century will undoubtedly be regarded as a golden age in mathematics, both for the emergence of brilliant new ideas and for the solution of long-standing problems (the two are, of course, not unrelated). One of its major triumphs was Andrew Wiles's 1994 proof of Fermat's last theorem, outstanding for over 350 years. Various attempts to prove the theorem had been made during the preceding three centuries, the most important being Kummer's in the nineteenth century (see above). In the 1920s, Harry Schultz Vandiver (1882–1973) established FLT for all primes $p < 157$ (recall that about 70 years earlier, Kummer had reached $p < 100$), and in 1954, he extended the result, using the SWAC calculating machine, to $p < 2521$. Using modern computers and advanced theoretical mathematics, the theorem was established for $p < 125{,}000$ in 1973 and for $p < 4{,}000{,}000$ in 1993.

But no matter how powerful and sophisticated, computations clearly cannot establish FLT for all exponents p. A theoretical departure was needed, and it materialized in the 1980s. Whereas previous attempts to prove the theorem were mainly algebraic, relying on ideas of Kummer and others, this approach was geometric, having its roots (in a sense) in Diophantus's work. The crucial breakthrough came in 1985, when Gerhard Frey (b. 1944) related FLT to elliptic curves. (An elliptic curve is a plane curve given by an equation of the form $y^2 = x^3 + ax^2 + bx + c$, with a, b, and c integers or rational numbers. Elliptic curves had, in effect, been studied by Diophantus and Fermat and investigated intensively by Euler and Jacobi; the curve represented by Bachet's equation $y^2 = x^3 + k$ is an important example.)

The association of elliptic curves with FLT was, according to experts, a most surprising and innovative link. Specifically, if $a^p + b^p = c^p$ holds for nonzero integers a, b, c, the associated elliptic curve, now known as the *Frey curve*, is $y^2 = x(x - a^p)(x + b^p)$. Frey conjectured that if such a, b, c did, in fact, exist, that is, if FLT failed, the resulting elliptic curve would be "badly behaved": It would be a counterexample to the Taniyama-Shimura conjecture (TSC). Put positively, Frey conjectured that if the TSC holds, then FLT is true. The outline of a possible proof of FLT now emerged: (1) Prove Frey's conjecture, namely that the TSC implies FLT; and (2) prove the TSC.

The TSC was formulated by Yutaka Taniyama (1927–58) in 1955 and refined in the 1960s by his colleague Goro Shimura (b. 1930). It says that every elliptic curve is modular. The notion of modularity is technically difficult to define, but the following statement from Harvard mathematician Barry Mazur (b. 1937) gives a sense of the scope and depth of the TSC: "It was a wonderful conjecture, but to begin with it was ignored because it was so ahead of its time. On the one hand you have the elliptic world, and on the other the modular world. Both these branches of mathematics had been studied intensively but separately. Then along comes the Taniyama-Shimura conjecture, which is the grand surmise that there's a bridge between the two completely different worlds. Mathematicians love to build bridges."

Enter Ken Ribet (dates unknown) of the University of California at Berkeley. In 1986, he proved Frey's conjecture that TSC implies FLT. It was a big event. Wiles was ecstatic on hearing of the proof: "I knew that moment that the course of my life was changing because this meant that to prove Fermat's Last Theorem all I had to do was to prove the Taniyama-Shimura conjecture. It meant that my childhood dream was now a respectable thing to work on. I just knew that I could never let that go." Work he did on it—for the next seven years. As he relates it: "I made progress in the first few years. I developed a coherent strategy. Basically, I restricted myself to my work and my family. I don't think I ever stopped working on it. It was on my mind all the time."

In 1993, Wiles was convinced that he had a proof of FLT, and he presented it in a series of three talks at a conference in Cambridge, although he did not reveal the goal of his lectures until the very end. He concluded the third lecture with the words: "And this proves Fermat's Last Theorem. I think I'll stop here." Mazur described the event: "I've never seen such a glorious lecture, full of such wonderful ideas, with such dramatic tension, and what a buildup. There was only one possible punch line." Specifically, what Wiles did was prove the TSC for an important class of elliptic curves, the "semistable" elliptic curves. This sufficed to prove FLT, for Ribet had earlier proved that if such curves are modular, then FLT holds.

Wiles's proof was very deep and technically demanding. Ram Murty (b. 1928), an authority in the field, described it thus: "By the end of the day, it was clear to experts around the world that nearly all of the noble and grand ideas that number theory had evolved over the past three and a half centuries since the time of Fermat were ingredients in the proof." So, in a sense—without detracting from Wiles's great achievement—the proof was a grand collaborative effort of dozens of mathematicians over several centuries.

Wiles's lectures at Cambridge in June 1993 were, however, not to be the end of this 350-year odyssey. The proof was very long and complex and required validation by experts. Many errors were found; most were easily and quickly corrected. One error, however, could not be fixed. Wiles worked for several months, without success, on repairing it, and in January 1994 sought the help of Cambridge mathematician and former student Richard Taylor (b. 1929). On September 19, they found the "vital fix." In October, two papers proving FLT (totaling over 120 pages) were published, one by Wiles, the other by Taylor and Wiles.

Many tributes poured in following the publication of the proof. Here are two, from eminent number theorists M. R. Murty and John Coates (dates unknown), respectively (the latter was Wiles's doctoral adviser at Cambridge): "Fermat's Last Theorem deserves a special place in the history of civilization. By its simplicity it has tantalized amateurs and professionals alike, and with remarkable fecundity led to the development of many areas of mathematics such as algebraic geometry, and more recently the theory of elliptic curves and representation theory. It is truly fitting that the proof crowns an edifice composed of the greatest insights of modern mathematics" (Murty). "In mathematical terms, the final proof is the equivalent of splitting the atom or finding the structure of DNA. A proof of Fermat's Last Theorem is a great intellectual triumph, and one shouldn't lose sight of the fact that it has revolutionized number theory in one fell swoop"(Coates).

The last word belongs to Wiles: "I had this very rare privilege of being able to pursue in my adult life what had been my childhood dream. I know it's a rare privilege, but if you can tackle something in adult life that means that much to you, then it's more rewarding than anything imaginable. Having solved this problem, there's certainly a sense of loss, but at the same time there is this tremendous sense of freedom. I was so obsessed by this problem that for eight years I was thinking about it all the time—when I woke up in the morning to when I went to sleep at night. That's a long time to think about one thing. That particular odyssey is over. My mind is at rest."

Bibliography

Adams, William W., and Larry J. Goldstein. *Introduction to Number Theory.* Upper Saddle River, N.J.: Prentice Hall, 1976.

Bashmakova, Isabella. *Diophantus and Diophantine Equations.* Translated from the Russian by A. Shenitzer. Washington, D.C.: Mathematical Association of America, 1997.

Dickson, Leonard E. *History of the Theory of Numbers.* 3 vols. New York: Chelsea, 1966.

Edwards, Harold M. *Fermat's Last Theorem: A Genetic Introduction to Algebraic Number Theory.* New York: Springer-Verlag, 1977.

Goldman, Jay R. *The Queen of Mathematics: A Historically Motivated Guide to Number Theory.* Wellesley, Mass.: A. K. Peters, 1998.

"Great Internet Mersenne Prime Search." www.utm.edu/research/primes.

Hardy, Godfrey H., and E. M. Wright. *An Introduction to the Theory of Numbers.* Oxford: Clarendon Press, 1938.

Ireland, Kenneth, and Michael Rosen. *A Classical Introduction to Modern Number Theory.* New York: Springer-Verlag, 1972.

Ore, Oystein. *Number Theory and Its History.* New York: McGraw-Hill, 1948.

Scharlau, Winfried, and Hans Opolka. *From Fermat to Minkowski: Lectures on the Theory of Numbers and Its Historical Development.* New York: Springer-Verlag, 1985.

Weil, André. *Number Theory: An Approach through History.* Boston: Birkhäuser, 1984.

Oceanography

Anita McConnell

Oceanography is the science of the sea. Early studies were limited to the physical aspects of depth, temperature, salinity, and currents, but it now encompasses the chemistry of seawater and of the seabed, the structural geology of the ocean basins, the diversity of plant and animal life within the water, and the atmosphere above it.

Seven-tenths of the globe is covered by water. The greatest extent and volume is contained in the deep oceans, their floors some 3 miles (4.8 kilometers) beneath the surface. The remainder is in the shallower seas filling, depressions in the continents themselves. In earlier centuries, a few peoples migrated over the vastness of the Pacific Ocean, while others crossed the Atlantic and Indian Oceans, but there was no interest in what lay below the surface waters where fish and other products were taken. In the fifteenth century, larger sailing ships ventured from Europe to the Americas, Africa, and Asia, but again their interest was confined to the shallow waters where they could safely anchor, avoiding rocks, reefs, and other hazards. Philosophers speculated on what lay beneath the surface; sailors recounted tales of the mermaids and sea creatures they had seen, and cartographers decorated the empty spaces on their maps with fearsome monsters emerging from the waters.

With the rise of scientific reasoning in the eighteenth century, serious thought was given to the nature of Earth as a whole—what forces beneath the surface gave rise to volcanoes, where springs and rivers came from, and why the ocean was salty. In 1660, Robert Boyle (1627–91), a natural philosopher with wide-ranging interests, undertook experiments and questioned seamen and travelers to distant regions. Boyle touched on what we would now regard as marine science in his *Tracts about the Cosmicall Quality of Things* (1670), a series of essays on the temperature and depth of the sea, and his *Observations on the Saltness of the Sea* (1673). Evidence from the fiery outpourings from volcanoes suggested that heat from the interior of Earth would warm the seawater in its lower layers, yet pearl fishers and divers who salvaged goods from wrecked ships testified that the water became colder with depth.

Boyle reasoned that since the water became deeper the farther one went from the shore, the deepest parts must be those most distant from land. But there were many well-known midocean islands that contradicted this hypothesis. Boyle also considered why the water should be salty when the seas and oceans were fed by large freshwater rivers. He was, moreover, one of the first to appreciate the increase in pressure with depth. Boyle spent much of his life in Oxford and London and was an early member of the Royal Society, chartered in 1662. From the beginning, its members were looking for ways to increase knowledge of the sea for the benefit of seafarers. A list of topics was drawn up, and in 1666, the society appealed to mariners and passengers sailing with the trading companies to Africa, the Far East, and the Caribbean, asking them to make the necessary experiments.

The lines of inquiry covered the depth of the sea and the nature of the seabed; the salinity, taste, and color of the water; its possible corrosive or preservative properties; the tides; and the extent to which wind stirred up the water. There were at this period no instruments or apparatus that could have reliably answered these questions. The only ones known to the society had been mentioned in books but never put to the test. Consequently, the Royal Society's curator of experiments, Robert Hooke (1635–1703), set about devising sounders (for depth measurements) and a water sampler. Hooke never went to sea, merely conducting a few trials in inshore waters. Much of the failure of his apparatus to perform as expected was due to the wooden parts becoming waterlogged at depth.

At around the same time, England was concerned to hold the North African port of Tangier, and this necessitated frequent voyages between England and the western Mediterranean through the Strait of Gibraltar. The vessels on this route were handicapped by the strong irregular currents that flowed through the strait, and attempts were made to discover whether they were tidal and whether their speed and direction varied from one side to the other. But since ships could not then anchor in deep water, the investigators had no stable platform from which to measure flow. This was to be a long-standing problem, not resolved until the late nineteenth century.

Chemistry of Seawater and Light in the Sea

During the seventeenth and eighteenth centuries, the chemical investigation of seawater was part of a general interest in mineral waters, increasingly valued for their curative properties; it was also a factor in trials for new and reliable methods of rendering seawater potable on board ship. It was recognized that the sea was salty throughout its depth, but there were clear differences between oceanic water and that from confined seas, where salinity was likely to be increased by evaporation or diminished by river inflow. It was generally accepted that the salts in the sea had been washed out of the rocks and transported by rivers, but there was less agreement over the way in which water cycled, whether underground from ocean floor to springs and rivers, or through the atmosphere by evaporation and rainfall.

Before about 1800, there were three ways of finding the amount of dissolved matter in mineral water and seawater. The simplest was to compare its weight with that of fresh water by means of the instrument known as a **hydrometer** or areometer, the result being given as **specific gravity**. Alternatively, a given volume of water could be evaporated to dryness and the solid residue weighed. This could then be further analyzed by taking off those fractions that were soluble in alcohol or oil, or the salts could be precipitated by adding certain chemicals. These combined with particular salts and caused them to settle out, at which point they could be taken off and weighed.

* **Hydrometer.** An instrument that measures the density or specific gravity of liquids.

* **Specific gravity.** The ratio of the density of some material at a specified temperature to a standard reference material such as water.

By the time the French chemist Antoine-Laurent Lavoisier (1743–94) and the Swede Torbern Bergman (1735–84) turned their attention to the analysis of waters, advances in chemical analysis and an agreed nomenclature made it easier for scientists to compare and discuss their results. In fact, none of the chemists who examined seawater were particularly interested in marine science—seawater was simply the most complex of the mineral waters available for study.

In 1772, Lavoisier gave the first analysis of seawater in which the salts were itemized by both weight and type. His technique was evaporation followed by treatment with alcohol and oil. The principal constituents were (in modern terms) sodium chloride (common salt) and sodium and magnesium sulfates, with small quantities of calcium and magnesium chloride, the latter responsible for the bitter taste. (The bitter taste of seawater, not found in common salt, had previously been thought to derive from bitumen.) Lavoisier also identified earthy **calcareous** matter, probably derived from land, and gypsum.

* **Calcareous.** Containing or resembling calcium carbonate.

Bergman employed evaporation and precipitation, weighing the precipitated salts. In 1784, he included seawater among his analyses of mineral waters, recognizing its chemical complexity and mentioning the organic matter that was present in surface water. Both Lavoisier and Bergman tried to obtain samples from various seas and oceans and from beneath the surface. Contemporary methods of collecting subsurface water were extremely unreliable, however, and it is likely that all the samples analyzed came from close to the surface.

From 1818, when the Edinburgh chemist John Murray (1786–1851) proposed that seawater could be measured more easily and accurately by precipitation of the acids and bases in the sample rather than by removing the salts themselves, analytical chemists favored this technique. By this time, global exploration coupled with a renewed interest in marine science demonstrated in Britain, France, Russia, and the United States had led to the development of marine chemistry as a discipline in its own right. Samples were collected from all oceans and from high latitudes to the equator—and with the improvement in apparatus, from depths considerably below the surface. The further development of analytical chemistry had revealed the presence of other salts and indeed other elements, such as iodine and various metals.

The Danish chemist Johann Georg Forchhammer (1794–1865), in his influential 1865 treatise "On the Composition of Sea Water in Different Parts of the Ocean," introduced the term *salinity*, which he defined as the proportion of the totality of salts in the water. He identified 27 elements, several not previously recognized in seawater. His technique was to determine gravimetrically the chlorine, sulfuric acid, soda, potash, lime, and magnesia (those components that made up the various salts) and the numerous other elements having no influence on the salinity. Although confirming that the relative proportions of the different salts were everywhere the same, Forchammer was able to distinguish differences in the salinity of various water masses, such as contrasts between the Gulf Stream and its surroundings, and he predicted the existence of a subsurface Antarctic current before it was discovered by the *Discovery* voyages during the 1930s.

When the global circumnavigation of HMS *Challenger* was concluded in 1876, the 77 water samples collected on the voyage were given to the German chemist William Dittmar (1833–94), who employed a refinement on Forchhammer's method. He defined salinity as the total volume of salts in parts per thousand, his values falling between 35.342 and 36.833. Dittmar coined the word *oceanography* to describe the work performed by *Challenger*'s scientists. The results from other expeditions generally confirmed the work of Forchhammer and Dittmar and established chemical oceanography as a discipline in its own right.

Throughout the nineteenth century, however, problems remained in relation to the nature of salts and the way in which the various elements were combined and held in solution. These issues were not resolved until the Swedish chemist Svante Arrhenius (1859–1927) stated that salts dissolved in seawater broke up into dissociated molecules, now known as *ions*, each carrying an electric charge. The ions could not, however, be isolated, for when compelled to assume the solid state, they combined to form electrically neutral molecules.

Once the concept of proportionality of salts was accepted, it was sufficient to determine the chlorine in a sample and to apply a coefficient to ascertain the quantity of other salts. This coefficient was itself the subject of dispute, values varying between 1.806 and 1.829. From its early days, the International Council for the Exploration of the Sea (ICES) provided sealed glass vials of standard seawater for comparison. By the 1960s, with the benefit of new electronic apparatus, scientists returned to the conductivity method that had previously been found wanting. Conductivity salinometers had a typical accuracy of 0.003 percent and the advantage of giving a measurement on the spot, and they became the recognized method of determination.

Marine Science on Polar Explorations and Circumnavigations, 1750–1850

By 1750, ships engaged in whaling and sealing were sailing into the icy polar waters in Arctic and Antarctic regions. French and English government support for scientific maritime exploration was stimulated in the 1760s by the opportunities offered by the upcoming transit of Venus across the face of the Sun, due to occur in 1769; the timing of this event, if observed from two widely separated points, would establish the distance between the Sun and Earth. Although scientists in both countries were eager to obtain data from the unexplored regions of the world's oceans, the state of technical development at the time did not match their ambitions. There were as yet no self-registering thermometers that would show the temperature at the depth to which they had been lowered. The alternative was to insulate the thermometer from the warmer surface water, either by cladding it in nonconducting materials, such as wool, or by sending it down in a container fitted with flap valves, so as to capture colder water from below, insulating the thermometer until it was hauled up on deck. The most popular example of the latter apparatus was devised by the naturalist Stephen Hales (1677–1761) and first tried in 1751. It consisted of a wooden bucket with flap valves top and bottom, but Hales, and other seamen who employed versions of this gauge, failed to realize that the wooden staves would be deformed by pressure.

James Six (1731–93) was led by his interest in meteorology to devise the first practical thermometer to register maximum and minimum temperatures. He went on to construct some examples of such a thermometer for use at sea, describing and illustrating them in his book, *The Construction and Use of a Thermometer* (1794). Six's thermometer consisted of a tube folded into three limbs; alcohol filled a central reservoir and the upper portion of one limb, with the two alcohol sectors separated by a mercury sector filling the U-bend. In each outer limb, light steel-in-glass indexes floated on the mercury. As the temperature rose and fell, these indexes were transported on the mercury surface, and afterward left behind, to register against the scales the extremes reached. They were reset with the help of a magnet. This thermometer was sent down within a protective case, but even so, had several disadvantages: The indexes were often dislodged as the instrument was hauled in through the waves, and the tube itself was compressed by deep-sea pressure, falsifying the reading. This effect was not always appreciated and gave rise to a belief in some circles that the temperature at depth never fell below 40°F (4.4°C). In the 1860s, a pressure-protected version was produced by re-forming the tube into a simple U-shape and sealing the bulb into an outer glass sheath.

One of the first to combine experience and theory was William Scoresby, Jr. (1789–1857), of Whitby, England. Scoresby's success as a whaler was in part due to his careful study of the water and ice in the regions between Greenland and Canada. He spent several winters at the University of Edinburgh, where one of his professors had encouraged him to apply scientific methods to the study of conditions in the Arctic. Scoresby discovered that warmer water often underlay the cold surface layers, and this led him to speculate that part of the Gulf Stream might travel into the Arctic. He examined the microscopic marine plant and animal life under his microscope and realized that these organisms were the source of the red and green streaks found on some ice and of the luminescence seen on the water's surface. He also saw that whales were more plentiful in greenish water, where this microscopic life sustained a food chain on which they fed. Scoresby's two-volume *Account of the Arctic Regions* (1820) was a remarkable summary of his observations on all aspects of this region and was widely read and appreciated in other countries.

Science and Technology: Ice in the Sea

Mariners who ventured into polar regions were aware that sea ice melted into fresh water. This raised questions about its origin: whether it had broken off from Greenland glaciers, or had actually formed in the open sea and then been built up by falling snow. The subject was hotly debated during the latter years of the eighteenth century, spurred on by reports from the margins of Antarctica, and, as often happens, both interpretations were broadly correct.

When seawater freezes, its salts do not fit into the crystalline structure of the ice and are largely excluded, increasing the salinity of the surrounding water, which then sinks and is replaced by warmer, less dense water, thus maintaining the freezing process. Wind and waves break up and recombine the thin sheets, which may accumulate as corrugated pack ice many feet thick. Icebergs originate from glaciers in both polar regions. Broken pack ice and icebergs may be carried over vast distances by wind, and since much of their bulk lies beneath the water, they are also transported by currents.

In 1958, the nuclear-powered submarine USS *Nautilus* dived beneath the Arctic ice near Point Barrow, Alaska. Avoiding places where the ice almost touched the seabed, *Nautilus* found a deep uncharted channel enabling her to cross the pole and emerge near Svalbard (a group of islands north of Norway) some 36 hours later. In subsequent years, submarines were able to scan the base of the Arctic ice by means of upward-looking radar. This disclosed the thickness and make-up of the pack, completing the picture of the overall upper surface derived from aerial photography and later by satellite.

In 1817, Scoresby reported that the sea ice had retreated, offering the possibility of searching once more for the hoped-for Northwest Passage to the Pacific. A naval expedition under Captain John Ross (1777–1856) sailed north in 1818 with instructions to record tides and currents and to sound and collect bottom samples with the apparatus provided in the hope that this would assist their exploration. When the ships entered an ice-covered inlet, they were to sample the water; if this was found to be fresh, the inlet was probably fed by a river and could be ignored, but if it was saline, the inlet might lead to the passage. The expedition returned within months, having failed to find the passage. A second expedition under Commander David Buchan (1780–c. 1835), which had sailed due north, soon encountered impenetrable ice. Despite these setbacks and the unreliability of the apparatus supplied, both expeditions made numerous useful measurements of surface and subsurface temperatures and specific gravity. Buchan was more successful in obtaining subsurface water samples. Ross must have protected his thermometers from the effects of pressure, for their readings agreed with measurements made on mud samples retrieved with the simple grab that had been made on board.

Two later expeditions under Captain William Edward Parry (1790–1855), carrying the same instructions and apparatus, made further progress into the ice-bound northern regions, overwintering in 1819–20, 1821–22, and 1824–25. Parry acknowledged the difficulty of trying to measure deep currents, which often ran contrary to those visible at the surface, and he also recognized the problems of sounding when it was impossible to detect the moment when the grab hit the seabed, so that the paid-out length of line exceeded the true depth. In these stormy regions, calm days were rare; even then, the slow process of sounding put the ships at great risk if the wind increased before the line was recovered. Parry was also frustrated by the poor performance of his Six thermometers. Lowered into freezing water, the glass rapidly contracted, causing the fluid to bypass the index and ruining the reading. His water bottles, however, which were small and of heavy iron, recovered the sample with little change of temperature. While Parry was probing from the eastern end, Lt. Frederick William Beechey (1796–1856) took HMS *Blossom* to the Bering Strait in 1825–28 to search for the Pacific end of the passage. Beechey took numerous readings of surface and subsurface water and made several deep temperature soundings, some down to the considerable depth of 850 fathoms (5100 feet or 1500 meters). He failed, however, to protect his thermometers from pressure, and his values are now considered far too high.

Other important high-latitude voyages in the 1830s also combined marine science with exploration. The third Russian circumnavigation, under Otto von Kotzebue (1787–1846), took an insulated water sampler that yielded reliable deep-sea temperatures and permitted the measurement of temperatures to 36°F (2.2°C), far lower than hitherto recorded. His work showed for the first time that the point of maximum density of salt water differed from that of fresh water. More data came from French expeditions under Captain Abel du Petit-Thouars (1793–1864) in *Vénus*, 1836–39, and J.-S.-C. Dumont d'Urville (1790–1842) in *Astrolabe* and *Zelée*, 1837–40; and from the American Exploring Expedition, 1838–42, under Charles Wilkes (1798–1877).

Antarctic marine research began in earnest with the voyage of James Clark Ross (1800–62), nephew of John Ross. His expedition, in HMSs *Erebus* and *Terror* (1839–42), was part of an international project to measure Earth's magnetic field. Ross, who had earlier stood at what he believed to be the north magnetic pole, hoped to reach the south magnetic pole, located somewhere on Antarctica. Again, his instructions were to examine tides, currents, water depth, and bottom sediments (see Figure 1). Ross was not, however, alerted to the need to protect his thermometers against pressure, and although he managed to get readings from depths down to 1200 fathoms (7200 feet or 2118 meters), he never recorded temperatures below 39.5°F (4.2°C).

Nansen's Voyage in *Fram*

Fridtjof Nansen (1861–1930) had already traveled across the Greenland ice cap before he decided to explore the North Atlantic and Arctic Oceans. Nansen was aware of the American voyage of George Washington De Long (1844–81) on the *Jeanette* in 1879. DeLong had intended to sail through the Bering Strait to Wrangell Island and from there travel over the ice to the North Pole. But *Jeanette* became stuck fast in the ice and over the next two years drifted north of Wrangell Island to sink off the New Siberian Islands. DeLong and most of the crew perished, but in 1884, articles from the *Jeanette* were found in the pack ice off the eastern coast of Greenland, having presumably drifted over the pole.

Figure 1. Deep-sea soundings, c. 1840; typical of those conducted by James Clark Ross in his expedition to Antarctica. From J. C. Ross, Voyage ... in the Southern and Antarctic Regions *(London, 1847). Courtesy of the Thomas Fisher Rare Book Library, University of Toronto.*

This gave Nansen the idea of building a ship that could be safely frozen into the pack, which would carry it toward the pole. *Fram* (*Forward*) was built with a hull shaped to resist being crushed when held in the ice. She sailed from Oslo on June 24, 1893, and was beset by ice near the New Siberian Islands. Two years passed before, in November 1895, *Fram* reached its northernmost point, 244 miles (394 kilometers) from the pole, when Nansen and a few colleagues set out to reach it on foot. *Fram* drifted on, to break free on August 13, 1896, near Svalbard (a group of islands north of Norway), having been transported 1028 miles (1658 kilometers). This voyage showed that there was no polar continent; indeed, some of the soundings (made through holes cut in the ice) reached nearly 10,000 feet (3000 meters). Temperature measurements disclosed a body of warmer, more saline water between 500 and 3000 feet (150 and 900 meters), which Nansen identified as Atlantic water that had sunk beneath the colder but fresher Arctic water. The permanent pack ice was clearly not of glacial origin but had formed directly on the ocean surface.

Matthew Fontaine Maury: His Oceanography and Marine Meteorology

Matthew Fontaine Maury (1806–73) occupies a large but controversial place in U.S. maritime history. As head of the Washington Naval Observatory from 1842 to 1861, he oversaw improvements in deep-sea sounding techniques and gathered a vast quantity of data on sea temperatures, winds, and currents across the oceans. But his well-known *The Physical Geography of the Sea* (1855), although influential in its day, has always been judged by the scientific community to be a flawed and misleading work.

Maury joined the U.S. Navy as a midshipman in 1828 and spent the next nine years at sea. At the Naval Observatory, he began collecting data on winds and weather and their relation to surface currents, drawing his data from old logbooks and instructing commanders of naval and merchant vessels to gather such information and return their observations to him. His *Wind and Current Chart of the North Atlantic* (1847) was followed by his *Abstract Log for the Use of American Navigators* (1848), reissued in subsequent years as *Explanations and Sailing Directions to Accompany the Wind and Current Charts*. Maury's work coincided with that of Britain's Admiral Robert Fitzroy (1805–65), who was issuing standard barometers and thermometers to commanders willing to undertake regular observations at sea. Maury bought a number of barometers made to the same pattern and had them replicated by U.S. manufacturers for widespread issue to U.S. ships. He was the most active participant in the international maritime meteorology convention held in Brussels in 1853, gathering honors from many European countries.

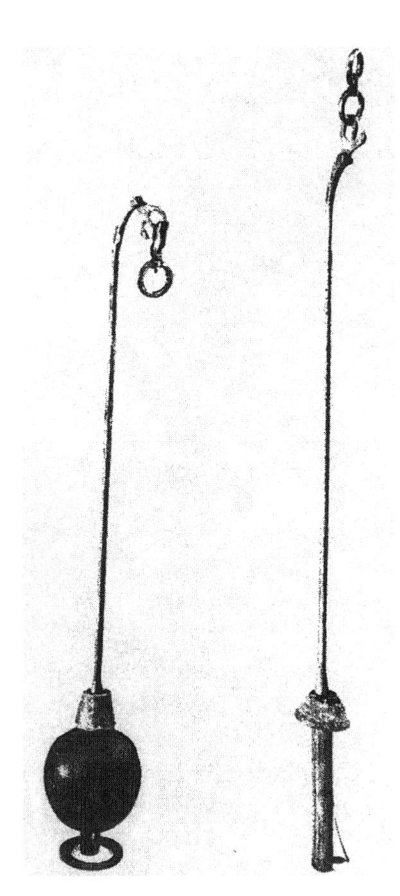

Figure 2. Brooke's detaching-weight sounder, with the weight slipped from the harness (left) and the valved sampler tube attached to the base of the rod (right). Courtesy of the Science Museum, London.

Maury's role in the development of deep-sea sounding techniques during the 1850s led to the invention in 1852 of the detaching-weight sounder by John Mercer Brooke (1826–1906) of the U.S. Naval Observatory (see Figure 2). Brooke's sounder consisted of a sounding rod weighted by a hollow cannon ball carried in a harness attached to a hinged arm at the top of the rod; when the apparatus struck the seabed, the arm dropped, slipping the harness and shot. The light tube and its sample were then hauled up. Versions of this sounder proved their worth in the charting of a track across the North Atlantic in preparation for the laying of a submarine cable in 1867. His interpretation of the sparse data led him famously to assert the existence of a "telegraph plateau" across which the cable might safely be laid, and he issued a contoured chart of the North Atlantic basin (see Figure 3). Given the small number of reliable soundings, this chart owed more to imagination than to fact.

Maury's publications allowed commanders to plot courses that considerably reduced their sailing time (that from New York to San Francisco, for example, was reduced from 180 days to about 130 days). But his fellow scientists scorned his claims that "the springs in the oceans which supply the sources of all the great rivers of the northern hemisphere are, for the most part, to be found where the S.E. trades blow, in the Pacific and Indian Oceans," and similarly, his declaration that "magnetism is a powerful agent in giving direction to the circulation of the atmosphere; and the question raised, if it be not concerned in the currents of the ocean also." These and other unprovable conjectures appeared throughout his writings. He ignored established physical knowledge and attributed the adverse judgments of his contemporaries to personal envy and malice. Among those most hostile to Maury's outpourings were

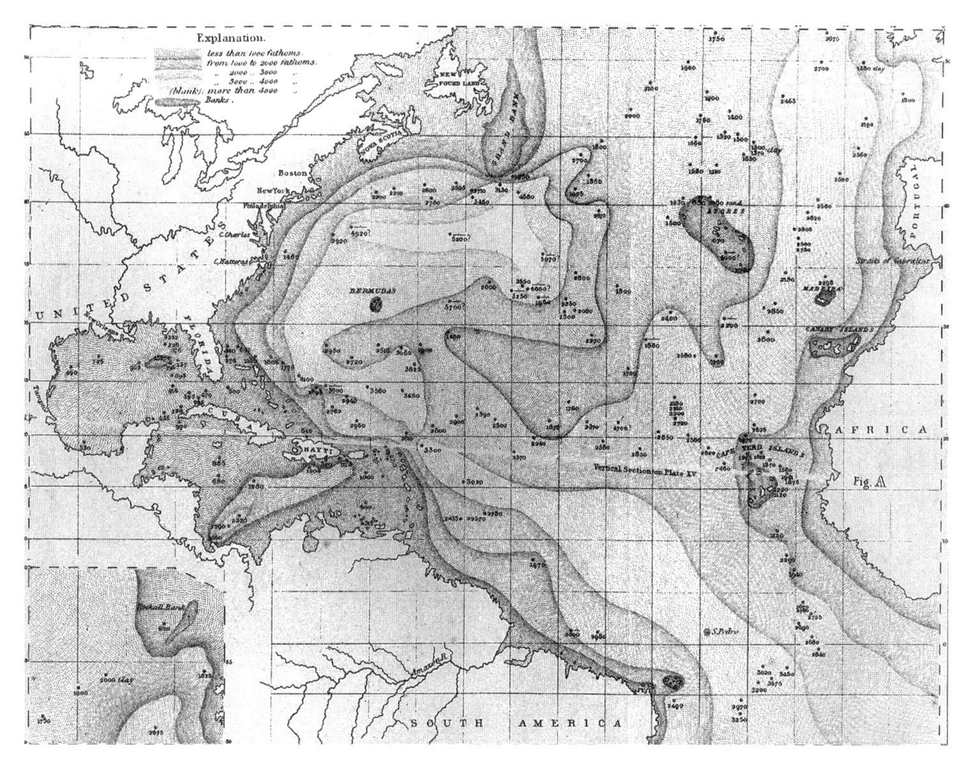

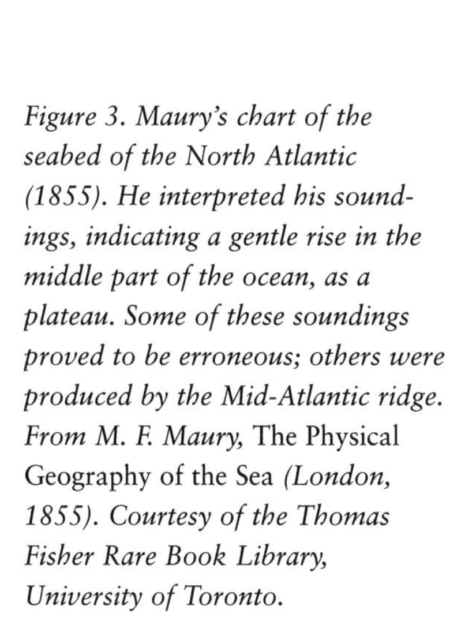

Figure 3. Maury's chart of the seabed of the North Atlantic (1855). He interpreted his soundings, indicating a gentle rise in the middle part of the ocean, as a plateau. Some of these soundings proved to be erroneous; others were produced by the Mid-Atlantic ridge. From M. F. Maury, The Physical Geography of the Sea *(London, 1855). Courtesy of the Thomas Fisher Rare Book Library, University of Toronto.*

Joseph Henry (1797–1878), secretary of the Smithsonian Institution, and Alexander Dallas Bache (1806–67), superintendent of the Coast Survey, who was directing detailed observations of the Gulf Stream.

Most of the ideas set out in *The Physical Geography of the Sea* had already been aired in journals laid before a wider readership and praised by literary and popular reviewers. The scientific community judged the book more severely. In revising his book, Maury generally paid little attention to his critics, although he vehemently defended his belief that oceanic circulation was driven principally by differences in salinity (ignoring the part played by winds), that saline water sank at the equator and fresher water moved toward it at the surface. Yet at the same time, he considered temperature to be the driving force of the Gulf Stream, which was an example of warm water moving toward the pole at the surface. He refused to admit wind as a driving force of currents, even though the correlation between the trade winds and the equatorial currents in the Atlantic had been known for nearly four centuries and indeed had been plotted on Maury's own charts. Maury also held to a wholly improbable concept of the circulation of the atmosphere. Among those rejecting this fantasy was William Ferrel (1817–91), who bent his efforts to the problems of motion in the air and oceans on a rotating Earth.

Upon the outbreak of the Civil War, Maury joined the Confederate States Navy. He was sent to England in 1862 to enlist aid for the South. He returned to the United States in 1868 and became professor of meteorology at the Virginia Military Institute, Lexington, where he died four years later.

Submarine Cables and the Benefits for Oceanography

Telegraphy, the overland transmission of signals by electric currents along a copper wire, became practical in the 1840s. Cables consisting of a copper core insulated by gutta-percha (a vegetable material resembling rubber) and armored by twisted iron strands were laid on land and under rivers, then across the English Channel and the

Baltic Sounds. Soon entrepreneurs were raising capital to lay cables across the Atlantic, linking the commercial centers of Europe and North America. The question then arose of how many miles of cable to manufacture and how strong it needed to be. Hitherto, few soundings had been made in the open Atlantic and the nature of the ocean floor was unknown. The first task, therefore, was to obtain closely spaced soundings, with samples of the seabed on which the cable would rest, and to sample the water at depth to see if chemical corrosion might threaten the life of the iron armoring on the cable.

By midcentury, ships on distant voyages were being issued with sounder samplers. Working from small boats (to avoid the strain on the rope exerted by the pitching of a large ship), crews deployed sounders in a variety of forms on hemp line 1 or 2 inches (2.5 to 5 centimeters) in diameter, each provided with some means of recovering a small sample of the seabed sediment. The considerable resistance of the thick rope made hauling in a lengthy business, and each sounding could take several hours, while submarine currents could sweep the line out of the vertical, without those at the surface realizing what was happening. The results of isolated operations that gave unexpected values were therefore suspect, and only a line of closely spaced soundings could give a reliable profile of the seabed. Another vital detail was to plot accurately the location of each sounding.

Various ships' captains and manufacturers had sought to overcome these difficulties. The brass way-wiser sounder, in which a small fan rotated as the sounder descended, its revolutions registering on a dial that could be read when it was hauled in, worked well in medium depths but was damaged by compression in deep water. Copper and iron wires were tried, but these could not then be manufactured in long lengths and always broke at the joints. Strong silk or twine passed through the water with less friction but was not strong enough to recover the sounder, and there was therefore no proof that bottom had been reached.

The first transatlantic cable surveys were made in 1856 by Lt. Berryman of the U.S. Navy. He sounded out and back using a thin line, by watching the speed at which the line ran out—the line was cut when the sounder was judged to have hit the seabed. Berryman merely located his soundings with reference to his midday position; not surprisingly, his two profiles differed considerably. Lt. Joseph Dayman of the Royal Navy made a third survey in 1857, sounding every 50 miles (80 kilometers). Dayman used several types of apparatus, including a sturdy version of Brooke's detaching-weight sounder. The Brooke sounder was deployed in various forms on surveys in the Mediterranean and Red Sea in connection with cables to India and on a northern transatlantic route via Iceland and Newfoundland.

By 1870, cable laying was at its peak; the cable companies were operating their own specialized steamships, which could hold station more easily than sailing ships and wind their gear in on steam-driven donkey engines. Yet the majority of cables failed because they were laid with inadequate knowledge of the seabed. Rough ground was discovered only when broken cables were recovered for repair; unsupported cables stretched until their insulation gave way; and strong tidal currents frayed the armoring wires. It became clear that the ocean basins had a topography as varied as the land, and that for commercial reasons, the cable companies needed to make rapid and closely spaced soundings before their cables were laid, if regular and reliable operation was to follow. This was achieved with the wire-sounding machine, invented by Sir William Thomson (later known as Lord Kelvin, 1824–1907), professor at Glasgow University.

Kelvin was financially involved in a number of telegraph companies and had devised many sorts of telegraph apparatus. He also owned a private steam yacht, and it was on board this vessel in 1872 that he built and tested the first efficient sounding machine, with wire replacing the conventional hemp rope. The paying-out, measuring, and winding-in reels; the accumulator, or spring, which eased the strain on the line as the ship pitched or rolled; and the steam-powered winch were now arranged within a frame, comprising a self-contained powered appliance. U.S. survey vessels adopted the Kelvin sounder, which underwent considerable improvement at the hands of Lt. Charles Dwight Sigsbee (1845–1923). Prince Albert I of Monaco (1848–1922) also adapted it for his own use, being the first to replace single-strand wire with fine wire rope. The compact and efficient Lucas sounding machine of 1887, designed by a cable company engineer, Francis Robert Lucas (d. 1931), enabled ships to sound within a few minutes, then steam to the next station as the wire was reeled in.

Mapping the Seabed

As cables traversed inland seas and the Atlantic and Indian Oceans and skirted around the North and South Pacific, there were great benefits for physical oceanography. Detailed profiles, although they spanned only centers of commercial importance, revealed the bottom topography and its sediments. When a faulty or broken cable was detected in the open ocean, the repair ship dragged backward and forward over the area where the break was believed to lie until the two broken ends were found, grappled, and brought to the surface for repair. These grappling charts produced a three-dimensional picture of the seabed. Similar detailed charts were prepared for areas where many cables ran up the continental shelf into a single terminal.

The cable electricians customarily monitored the speed of signal transmission, and as this was dependent on temperature, their logs provided continuous records of temperature at the seabed. Although the various companies were in commercial competition, much of this oceanographic information was made freely available. Some companies invited scientists on board to undertake their own research. Thus John Young Buchanan (1844–1925), the chemist who had been on board HMS *Challenger*, made detailed studies of small regions of the seabed. Some of his findings were of value to the cable company; others, such as his work off the west coast of Africa, were in areas where there were no plans to lay cables. Yet here Buchanan discovered the structure of underwater canyons, which cut through the flanks of continental shelves at the mouths of large rivers.

The first world map of the ocean floor was proposed in 1899; funded by Prince Albert, the 24 sheets of the General Bathymetric Chart of the Oceans (GEBCO) came off the press in 1905. This chart was based on individual soundings. During both world wars, marine science continued apace, but under cover of military secrecy. *Sonar* apparatus developed during World War I afterward revealed the true ruggedness of the ocean floor and led to the discovery of ocean ridges other than those previously recognized from Atlantic soundings. Radar, developed during World War II, was adopted for scanning wide swathes of the seabed, giving a view of the ocean's structural geology. Studies of the velocity of sound through water, initiated for submarine detection, led to a greater understanding of the complexity of temperature and salinity distribution within the water column and the implications for the global circulation. Since the 1960s, side-scan sonar devices, which gather images of wide swathes of the seabed, have provided fine detail for certain areas, and submersibles have photographed a very few parts of the ocean ridges and abyssal troughs. Notwithstanding these advances, the oceans are so vast in extent that they remain largely unexplored.

Later Scientific Circumnavigations

By the middle years of the nineteenth century, ships of many nations were contributing toward a real science of the sea. As apparatus became more reliable, and theories of a global circulation of water took shape, Germany, Sweden, and the United States began to formulate plans for major scientific voyages. The Austro-Hungarian government (whose territory then extended to the Adriatic) sent the *Novara* around the world in 1857–59 and published the results in 24 volumes in 1862. In Britain, the government and the Royal Society planned a voyage that would visit several oceans, undertake physical and chemical measurements throughout the water column, and measure tides and currents, meteorology, and terrestrial magnetism. A steam-assisted corvette, HMS *Challenger*, was fitted with laboratories and storerooms to hold collected specimens, as well as the many miles of hemp rope for sounding. For the first time, there was a civilian scientific staff, under the zoologist Professor Charles Wyville Thomson (1830–82), assisted by the chemist John Young Buchanan and naturalists Henry Nottidge Moseley (1844–91), John Murray (1841–1914), and Rudolph von Willemöes-Suhm (1847–75), who died during the voyage. The ship sailed in December 1872, returning in May 1876, having traversed every ocean except the Arctic.

Challenger took some newly invented items of apparatus, including a Kelvin sounding machine, but the crew were reluctant to use this, and all *Challenger*'s soundings were made with hemp rope, which undoubtedly slowed down the proceedings. Buchanan improved the reliability of water bottles for collecting samples at the seabed and in midwater; the midwater bottle, known as the *Challenger stopcock water bottle*, was widely used thereafter. Pressure-protected Miller-Casella thermometers (versions of the Six thermometer) were used for most of the observations but gave anomalous results in the Antarctic, where cold surface water overlay warmer levels. Current measurements were restricted to shallow levels. Two new devices were sent out to meet the ship: the heavy Baillie sounder, invented in 1873, retrieved seabed samples in deep water, and the reversing thermometer, devised and manufactured by Negretti and Zambra in 1874, was able to register temperatures at a preselected depth, irrespective of the surrounding temperatures (see Figure 4). The reversing technique was subsequently adopted generally as a reliable method of operating thermometers and water bottles at depth.

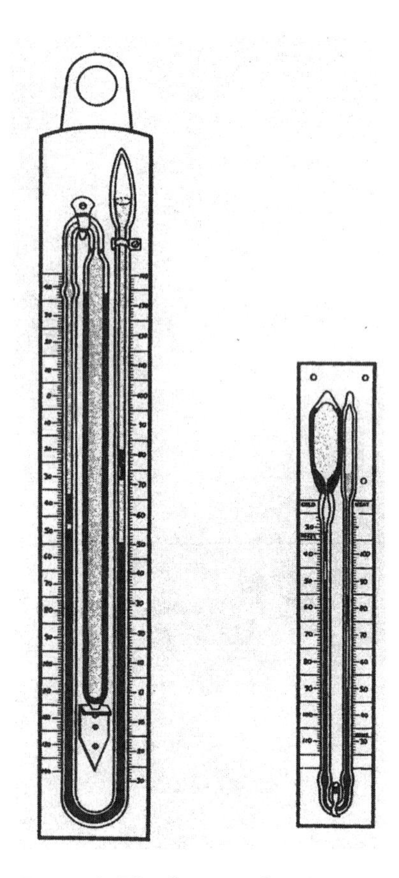

Figure 4. The deep-sea thermometer, designed by the firm of Negretti and Zambra (on the left), which negated the effect of pressure by enclosing the bulb in an outer jacket of glass. This device was soon supplanted by the two-limb version designed by Miller and Casella (on the right). Courtesy of the Science Museum, London.

Challenger's naturalists made regular sweeps with surface and midwater nets and dredged across the bottom, bringing to light a host of new creatures large and small, from the surface to the abyssal depths. To their surprise, the deep ocean floor was masked with fine clay and peppered with manganese nodules, another discovery new to science. Despite working with antiquated and somewhat imprecise apparatus, with observing stations often hundreds of miles apart, *Challenger*'s voyage set boundaries to the physical parameters of the ocean. But so much material was collected, and so many measurements made, that publication of the reports was not completed until 1895.

In 1964, Georg Wüst (b. 1890) classified the major deep-sea expeditions into four periods: the eras of exploration (1873–1914), of national systematic and dynamic ocean surveys (1925–40), of new marine geological, geophysical, biological, and physical methods (1947–56), and of international cooperation (post-1957). He identified a considerable political motivation in the *Challenger* voyage, the German circumnavigations in *Gazelle* (1874–76) and *Planet* (1906–07), the Antarctic voyage of the *Deutschland* in 1911–12, and the Dutch *Siboga* expedition (1899–1900). This is not to deny their scientific value, which was enhanced by the reasonably prompt publication of the results. Between the two world wars, there was a shift toward regular intensive research in home waters or around colonial territories; thus the Italians worked in the Mediterranean, the Dutch in the Netherlands East Indies, RRS *Discovery* circled Antarctica on behalf of Britain's whaling industry, John Murray took the Egyptian research ship *Mabahiss* to the Indian Ocean in 1933–34, and the German *Meteor* (1925–27) showed the flag in the South Atlantic. Investigations were now directed toward fisheries research, with its need to determine the fine seasonal and annual variations in water quality, and led to the development of vastly improved thermometers and water bottles.

The period of international cooperation has brought scientists of many nations together, either on a single ship or sharing research projects between nations. Observations are made with apparatus deployed from the ship or from free-floating or buoyed instruments that transmit data when interrogated by satellites passing overhead. Satellites keep a constant watch on the sea, measuring surface temperature, wave heights, and surface topography, and assess the quantity of biological life from discoloration of the water.

Oceanography for Fisheries Research

Until the late nineteenth century, the saying that there were "more fish in the sea than ever came out of it" held broadly true. But as sail gave way to steam and fishermen trawled more frequently across the spawning grounds, fewer young stock survived to maturity, and the harvest of the sea began visibly to decline. Yet individual governments were unable to do much about this, for the fish were no respecters of national boundaries. Species that spawned in the waters of one country often migrated to a considerable distance to feed and could sustain fisheries in both zones.

The base of the marine food chain was the microscopic floating plant and animal life known as *plankton*. Investigations into the chemical interactions between seawater and the plankton that led to its localized abundance and the reason for its annual "bloom" were taken up by Victor Hensen (1835–1924) of the Kiel Commission laboratory, established in 1870. To this end, annual cruises were undertaken and nets devised to provide measures of the abundance per unit volume of water. The word *plankton* was coined for Hensen's important monograph *Ueber die Bestimmung des Planktons. . .* (1887). The Plankton Expedition in *National* from July to November 1889, which undertook investigations in the North and South Atlantic, showed that plankton was most plentiful in cold northern regions, refuting the belief that tropical seas were the oceanic equivalent of the productive tropical rain forests. Under Karl Brandt (1854–1931), continuing studies showed the importance of the nitrogen cycle in determining the time and location of plankton abundance.

Other government laboratories were soon established: the Commission for Scientific Research in German Seas (1871); the United States Fish Commission (1872) with its marine laboratory at Woods Hole, Massachusetts (1885); the Stazione Zoologica founded by the German Anton Dohrn (1840–1909) at Naples (1873), which hosted scientists from many nations; and the Marine Biological Association at Plymouth, England (1888). Although these organizations sought to make their findings available by rapid publication, the first truly international body was the International Council for the Exploration of the Sea (ICES), founded in 1902 and ultimately encompassing the Atlantic nations from Portugal to Scandinavia, including Canada and the United States.

Two small states whose populations depended largely on fishing were fortunate in having sovereigns with a passionate interest in marine science; toward the close of the nineteenth century, Prince Albert of Monaco and King Carlos of

Portugal (1863–1908) both undertook research in their own ships with the aim of locating new areas for fishing. From 1884, Prince Albert commanded annual cruises involving meticulous investigation of the physical and biological environment of his chosen area, which over the years covered much of the North Atlantic. His engineers improved several of the physical instruments in use elsewhere. Nets and traps that could collect all forms of life, from the surface to the seabed at depths greater than 20,000 feet (6000 meters), were devised. These specimens were brought back for examination, and the reports of each cruise published in full. King Carlos's first cruise was in 1897; working around the Portuguese coast, his program was modeled on that of his friend Prince Albert but was cut short by his assassination in 1908.

The North Sea nations were concerned principally with cod and herring, both of which spawned off the Norwegian coast before moving to more distant feeding grounds. These species sustained important onshore industries, as they were salted or dried for export into Europe, and poor catches brought much hardship to the coastal populations. By the closing years of the nineteenth century, marine scientists in the Scandinavian countries, Germany, and Scotland were exchanging information gathered on their annual cruises.

ICES established a central laboratory in Christiana (now Oslo) in Norway to gather basic physical and chemical data. Research focused on the migration routes of cod and herring, relating this to water temperatures and identifying where to prohibit trawling over the seabed where flatfish such as plaice, sole, and haddock laid their eggs. The director was Fridtjof Nansen, famous for his polar explorations. To ensure that the data collected were truly comparable, water bottles, thermometers, and current meters were designed and manufactured locally for ships taking part in the programs. The insulated water bottle, devised some years earlier by Otto Pettersson (1848–1941), brought new accuracy to deep-sea temperature measurements. The insulation took the form of concentric cylinders of brass and celluloid, which prevented heat transmission. An ingenious device allowed it to descend open; at the required depth, the lid closed and locked. A long-stem thermometer passing through the lid allowed the temperature to be read as soon as the bottle was hauled up, without needing to open it. ICES also developed reversing bottles paired with thermometers, so that the temperature was registered at the instant of reversal. The cruise of the Norwegian fisheries research vessel *Michael Sars* in 1910 under Johan Hjort (1869–1948), director of fisheries, extended from Europe to Newfoundland and picked up interesting variations in the depth of maximum plankton abundance. A narrative of the cruise, which has since become a classic text, was soon available: John Murray and Johan Hjort, *The Depths of the Ocean: A General Account of the Modern Science of Oceanography Based Largely on the Scientific Researches of the Norwegian Steamer* Michael Sars *in the North Atlantic* (1912).

American scientists were more dependent on private benefactors. The U.S. Fish Commission under Spencer Fullerton Baird (1823–87) had its own purpose-built steamer, *Albatross*, launched in 1882. After Baird's death, the vessel was refitted for Alexander Agassiz (1835–1910), who had taken over from his father as director of the Museum of Comparative Zoology at Harvard. Agassiz had made a fortune from copper mines, much of which he spent on oceanographic expeditions and the publication of their results. He was the first to employ steel rope for deep-sea dredging, and the apparatus he devised for use on USS *Blake*, and later on *Albatross*, became known as the *Agassiz* or *Blake dredge*. After his death, other zoologists and philanthropists were inspired by Scandinavian methods to follow similar programs of investigation. The Fish Commission also operated small vessels on the Atlantic

Science and Society: U.S. Government Support for Coastal Research

The federal government acknowledged responsibility for a survey of the sea coast early in U.S. history, but didn't formally institute coastal research until over 30 years after the country's birth. The U.S. Coast Survey was enacted by Congress in 1807, with the immigrant Swiss surveyor Ferdinand Rudolph Hassler (1770–1843) as superintendent. Instruments were procured, but although the War of 1812 brought home the need for accurate charts, little progress was made and the survey was suspended in 1818. In 1831, the U.S. Navy established the Depot of Charts and Instruments and began producing nautical charts. The Coast Survey was reestablished in 1832, again under Hassler. Coasts and harbors were charted, tide gauges installed, and surveys extended to the Gulf Stream. Under the superintendence of Alexander Dallas Bache, marine investigations extended to the Pacific coast. In 1878, the organization transferred to the Commerce Department as the U.S. Coast and Geodetic Survey.

shore, one of its programs being the Gulf of Maine investigations (1912–24) under Henry Bryant Bigelow (1879–1967).

On the Pacific coast, the little Marine Biological Association of San Diego was transformed when in 1903 newspaper mogul E. W. Scripps founded the Scripps Institution of Biological Research, the oldest and best endowed of its kind in the United States. Its research vessel launched in 1907 was named the *Alexander Agassiz.* On the Atlantic coast, Woods Hole Oceanographic Institution, founded in 1930 on the site of a small U.S. Fish Commission laboratory, benefited from plans put forward by the National Academy of Sciences and made possible by generous funding from the Rockefeller Foundation. As first director at Woods Hole, Bigelow was able to procure the *Atlantis*, a research ship capable of working on the high seas in all weathers. The program was varied, but annual cruises were made to explore the Gulf Stream and the physical characteristics of the western North Atlantic, and continued Bigelow's earlier work in the Gulf of Maine, concentrating on the Georges Bank, a rich fishery where environmental conditions were hardly known.

Ocean Currents

Currents in the open ocean may take the form of narrow streams moving faster than their surroundings. The Gulf Stream, Kuroshio Current, and Equatorial Current are among the major features of this kind. To restore the balance of water, countercurrents may flow beneath these streams. In shallow water, slow or variable currents may result from surface winds, differences in temperature or salinity, or the ebb and flow of tides.

From the fifteenth century, Portuguese explorers heading for the southern tip of Africa knew of the trade winds and surface currents that compelled them to swing westward to the Brazilian coast before crossing back to the African shore. By 1519, seamen had learned to sail to America with the Equatorial Current and back along the Gulf Stream as far as Cape Hatteras before turning eastward to Iberian ports. On venturing into Indian waters, they learned from Arabian pilots the regime of the monsoon winds and currents that controlled the times of year when it was possible to cross the Arabian Gulf. However, although such permanent and seasonal currents were recognized in certain regions of the oceans, their true direction and velocity could not be measured from a ship that was itself at the mercy of combined wind and current. Only in the late nineteenth century, when it became possible to anchor in deep water, thus providing a stable platform for such measurements, could scientists seek to explain the origins of currents and the global circulation of the water itself.

James Rennell (1742–1830), surveyor for the East India Company, was among the first to chart currents and consider their causes and effects. Rennell retired in 1777 and on his way home from India encountered the peculiar patterns of ocean currents off the coast of southern Africa. Over the remaining 50 years of his life, which he devoted to a variety of geographical investigations, Rennell studied the intermittent currents off the southwestern coast of England, which often carried British ships unawares toward the dangerous Cornish coast. He developed in detail, albeit imperfectly, methods of extracting information from ships' journals. His ideas were published posthumously as *An Investigation of the Currents of the Atlantic Ocean and Those Which Prevail between the Indian Ocean and the Atlantic* (1832).

The Gulf Stream, which emerges from the Gulf of Mexico through the Florida Straits and snakes north parallel to the American shoreline before turning east to wash up against Scandinavia, was recognized from the earliest times because it carried perceptibly warmer water into colder regions and because whales were known to congregate on its cold western flank. Benjamin Franklin (1706–90), best known as a natural philosopher and American statesman, became postmaster general in 1775. He was concerned because mail packets took two weeks longer to cross the Atlantic than the merchant ships, whose captains laid course to avoid sailing against the current. Franklin attributed the Gulf Stream to the effect of the trade winds, which piled up the water on the American coast so that it ran downhill in the form of the Gulf Stream. In this he was wrong, but he pioneered the use of the thermometer to define the boundaries of the flow and attempted to collect water from depths down to 100 feet (30 meters). Captain William Strickland's (dates unknown) temperature investigations at the turn of the century showed that one arm of the Gulf Stream washing the northern coast of Europe was responsible for the milder climate of that region compared with the equivalent latitudes on the American side. By the early nineteenth century, the transitory nature of the Gulf Stream had been appreciated. Unlike the steady equatorial currents, it was seen to consist of an undulating flow, which from time to time cast off giant eddies enclosing cold patches in their centres. A deeper countercurrent was discovered by deploying the neutrally buoyant floats first devised by the English oceanographer John Swallow (1923–94). These floats traveled vast distances within the body of the flow, transmitting their positions by radio to a mother ship.

Many captains on exploring voyages were instructed to cast overboard drift bottles or floats containing messages asking finders to report when and where they had come ashore. A "Bottle Chart" depicting their tracks was published in the British *Nautical Magazine* in 1856. Prince Albert of Monaco made extensive use of this method. Beginning in 1880, he released a great number of barrels, bottles, and specially constructed floats from the Azores to Newfoundland, showing that the Gulf Stream divided as it turned east, one branch heading toward Norway, the other toward France. The U.S. Coast Survey extended this idea by casting weighted drifters and marked seashells over the inshore seabed of the northeastern United States to discover the residual bottom currents as distinct from the daily ebb and flow of the tide.

The first successful deep-water measurements were made from the USCS schooner *Drifter*, engaged on a continuous program of research on the Gulf Stream, taking water samples and soundings. *Drifter* anchored in 800 feet (250 meters), but to gain access to deeper levels, the USCS steamer *Blake* was provided with strengthened gear to resist the enormous strain put on the taut anchor cable as it held the ship against the force of waves and current, eventually anchoring in over 6000 feet (over 1800 meters). In 1885, measurements were obtained with a current meter devised in 1876 by Lieutenant John Pillsbury (1846–1919) (see Figure 5). This consisted of a heavy frame supporting a set of rotating cups with a counter, and a vane connected to a compass bowl. The apparatus was lowered down a jackstay linked to the anchor chain and set running for a period. On hauling up, the direction of flow was given by the angle between the compass

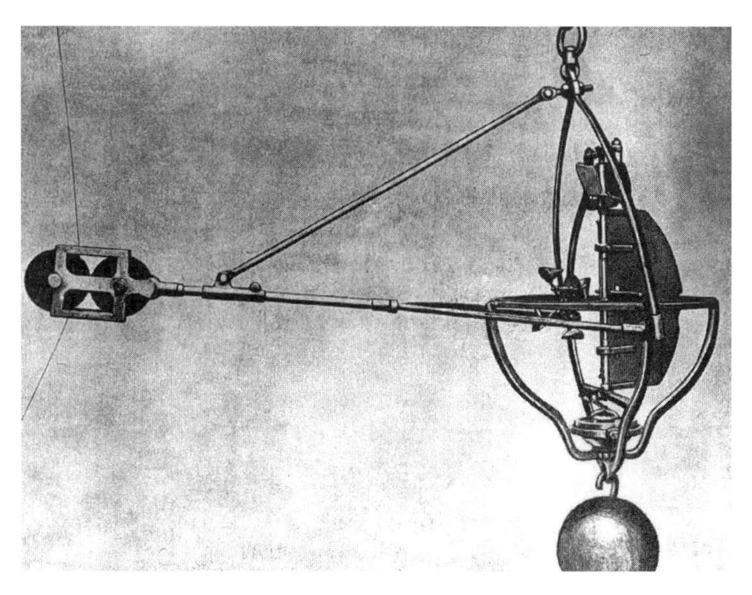

Figure 5. Photograph of the Pillsbury current meter, with the extended arm running down a cable. From Coast and Geodetic Survey Report *(Washington, D.C., 1890).*

needle and magnetic north, and the velocity was calculated from the number of revolutions shown on the counter. The apparatus was modified for greater efficiency, and by the 1890s, the Royal Navy was also using Pillsbury meters in home waters and on surveys of the Equatorial Current.

In regions where currents are slow or oscillating, rotating-cup meters are ineffective. Working in shallow polar seas, Fridtjof Nansen devised a form of pendulum meter that was suspended in a tripod lowered to the seabed, but this soon gave way to one of the most successful forms of lightweight meter, devised by the Swede Vagn Walfrid Ekman (1874–1954). Tested on the first *Michael Sars* cruise in 1903, it was available for ICES the following year. In use, the meter hung on ball bearings in its frame and was turned into the current by a light tail fin. Once it was set running by messenger, a propeller driving a counter provided a record of velocity, and the direction that it had taken up was registered by the regular delivery of balls from a hopper, each ball falling into a grooved compass needle and dropping into one of 36 compartments forming the compass bowl. After a measured length of time, the meter was locked and hauled up. The distribution of balls in the compass bowl showed whether the current was constant or fluctuating; Ekman later refined the meter by numbering the balls, so that he could discover the characteristic motion of water at each state of tide or through the seasons.

In the late twentieth century, other methods became possible. With the miniaturization of radio transmitters, one or more current meters, together with other apparatus, could be hung from weighted buoys and left to report back to a mother ship. In the 1980s, Walter Munk (dates unknown) of Scripps Institute and Carl Wunsch (dates unknown) of Massachusetts Institute of Technology devised ocean acoustic tomography, involving the transmission of low-frequency sound across vast distances of ocean to a bank of receivers. Increases in temperature, salinity, and pressure all increase the velocity of these sound waves. The first demonstration in 1981 covered a 186-mile (300-kilometer) area southwest of Bermuda, detecting a large cold eddy that moved across the area. Subsequent tests were successful over larger distances, and in 1987, a seven-transceiver system was set up off New England to study the Gulf Stream. The launch of the short-lived Seasat in 1978 showed the value of satellite photography. False-color images of surface temperature yielded sequences demonstrating the dynamic nature of the Gulf Stream and other surface currents.

Theories of Coral Reef Formation

Coral reefs have been recognized since classical times as a distinct geographical feature of warm sunlit seas. Close to the surface, their rocky nature presented a hazard to shipping, yet they were also a source of the precious red and white coral, highly esteemed for decoration and as a lucky charm. Until the early nineteenth century, when overfishing made it unprofitable, Mediterranean coral was taken from small reefs and rocks off the coasts of France, Italy, and North Africa by lowering nets to entangle and raise branches of living coral. Scientists ashore generally saw only dead, often polished, stems, and they were unable to decide on its nature. The Italian naturalist Luigi Ferdinando Marsili (1658–1730), engaged on a study of the

Light in the Sea

Sunlight penetrates a few hundred meters beneath the surface, depending on the matter suspended in the water. The plankton, and the fish that feed on it, mostly inhabit these sunlit levels, so the depth of penetration forms an important part of fisheries research. Before photography, this penetration was estimated by lowering a white disk into the water and measuring the depth at which it became invisible. Around 1880, with the arrival of silver-bromide gelatin film, underwater photometers were designed to measure the penetration of light, not just from above, but also by internal reflection within the water. Two Swiss scientists, Hermann Fol (1845–92) and Edouard Sarasin (1843–1917), constructed a photometer consisting of a gelatin plate that could be exposed underwater for a set time. Tried out in Lake Geneva, it was employed in 1885 in the Mediterranean at depths between 656 and 1312 feet (200 and 400 meters). Photometers that could be commanded remotely to make multiple records or to record light from all directions and from below were then devised. At the Plymouth Marine Laboratory in England, trials with photocells in the 1920s and subsequently with the simpler selenium rectifier photocells led to their adoption as an ICES standard in 1938.

Mediterranean sea off the coast of France, went to sea with the fishermen and examined coral as it came from the water. In 1706, writing to the Academy of Sciences in Paris, he described what he believed to be the flowers of coral. Marsili's subsequent microscopical examination and chemical analyses of various types of coral were devoted to learning more about this strange growth, which he took to be a plant. His results were published in the first book on marine science, his *Histoire physique de la mer* (1725). Coral's animal nature was established by Jean-André Peyssonnel (1694–1759), who studied corals off the coast of Algeria and in the Caribbean, where he worked as a military doctor. Peyssonnel's remoteness from Paris cost him much of the honor due to him, which fell to Bernard de Jussieu (1699–1777), whose account was published by the Academy of Sciences in 1742. Thereafter, coral was firmly established as a small animal—a polyp—living in a stony habitat of its own making. The naturalist John Ellis (1710–76) extended his studies to the coldwater **corallines** and in the 1750s and 1760s published a series of masterly descriptive papers on these creatures.

* **Corallines.** Animals that resemble coral.

Precious coral was also taken in the Red Sea and Indian Ocean, but as more ships ventured into the Pacific Ocean, numerous massive reefs were discovered far from land and adjacent to deep water. Reefs take three forms: fringing reefs, which surround a volcanic island, presenting a shelf at low tide; atolls, which consist of circular reefs, or segments of reef, with no sign of land; and barrier reefs, off the coast of a substantial landmass, the most impressive being the Australian Great Barrier Reef. Where nutrients and the physical conditions of temperature and light are balanced by erosion by wave action and the depredations of coral-eating fish, these reefs are essentially steady-state features.

As the charts multiplied, scientists began to theorize about the origins of these reefs and whether they represented a chronological succession from fringing reefs to the atolls remote from land. The French naturalists J.-R.-C. Quoy (1790–1869) and J. P. Gaimard (1796–1858), on the circumnavigation of *Uranie* and *Physicienne* in 1817, made the important observation that reef-building corals live only in shallow sunlit water, thus removing any prospect that they might have grown up from the deep seabed. Samuel Stutchbury (1798–1859), surgeon and zoologist to a commercial pearl-fishing expedition to the Pacific in 1825, suggested that coral atolls had grown on the rims of extinct volcanoes, their craters marked by shallow lagoons. Stutchbury, who had seen fossil reefs in rocks well above sea level, thought that uplift could be accounted for by volcanic action. The geologist Charles Lyell (1797–1875) at first supported Stutchbury but was later converted to the opinions of Charles Darwin (1809–82).

Darwin, during his voyage as naturalist on HMS *Beagle*, saw clear evidence of changing levels of land on the west coast of South America. Before he even saw his first coral reef, at Cocos Keeling, he was convinced that atolls began life as reefs fringing a subsiding volcanic island, with the reef building up at the same rate as the cone subsided. He proposed that the shallow lagoons found within many circular atolls arose from the infilling by sediments and a weak growth of coral largely deprived of nutrients. He also accounted for the Great Barrier Reef as a **subsidence** feature. Darwin's book *The Structure and Distribution of Coral Reefs* (1842) was enthusiastically received, although there was at this time no known geological mechanism for such subsidence, and he was mistaken as to the origin of the lagoons.

* **Subsidence.** The sinking of a large area of Earth's crust.

The U.S. geologist W. M. Davis (1850–1934) attacked the coral reef problem in the 1920s, in the light of recognition of the glacial periods with their substantial changes in sea level and temperature. He traveled the world visiting most of the known coral features and concluded that many of the reefs fringing high islands had been eroded while exposed to the weather, but this was not the case for most of the open-ocean islands. Davis eventually came to accept Darwin's concept, which he considered was supported by glacial history, of upgrowing reefs on intermittently subsiding foundations. Even so, he was unable to account for the shallow lagoons inside atoll and barrier reefs. Darwin's theory was once more overturned by marine naturalist John Murray (1841–1914), in 1880 recently returned from the circumnavigation of HMS *Challenger*, who argued that barrier reefs and atolls rested on sediments deposited on submarine volcanoes. Their circular shape arose from the stronger growth of coral on the nutrient-rich outer flanks of the cones. Such atolls could appear without any rise or subsidence in the underlying volcano. The elderly Darwin, still receptive to ideas concerning coral reefs, was among those who acknowledged that the only way to resolve the problem of origin was to sink a borehole down to the foundation of the reef.

In 1896–99, the British Funafuti Expedition drilled on the atoll of Funafuti in the Ellice Islands (modern Tuvalu) in the Pacific Ocean. The deepest bore, penetrating to 1114 feet (340 meters), encountered only coralline limestone. On one side, this was claimed as evidence of subsidence; on the other, that the drill had merely passed through the deep accumulation of debris on the outer reef margin. The first evidence of deep volcanic rock came after World War II from seismic soundings, confirmed by drilling to depths of over 4000 feet (1220 meters) at Eniwetok Atoll in 1951.

The origins and structure of barrier reefs were investigated over a number of years by the Great Barrier Reef Committee, founded in 1922. In Australia, borings showed relatively thin layers of reef limestone on quartzite sand and were interpreted as recent shallow subsidence features. Elsewhere in the Pacific, and off Florida, similar recent reefs have formed on shallow volcanic or sedimentary foundations close to shore. Several nations subsequently established field stations where the physical and biological environment of coral reefs could be studied over time.

Tidal Investigations

The range between high and low tide varies enormously around the world and even from place to place along the same coast; it also varies seasonally. Tide measurement was an ancient science, practiced in India, China, and the Middle East, for tidal range was a matter of vital importance to coastal shipping and to civil engineers constructing docks and harbors. At their simplest, tide gauges are merely graduated staffs, or a series of notches cut into a harbor wall. But wind and waves make it difficult to judge the level against such exposed marks, and it is preferable to set the staff within a well filled by the sea so that the level rises and falls smoothly with the tide, however rough the weather.

While tides bear a close relationship to the Moon's passage overhead, they are clearly influenced in both height and timing by wind and weather. In the seventeenth century, their fundamental cause and the major influences on their variability were still matters of debate. The Royal Society enlisted the help of people living near the sea in Britain and overseas, urging them to keep continuous records of the daily times of high and low tide throughout the year and to note any unusual events. In 1668, John Winthrop wrote from New England promising to check sailors' reports of a remarkable tidal range observed in the Bay of Fundy, Nova Scotia. This turned out to be the largest in the world, having at its northern end a maximum range of 56 feet (17 meters).

The search for an explanation of tides could also be conducted theoretically. Three causes had their supporters: Simon Stevin (1548–1620), William Gilbert (1544–1603), and Johannes Kepler (1571–1630) favored the magnetic or attractive hypothesis; Galileo Galilei (1564–1642) argued from Earth's motion; and René Descartes (1596–1650) held that Earth and atmosphere were compressed by the Moon's passage overhead. Mathematicians debated and elaborated on these ideas, but no hypothesis seemed to account for all known situations. In his *Principia* (1687), Isaac Newton (1642–1727) offered the first clear demonstration that tides were caused by the gravitational attraction of the Sun and Moon on both the solid Earth and its watery envelope. The highest spring tides occurred when Sun and Moon were both overhead, or on opposite sides of Earth, the minimal **neap tides** when one body was overhead and the other on the horizon. Newton went on to explain the influence of latitude and the Sun's annual progression north and south of the equator.

* **Neap tide.** A tide occuring every two weeks during the first and last quarters of the Moon.

The shape of the coast and of the adjacent seafloor was also seen to modify tides, and the height and the strength of tidal streams was sometimes noted on seventeenth-century hydrographical charts. Edmund Halley (c. 1656–1724) undertook a three-month voyage during the summer of 1701 to chart the tides and streams of the English Channel. Naval vessels sent on scientific expeditions that involved setting up temporary observatories in distant locations, notably for the two transits of Venus in 1761 and 1769, generally recorded the tides while they were at anchor.

The tedium of observing at frequent intervals led to the invention of recording gauges. The instrument devised and installed in London Dock by Henry Robinson Palmer (1795–1844) in 1831 was soon replicated elsewhere. Water entered a shaft, and a copper float riding on the surface was attached over a pulley to a counterweight. Revolutions of the pulley were transmitted through gears and shafts to a pencil that drew a scaled-down trace of water height on a chart driven by clockwork. In 1833, William Whewell (1794–1866) used port and expedition records to draw the first chart of cotidal lines (lines joining places that shared the same times of high and low water).

Measurement of tides in the open sea was solved in an entirely different manner. At high tide, the additional weight of water increases pressure at the seabed, and in 1887, Louis Favé (1853–1922) devised a gauge resembling an **aneroid barometer** for the French Hydrographic Service. The pressure sensor drove a marker pen, scribing onto a smoked glass disk rotated by an eight-day clockwork drive.

* **Aneroid barometer.** A barometer that utilizes a capsule containing no liquid.

Enclosed in a watertight case and buoyed for easy recovery, the gauge could operate down to 656 feet (200 meters) and record a range up to 50 feet (15 meters).

Several month's observations at a given place are essential to apply the mathematical formula for predicting future tides throughout the year. Mechanical calculators built for the purpose speeded up analysis and prediction. Once the various astronomical components were identified, they could be recombined and extrapolated for any given place. Nevertheless, the process was painfully slow by present standards: Lord Kelvin's elaborate tide-predicting machine of 1872 took four hours to run off one year's tables for a single port. Similar mechanical predictors did good service in ports throughout the world over the next half-century.

Recurrent Systems

The world's oceans have undergone gradual long-term changes, within which cyclical events have been detected. Glacial episodes lasting for many millions of years appear to depend on a combination of factors, among which continental drift changing the very shape of the oceans and deflecting warm and cold currents may be fundamental. On a seasonal time scale, annual monsoons affect the Indian Ocean and adjacent seas; hour by hour, sea levels rise and fall with the tides, with consequent reversals of currents.

The recurrent cyclic event known as El Niño, and more recently as the Southern Oscillation, is now seen to be global in its effects and is attracting much intensive research. El Niño was the name given to the warm southward current that appears off the barren coasts of Ecuador and Peru around December (coinciding with the advent of El Niño, the Christ Child), displacing the usual cold northward flow. Every few years the current is exceptionally warm and intense. Then rainfall in the coastal deserts of South America brings better pasturage for grazing animals, but can also lead to devastating floods. The warm current disturbs the marine ecology; plankton vanishes, fish and seabirds decline in numbers, and tropical species may appear. The first historical references to the system of currents and winds on the Pacific coast of South America date from 1535; their regularity became better known as Spanish voyages, and settlement in the region multiplied. The South American rainy season was also compared with other regions, and in his *Tratado de los tres elementos,* Tomás Lopez Medel (1509–82) explained that rain in the Caribbean coincided with droughts elsewhere, and vice versa.

Spanish expeditions working around South America in the eighteenth century were more interested in biology ashore and largely ignored the oceans. Interest was renewed only in the early twentieth century. American workers in the region considered the 1925 El Niño to have been the most intense since 1891. Between Panama and the Galápagos Islands, an extended mass of floating debris, aligned northeast to southwest, marked the position of the convergence of opposing currents, with unusually large numbers of fish, marine mammals, and birds. After World War II, Ecuadorian fisheries research showed that a strong El Niño could lower marine fertility by one-fifth.

From the 1950s, meteorologists realized that the reversing periods of easterly and westerly winds were part of the global atmospheric circulation. Unusually strong westward winds led to heavy rainfall and flooding in Indonesia and northern Australia, while the following year was marked by droughts elsewhere in the world. Surface currents are dictated by the prevailing winds, and as intense trade winds blow westward along the Equator, warm water piles up on the Australian side, whereas under the eastward winds leading to El Niño the 20° thermocline (the layer of rapid change in temperature between two water masses) is nearly horizontal across the Pacific. Atmosphere and ocean are therefore partners in the alternating states, being both cause and effect, although the atmosphere responds faster, the ocean taking a long time to adjust to changing winds. The oscillation may, however, be self-damping, for the evidence from coral reefs indicates that the effect is more prominent in some decades than in others.

Recent Developments

Oceanography has always been "big science," and much of the research is now done through international projects, often running for several years. Research vessels, with their costly laboratories, remain at sea for much of the time, with individual scientists being ferried out for the duration of their particular projects. The development of sophisticated onboard apparatus and the use of unmanned submersibles and satellites makes it possible to generate vast quantities of data, which are analyzed and processed rapidly by computer.

With a greater understanding of the global exchange of heat and momentum between air and ocean, the focus is now on the ocean-atmosphere system in its entirety. Working with their land-based colleagues, scientists can monitor global warming, the melting of ice fields, and the consequent rise in sea level. They can also probe the seabed, exploring for mineral resources and recording seismic activity. Oceanographers are likely to be fully employed for a long time to come, as their science continues to meet economic, environmental, and defense needs.

Bibliography

Deacon, Margaret. *Scientists and the Sea, 1650–1900: A Study of Marine Science*. 2nd ed. Aldershot, England: Ashgate Publishing, 1997.

Maury, Matthew Fontaine. *Physical Geography of the Sea*. New York: Harper, 1855. Reprint: Cambridge, Mass.: Belknap Press of Harvard University Press, 1963.

———. *Wind and Current Chart of the North Atlantic*. 3rd ed. Washington: U.S. Hydrographic Office, 1852.

McConnell, Anita. *No Sea Too Deep: The History of Oceanographic Instruments*. Bristol, Gloucestershire, England: Adam Hilger, 1982.

Mills, Eric L. *Biological Oceanography: An Early History, 1870–1960*. Ithaca: Cornell University Press, 1989.

Murray, John, and Johan Hjort. *The Depths of the Ocean: A General Account of the Modern Science of Oceanography Based Largely on the Scientific Researches of the Norwegian Steamer Michael Sars in the North Atlantic*. London: Macmillan, 1912.

Ocean Sciences: Their History and Relation to Man: Proceedings of the 4th International Congress on the History of Oceanography, Hamburg, September 23–29, 1987. Edited by Walter Lenz and Margaret Deacon. *Deutsche Hydrographische Zeitschrift*, Ergänzungsheft Reihe B, Nr. 22 (1990).

Oceanography: the Past Proceedings of the 3rd International Congress on the History of Oceanography, 1980. Edited by Mary Sears and Daniel Merriman. New York: Springer-Verlag, 1980.

Premier congrès international d'histoire de l'océanographie (Monaco, 1966). Monaco: Bulletin de l'Institut Ocèanographique, Numéro spécial 2. 2 vols + Index vol., 1968.

Rennell, James. *An Investigation of the Currents of the Atlantic Ocean and Those Which Prevail between the Indian Ocean and the Atlantic*. Edited by J. Purdy. London: Rivington, 1832.

Scoresby, William, Jr. *Account of the Arctic Regions*. Edinburgh: Constable, 1820. Reprint: Newton Abbot, Devonshire, England: David & Charles, 1969.

———. *Journal of a Voyage to the Northern Whale Fishery Made in the Summer of the Year 1822*. Edinburgh: Printed for A. Constable, 1823.

2nd International Congress on the History of Oceanography. *Challenger* Expedition Centenary, 1972. *Proceedings of the Royal Society of Edinburgh,* Sec. B (Biology) 72 and 73 (1971–72).

Thurman, Harold V. *Essentials of Oceanography.* 4th ed. New York: Macmillan, 1993.

Wüst, Georg. "The Major Deep-Sea Expeditions and Research Vessels, 1873–1960: A Contribution to the History of Oceanography." *Progress in Oceanography* 2 (1964): 1-52.

Optics and Light

Sungook Hong

The nature of light and colors, the mechanisms of human vision and color perception, the role of the ethereal medium in the transmission of light, and optical phenomena such as reflection and refraction have had scientific, philosophical, and religious significance through the ages. In this article, we survey the history of the study of optics and light, with a particular emphasis on the modern period, which is full of highly interesting interactions between scientific theory, experimental practices, and instruments.

Pre-Eighteenth-Century Roots

LENSES, SPECTACLES, AND THE CAMERA OBSCURA

People have long been fascinated with optical phenomena. Mirrors and burning lenses date at least to 1500 B.C. A pair of crude lenses found in Crete may date to 2000 B.C. The early lenses of the Greeks were globular rather than lenticular. The first written record of the lens is found in the writings of the Greek philosopher Aristophanes (c. 450–388 B.C.), who spoke in 424 B.C. of a glass globe filled with water. Seneca (4 B.C.–A.D. 65) later stated that this globe of Aristophanes could be used to read small letters. The legend also says that Archimedes (c. 287–212 B.C.) used a huge burning mirror (a concave mirror) to burn the invading Roman fleet.

In the Islamic world, the magnifying property of a spherical glass was well known. It was mentioned by Alhazen (Ibn al-Haytham, c. 965–1040), a famous Arabic scholar. However, these globular lenses and spherical glasses were not used as an aid to vision, like modern eyeglasses. This is no doubt due to the fact that the lenses and spherical glasses were irregular. Thus their distorting effect would have been much greater than their magnifying effect.

The use of lenses as an aid to vision was first clearly expressed by Roger Bacon (c. 1220–92), a scholastic philosopher in medieval Oxford. In his *Opus Majus*, he noted that "if the letters of a book or any minute objects, be viewed through a lesser segment of a sphere of glass or crystal, whose plane base is laid upon them, they will appear far better and larger. . . . And thus from an incredible distance we may read the smallest letters, and may number the smallest particles of dust and sand. . . ."

Bacon's optical work was conceived of as a dangerous sorcery by religious priests. He was imprisoned for 10 years and was released only in 1294, one year before his death. Although Bacon used a lens to magnify letters, he did not invent spectacles—with two lenses, one for each eye, fixed within a solid frame. They were invented, instead, by Alessandro della Spina (d. 1313), a Dominican monk of Pisa, some time between 1285 and 1299. Spectacles provided an important material resource (i.e., two lenses) and conceptual resource (i.e., the idea of magnification) for telescopes.

Another well-known optical phenomenon was the creation of an inverted image by using a pinhole on a screen. This had been known in China since the fifth century B.C. In Greece, Aristotle knew that a smaller pinhole produced a sharper image. The device was known to Alhazen and Roger Bacon. It evolved into the modern form of the camera obscura, a dark chamber with a small hole open to the outside. The first detailed description of the camera obscura was made by Giambattista della Porta (1535–1615), who described it in detail in his *Magia naturalis* (Natural Magic) in 1558. He suggested that it be used to assist in the prac-

Science and the Arts: The Camera Obscura as an Artistic Device

In the seventeenth and eighteenth centuries, the camera obscura became standard equipment for artists. An artist could draw an object very precisely by tracing the image of an object projected on the glass screen of a camera.

In his *Essay on Painting* (1764), Count Francesco Algarotti (1712–64) wrote: "The best modern painters among the Italians have availed themselves greatly of this contrivance [i.e., a *camera obscura*]; nor is it possible they should have otherwise represented things so much to life." *Rees's Cyclopaedia* (1819) lists drawing as one of the principal uses of the camera: "By means of this instrument, a person however unacquainted with drawing may delineate objects with great facility and correctness; and to the skillful artist it will be found indispensably useful in comparing his sketches with the perfect representations given in the camera. . . . "

In 1807, the English scientist William Hyde Wollaston (1766–1828) invented another mechanical device to aid drawings, which he called the *camera lucida*. A glass prism in this device enabled one to see the object and one's drawing together as if they were in the same plane.

The camera obscura did not simply reproduce natural objects. Paintings by the Dutch painter Jan Vermeer were probably aided by a camera obscura. Instead of capturing and reproducing nature mechanically, Vermeer's paintings show a very tight use of space, unexpected point of view, and several points of focus. These paintings are perhaps the first instance in art of "camera vision." They demonstrate, however, that camera vision did not so much mechanically reproduce nature as create an entirely new way of visualizing the world.

* **Refraction.** The bending of a beam of light as it moves between two mediums.

tice of drawing. Della Porta also wrote a major work on optical **refraction,** *De refractione, optices parte,* published in 1593.

Johannes Kepler (1571–1630) first devised a semiportable camera obscura in the seventeenth century. The astronomer Johann Hevelius (1611–87) also used it to investigate the Sun. A convex lens was inserted in the pinhole, and a mirror was employed to turn around the inverted image. In the early modern period, the camera obscura had two major uses. First, it was used as an artificial eye with which to investigate the nature of human vision, and second, it was used as an aid for the practice of drawing.

KEPLER'S NEW THEORY OF VISION

There were four different theories of vision in ancient Greece. First, the atomists believed that atoms progressed from the material object to the human eyes, creating the sensation of vision (the intromission theory). Second, Euclid (fl. 295 B.C.) and Ptolemy (c. A.D. 100–c. 170) claimed that visual rays are emitted from the eyes to the object (the extramission theory). Third, Plato (427–348/7 B.C.) proposed a hybrid theory combining the intromission and extramission theories. Finally, Aristotle (384–322 B.C.) proposed a theory in which the forms of an object were first impressed in the air that subsequently impressed the human eye.

Optics, as well as the theory of vision, were well developed in the Arabic world. Al Kindi (A.D. c. 801–c. 870) supported the extramission theory of Euclid and criticized the rest. However, influential Arabic philosophers such as Avicenna (Ibn Sīnā, A.D. 980–1037) and Averroës (Ibn Rushd, 1126–98) supported the Aristotelian theory of vision. It was, however, Alhazen who made a lasting contribution to optics. He convincingly criticized the extramission theory and proposed an intromission theory of his own. Before Alhazen, the intromission theory was either atomistic (the progression of atoms) or Aristotelian (the progression of forms). Alhazen's originality lay in his proposition that every point of an object should be regarded as a point source from which rays emanate in all directions. He also suggested that an image of the object is created inside the human eye (in the glacial humor), and he distinguished the sensation of the image in the eye from the perception of it in the brain. He posited a one-to-one correspondence between the image and the object.

How did Alhazen establish a one-to-one correspondence between the object and image if rays are emitted in all directions from every point of an object? He reasoned that although an infinite number of rays are emitted from any point, only a single ray can enter the eye at a right angle, and this ray alone is not refracted. This unrefracted ray is stronger than other refracted rays and is therefore responsible for the formation of the image of that point inside the eye. The image of the entire object consists of the aggregation of all the points, each of which is created by unrefracted rays.

Alhazen's theory synthesized the pointlike emission of rays, the principle of one-to-one correspondence between object and image, and the geometrical and physical considerations of vision. Yet it had the fundamental weakness that it ignored all the refracted rays and considered only an unrefracted ray to be responsible for image formation. Although this problem was not solved until Kepler, Alhazen's theory became very influential in medieval Europe, exercising an impact on Roger Bacon, Witelo (Vitello, c. 1230–75), John Pecham (c. 1230–92), Francesco Maurolico (1494–1575), and Giambattista della Porta.

Johannes Kepler became an assistant to Tycho Brahe (1546–1601) in 1600. While wondering about a curious optical anomaly exhibited by the camera obscura, Brahe asked Kepler to provide an explanation for it. This led Kepler to the field of optics and vision. In particular, the formation of an inverted image on the back screen of the camera obscura particularly impressed him. After explaining the anomaly of the camera obscura that had puzzled Brahe, Kepler turned his attention to human vision. He found medieval theories of vision to be confusing and conflicting. In particular, he did not accept Alhazen's theory that all the refracted rays could be ignored in image formation. How was it possible to believe that only rays entering perpendicularly contribute to the formation of the image, while slightly refracted rays do nothing? But if all the refracted rays were counted, why wouldn't we see thousands of different images of a single point?

While he pondered this problem, he became aware of a new anatomical work by Felix Platter (1536–1614) in which the retina, not the humor, was identified as the site of vision. By comparing the action of a human eye with that of a convex lens and by considering the retina to be the site of vision where all the refracted rays converged, Kepler was able to articulate what is essentially a modern theory of vision, in which an inverted (and reversed) image of an object is formed on the retina. Thousands of refracted rays are simply focused by the action of the lens on a single point on the retina. Kepler's theory, publicized in *Ad vitellionem paralipomena* in 1604, beautifully explained the problem that had puzzled scientists and natural philosophers for more than 2000 years. The formation of the image on the retina was later verified experimentally by Christopher Scheiner (1573–1650).

GALILEO'S TELESCOPE

There were two different kinds of optical telescope: reflecting and refracting. In the reflecting telescope, the objective (i.e., the primary and largest lens or mirror) is a mirror, whereas in the refracting telescope it is a lens. The principle of the reflecting telescope was known in the later sixteenth century. Leonard Digges (c. 1520–59) at Oxford remarked in his *Pantometria* (1571) that "by concave and convex mirrors of circular and parabolic forms, or by pairs of them at due angles . . . any part of it [the whole region] may be augmented so that a small object may be discerned as plainly as if it were close to the observer, though it may be as far distant as the eye can describe." The invention of refracting telescopes was not far away. In the 1589 edition of *Magia naturalis*, Della Porta speculated that "by means of a concave glass you will see distant objects small but clear; with a convex glass, near objects magnified but dim. If you know how to combine them exactly you will see both distant and near objects larger than they would otherwise appear and very distinct."

It is known that the refracting telescope was available by 1608, the year that is usually cited as the year of its invention. Two craftsmen laid competing claims to its invention, however: Hans Lippershey (c. 1570–c. 1619) and Zacharias Jansen (1588–c. 1631). Both were spectacle makers in Middleburg in the province of Zeeland in the Netherlands. Lippershey was the first to apply for a patent, but because of competing claims from others, including Jansen, the patent was not granted to Lippershey. Because of this decision, the telescope began quickly to spread all over Europe.

Galileo Galilei (1564–1642) heard the news that someone in the Netherlands had constructed a telescope with which one could view visible objects at a great distance quite distinctly. Within a couple of weeks of hearing this news, Galileo constructed his own telescope by carefully studying the principles of refraction in the telescope's lenses. He showed his telescope to military officials in Venice, and in return he was promised an increase in his salary for the professorship in Padua. Although Galileo never claimed that he was first to invent the telescope, he did not fail to stress that his invention was not triggered by a mere chance, but was made possible by deep reasoning.

Galileo's telescope used a convex lens as the object lens, and a concave lens as the eyepiece. It had the merit of having an upright image with only two lenses (with convex lenses only, one needs three to have an upright image), but its magnification was limited. Although Galileo's first telescope had a magnification of three times, his second and third telescopes could obtain a magnification of 20 and 30 times, respectively. These telescopes were good enough to reveal for the first time the new marvels of the universe.

In January 1610, Galileo discovered the four moons of Jupiter, named them the Medicean stars, and dedicated them to Grand Duke Cosimo II de Medici. He also showed that the surface of the Moon was not smooth, as Aristotelian philosophers had imagined, but coarse, with mountains and valleys. However, even the power of the telescope was insufficient to magnify the size of fixed stars, which he interpreted as being due to their remote distance. He reported these new heavenly discoveries, which provided some moderate support for Copernican heliocentrism, in *Sidereus nuncius*, which was published in March 1610 [see *Astronomy and Cosmology: Pre-Eighteenth Century*].

In December 1610, Galileo observed through his telescope the variation of the phase of Venus. This new evidence forcefully supported the Copernican system. It was terribly difficult, although not impossible, to explain the phase of Venus in terms of the Aristotelian-Ptolemaic system. In June 1611, he discovered black spots on the surface of the Sun and observed that they were rotating. The Sun's rotation was a heavy blow to the Aristotelian-Ptolemaic theory, which had claimed that the supralunar world (i.e., the world beyond Earth and the Moon) was perfect and motionless (except for the rotation of crystalline spheres around Earth).

Christopher Scheiner, who had discovered the Sun's spots independently of Galileo, explained the rotation of the Sun's spots in terms of the rotation of little planets around the Sun. As Scheiner's incorrect opinion was favored by the church, Galileo had to challenge him. In so doing, he reluctantly became involved in a dispute with Scheiner. This debate in turn highlighted his Copernican stance. In 1616, an injunction from Rome warned him that he should neither defend nor hold the Copernican hypothesis. To the church, Galileo appeared to have violated this injunction by publishing his *Dialogue* in 1632, for which he was forced to stand in trial in 1633.

KEPLER AND THE THEORY OF ASTRONOMICAL TELESCOPES

The term *telescope* was coined at Prince Frederico Cesi's (1585–1630) banquet in Rome in 1611, where Galileo was invited to demonstrate his new instrument to Cesi and his powerful guests. The word rapidly replaced other terms, such as Galileo's *perspicillum* or Kepler's *instrumentum*.

The optical theory of the telescope was proposed by Kepler in his *Dioptrice* (1611). Having explained the converging and diverging effect of light rays through lenses, Kepler described for the first time an effect called *spherical aberration* that was detrimental to telescopes (see Figure 1). Spherical aberration is caused by the failure to create a focus on a single point, because rays entering the lens' center and rays entering the periphery are refracted differently. Kepler not only reported this, but also hypothesized that it could be remedied if a hyperboloidal lens was used (a hyperboloid is a surface generated by rotating a hyperbola about its main axes). He arrived at this novel idea by examining the human eye and finding the shape of the eye's lens to be hyperboloidal. Human eyes proved to be an excellent model because they did not suffer from any kind of spherical aberration. Also, Kepler for the first time proposed the use of a telescope with two convex lenses. It had the advantage of providing a wider view than did the Galilean telescope, which had two lenses, one convex and one concave. However, Kepler did not construct this new telescope. He confessed that "for observations his eye was dull and for mechanical operations his hand was awkward." The convex eyepiece was mentioned again by Scheiner in 1630. The first working model of the Keplerian telescope was made as late as 1645.

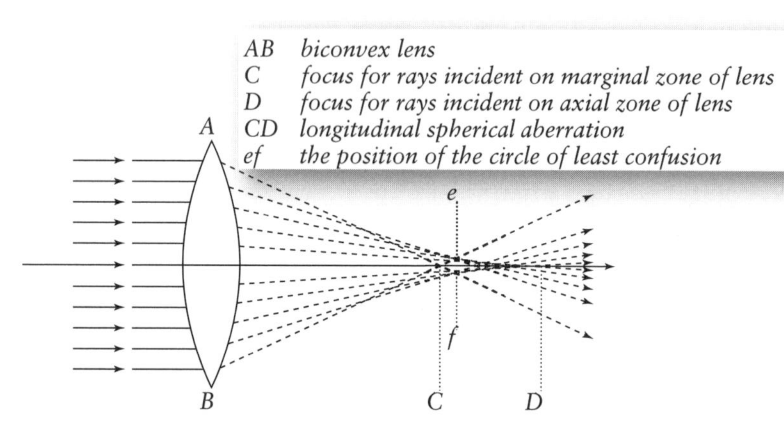

Figure 1. Illustration of spherical aberration.

Another defect of early telescopes was chromatic aberration, which was caused by the unequal refraction of rays of different colors (see Figure 2). That is, even when a single beam of white light passed through a lens, its blue and red rays refracted to different degrees, failing to focus on a single point. This usually created a colored fringe around the edge of an object seen through a telescope. Because spherical and chromatic aberration appeared simultaneously in uncorrected telescopes, it was not easy to differentiate them. René Descartes (1596–1650), for instance, who proposed a method to correct spherical aberration, did not realize that it could be distinguished from chromatic aberration. The mechanism of chromatic aberration was fully understood only when Isaac Newton (1642–1727) proposed his new theory of

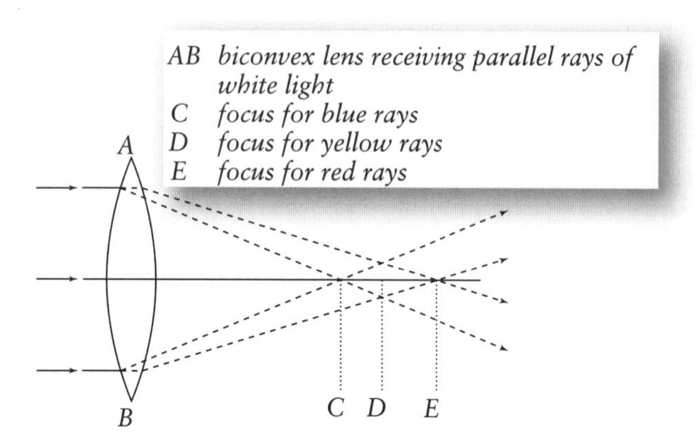

Figure 2. Illustration of chromatic aberration.

color, according to which a white ray consists of different rays that are refracted by differing amounts.

One simple way to avoid chromatic aberration was to use mirrors instead of lenses, because mirrors involve only reflections. In 1616, Niccolò Zucchi (1586–1670) proposed the first reflecting telescope to eliminate the chromatic effect. Zucchi's telescope, however, had a problem of blocking images. Marin Mersenne (1588–1648) overcame this problem by ingeniously combining a large concave mirror with a small one. However, largely because of the difficulty in securing good-quality large concave mirrors, Descartes persuaded Mersenne to abandon this idea. Although Mersenne decided not to build this reflecting telescope, he mentioned the idea in his *L'harmonie universelle* in 1636.

LAW OF REFRACTION: SNELL, DESCARTES, AND FERMAT

Reflection and refraction had been known since antiquity. Ptolemy stated the law of reflection, that is, the equality between the angle of incidence and that of reflection, and came fairly close to the law of refraction. On the observational ground that the angle of refraction was always smaller than the angle of incidence, Ptolemy concluded that the former was less than the latter in a given proportion. Alhazen found this to be wrong. He asserted that the angle of refraction was not simply proportional to the angle of incidence but did not provide empirical justification. Witelo gave a table of incident and refracted angles, both empirically observed, which supported Alhazen's assertion.

When Kepler was developing the theory of telescopes, he realized that spherical aberration was linked to the nature of refraction. He made the important discovery that refraction was dependent on the density, not the nature, of the medium, but he failed to obtain the law of refraction. To deal with the theory of telescopes, he reasonably assumed that the glass would have a **refractive index** of 1.5 relative to air, and further that the angle of refraction was proportional to the angle of incidence. With these two (incorrect) assumptions, he was able to handle most telescopic problems.

* **Refractive index.** For a ray of light undergoing refraction when traveling from one medium to another, the ratio of the sine of the angle of incidence to the sine of the angle of refraction.

Willebrord Snell (1580–1626) discovered the correct law of refraction in 1621, although he died in 1626 without publishing his law. In 1632, Jacobus Golius (dates unknown) discovered the sine law of refraction in a note in Snell's manuscripts (the note was subsequently lost). Its date was given as December 22, 1621. In 1632, René Descartes wrote to a friend that he had discovered the law of refraction in 1626–27. Descartes first published the sine law in *La dioptrique* in 1637.

What has been in debate is whether Descartes had known Snell's law before Snell's death in 1626, or between 1626 and 1632. Christiaan Huygens (1629–95) suspected in his *Dioptrica* (begun in 1653 and published posthumously in 1703) that Descartes had perhaps derived the sine law from Snell's manuscripts. Isaac Vossius (1618–89), who had also seen Snell's manuscripts, stated that "Cartesius [Descartes] got his law from Snell, and in his usual way, concealed it." Although not improbable, there is no textual evidence that Descartes obtained the idea of the sine law of refraction from Snell's note.

To appraise Descartes's law of refraction, we need to first look at the nature of light in Descartes's natural philosophy. Descartes furnished several different mechanistic models for light. His universe was filled with matter, and because of this, he considered rays of light to be similar to lines of pressure through the medium, which started from a luminous body and terminated on the surface of opaque matters. According to this idea, light was something like the transmission of energy or pressure rather than the transmission of real objects. However, elsewhere in his writings, particularly where he discussed the production of colors, Descartes seemed to conceive of light in terms of the movement of small rotating particles.

Descartes's derivation of the sine law of refraction in his *La Dioptrice* had a tremendous influence upon future studies in physics. It was one of the first instances of the successful application of mathematical reasoning to physical problems. It was grounded upon the assumption that the speed of light was faster in a denser medium than in a rarer medium. Descartes (incorrectly) thought that as light could only be produced in matter, light was more easily produced in a denser medium (and thus faster). This is not only contradictory to what is known today, but also even at that time, it was thought to be in conflict with a common belief that light was faster in a rarer medium, a belief that had been widely held since the time of Aristotle.

Descartes discussed refraction in detail using the model of a tennis ball moving through different media. His mathematical derivation was based upon two assumptions. His first assumption was concerned with the speed of light, and can be expressed as $v_r = nv_i$, where v_r and v_i are the velocities of refracted and incoming light, respectively, and n is a constant. His second assumption was that the parallel (or horizontal) component of the speed was conserved before and after the refraction. That is, $v_i \sin i = v_r \sin r$, where i and r are angles of incidence and refraction, respectively. Combining these two assumptions, Descartes obtained his sine law of refraction.

$$\frac{\sin i}{\sin r} = \frac{v_r}{v_i} = n \text{ (constant)}$$

The above formula "sin i/sin r is equal to a constant" is correct, but sin i/sin r = v_r/v_i is not. As is now known, the correct relationship is sin i/sin r = v_i/v_r.

Besides the troublesome relationship between the speeds of incident and refracted light, this law posed another puzzle. Descartes emphasized several times that the propagation of light was instantaneous. If so, how could the instantaneous propagation of light be made compatible with the different velocities in different media? When asked about this, Descartes explained that light was a tendency of motion (which is instantaneous), not an actual motion (which takes time). He thought that the speed of light, which was analogous to the speed of a tennis ball, was the speed of the "force" of light, not of light itself.

Despite being confusing in places, Descartes's optical studies were very influential. He applied the law of refraction to the rainbow and calculated the range of angles within which the rainbow could be seen. He also applied the sine law to lens grinding. Because he knew that any spherical lens would cause spherical aberration, he tried to construct a lens-grinding machine with which to produce nonspherical lenses. That this attempt failed is hardly startling, since nonspherical lenses were manufactured only in the middle of the twentieth century.

Pierre de Fermat (1608–65), unhappy with Descartes's method of deriving the sine law, engaged in a debate with him beginning in 1637. In the 1650s, he developed a derivation of his own. Fermat's derivation of the sine law was based on the principle of least time, which postulated that nature's action be performed in its most simple and economical form. He was convinced of this principle in 1657 and applied it to refraction. He reasoned that when light traveled between two points across the boundary of two media, it would take more time if it traveled in a straight line connecting the two points than if it traveled along the refracted path. This reasoning was effective only if one assumed that, contrary to Descartes's reasoning, the velocity of light was faster in a rarer medium than in a denser medium. By using these two assumptions, Fermat derived the sine law of refraction in 1662: sin i/sin r = n, where *n* is constant. However, in Fermat's law of refraction, the sine of an angle was directly proportional to the velocity: $\sin i/\sin r = v_i/v_r$. He confirmed that the sine ratio of incident to refracted angles was constant. In this limited sense, Fermat admitted that Descartes was right. But he also proved that Descartes's demonstration was "faulty and full of paralogisms."

Whether $\sin i/\sin r = v_r/v_i$ or $\sin i/\sin r = v_i/v_r$ was true remained undecided. Robert Hooke (1635–1702) followed Descartes, although Hooke's light was a pulse or wave. Newton's light was a corpuscle, but Newton's law of refraction was the same as that of Descartes. Gottfried Wilhelm Leibniz (1646–1716) criticized both Descartes and Newton, but he also concluded that light must travel faster in a denser medium. Huygens, however, agreed with Fermat, criticizing Descartes's hypothesis that light was faster in a denser medium. The velocity of light in space was first measured in 1676 by Ole Römer (1644–1710) using astronomical observation. The comparison between the velocities of light in air and water was first performed by Armand Fizeau (1819–96) and Jean Bernard Léon Foucault (1819–68) in 1850. The result for the first time proved experimentally that Fermat and Huygens were correct.

DIFFRACTION, INTERFERENCE, AND DOUBLE REFRACTION

Descartes solved the puzzle of refraction in the 1630s and provided an explanation of colors in mechanical terms. In his conception, colors were produced by the different rotational velocities of small particles consisting of light. Descartes was a system builder. He built a grand system of the world based on mechanical philosophy, in which **epistemology**, physics (including optics and mechanics), cosmology, and physiology were all tightly integrated.

* **Epistemology**. A branch of philosophy that is concerned primarily with the nature of knowledge.

Three startling new optical discoveries made in the 1660s had a fundamental impact upon the course of optics in the seventeenth century. In 1665, Francesco Maria Grimaldi's (1618–63) *Physico-mathesis de lumine* was published posthumously. In it, he reported the experimental observation that light did not travel in a straight line. He found that when light passed a sharp object, it seemed to bend inside, creating a colored band around the edge of the shadow. This was the phenomenon of **diffraction** (or inflection, in Newton's term). Since Grimaldi viewed light as a fluid substance, he attempted to explain diffraction in terms of his fluid theory of light in combination with Descartes's modification theory of colors.

* **Diffraction.** The bending of light around obstacles.

In the same year, Hooke published his widely read *Micrographia*. In the optical chapter of *Micrographia*, he reported for the first time the phenomenon of interference in a thin film. Hooke discovered that the color appeared only in a certain thickness of the film and that a change in the film's thickness altered the color pattern. If incident rays of light were partially reflected at both inner and outer surfaces of the thin film, he reasoned, the mixture of these two partially reflected beams constituted the colors of the thin film. Hooke pictured light as a pulse in a transparent and homogeneous medium and had an elementary notion of the spherical **wavefront**.

* **Wavefront**. The surface composed at any instant of all the points just reached by a vibrational disturbance in its propagation through a medium.

In 1669, Erasmus Bartholin (1625–98) reported a strange effect in the crystal Iceland spar. An incident ray of light was refracted in two different ways in this crystal. One was an ordinary refraction, but the other was extraordinary, for this refracted beam changed its angle of refraction as the crystal was rotated. This phenomenon, which was dubbed *double refraction*, raised a great deal of interest. In 1679, Huygens provided a theoretical explanation for it by assuming that light was a wave process.

NEWTON'S OPTICAL RESEARCHES IN THE 1660s AND EARLY 1670s

When Newton began experimenting on optics, Descartes's theory of light and Robert Boyle's (1627–91) theory of color were available to him. Descartes had proposed that light was a certain motion, or an action in the luminous body, transmitted instantaneously. Boyle had suggested that the color of a substance was created by the interaction between white light and a certain disposition on the substance's surface. Although Boyle performed a variety of experiments to reveal the secret of color, he could not completely solve the mystery involved in the production of colors by refraction and confessed that refraction was "one of the abstrusest things that I have met with in Physicks."

Newton did not accept Descartes's doctrines. From the beginning he believed that light consisted of corpuscles with a finite velocity. Newton accepted the general mechanical explanation of colors, but rejected the idea that colors were nothing but the modifications of white light (achieved by mixing white with dark light). Newton later recalled that he had begun to study optics with a triangular prism in January 1666 while staying at home in the countryside because of the outbreak of plague in the summer of 1665. His manuscripts indicate, however, that it is plausible that he had performed some rudimentary optical experiments in Cambridge before the outbreak of the plague. The ambitious young Newton took up Boyle's lead. He focused on the prism's production of colors by refraction. He wanted to make the science of colors "as certain as any part of optics."

From his experiments with a prism performed between 1664 and 1666, Newton made several important discoveries that he later incorporated into his first paper, "New Theory about Light and Colors," published in the *Philosophical Transactions of the Royal Society* in 1672. During these early years, he noted that "no colors arise out of the mixture of pure black and white," criticizing the modification theory of colors of Descartes, which in fact dated back to Aristotle. Newton also experimented on the mixture of colors produced by refraction, finding that "blue from a prism falling upon red gives green."

The most significant fact that Newton established was that "the rays which make blue are refracted more than the rays which make red." He first established this by looking at a colored (half red and half blue) thread through a prism. To his surprise, he observed the blue part above the red part, which made the thread appear discontinuous. He drew from this observation a radical conclusion, that blue rays were traveling more slowly than red rays and therefore were more bent by refraction. This conclusion later became the first theorem in his *Opticks* (1704): "Lights which differ in Color differ also in Degrees of Refrangibility." Newton's other early experiment supporting the different refrangibilities (degrees of refraction) of rays of light was the creation of the rectangular spectrum by having a circular ray of light pass through a prism.

In 1668, Newton performed the "crucial experiment" with two prisms (see Figure 3). This experiment demonstrated two things: first, blue rays were more refracted by the second prism than red rays, and second, neither blue nor red rays in the first spectrum could produce any other colors when they passed through the second prism. These early experiments together pointed to the following three fundamental conclusions: (1) a pure-colored ray does not split any further, (2) different-colored rays have different refrangibilities; and (3) white light is a mixture of colored rays. These fundamental doctrines of the Newtonian optics had been firmly established by late 1668. They were first

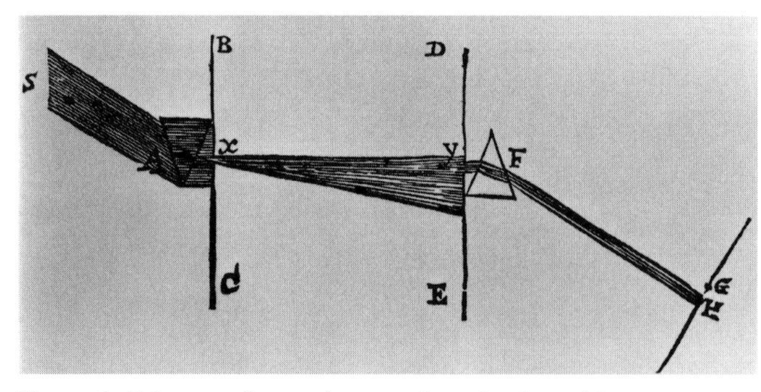

Figure 3. Diagram from a letter written by Isaac Newton (1642–1727) to the Royal Society (1672), illustrating his two-prism experiment. BC *and* DE *are boards with holes indicated by* x *and* y, *and the two prisms are indicated by* A *and* F. © *SPL, Photo Researchers, Inc.*

Scientific Method: Crucial Experiments

The concept of the crucial experiment (*experimentum crucis*) was originated by Francis Bacon (1561–1626). However, he did not coin the term itself. In his *Novum organon*, he referred to "crucial instances" (*instantiae crucis*). He maintained that scientists must "judge about natural causes by means of crucial instances and luciferous experiments and not solely by probable reasons." The word *crux* came from "crossroad," and a crucial instance to Bacon was analogous to a signpost at the crossroad. Robert Hooke first used the phrase *experimentum crucis* in his *Micrographia* in 1665. Hooke regarded the interference phenomenon exhibited in a thin film as the crucial experiment that could falsify Descartes's explanation of colors.

Newton employed the same phrase in his first paper in 1672. He was referring to the experiment with two prisms to demonstrate the unequal refrangibility of rays of light. From this experiment, he concluded that "Light consists of Rays differently refrangible." Newton's crucial experiment was criticized by several scientists. Robert Hooke was one of them. Hooke thought that "the same phenomenon will be saved by my hypothesis as well as by his without any manner of difficulty." To Hooke, Newton's crucial experiment was not crucial at all, although Newton replied to Hooke that "on this [the *experimentum crucis*] I chose to lay the whole stress of my discourse." It was perhaps because of this confusion that Newton did not use the phrase "crucial experiment" in his *Opticks* in 1704. In 1716, the Newtonian John T. Desagulier (Theophilus, 1683–1744) demonstrated Newton's crucial experiment, but at that time, he replaced Newton's original two-prism experiment with his own versions of prism-lens experiments, in which dispersed rays of light by a prism are made to converge into white light by means of a lens.

published in his 1672 paper, where he clearly asserted that "light itself is a heterogeneous mixture of differently refrangible rays" and declared that "a specific refrangibility is in one-to-one correspondence with each colour: so the science of colors is mathematical."

Criticism from Hooke, Ignace Gaston Pardies (1636–73), and Huygens immediately followed, but their criticism originated more from a misunderstanding of Newton than from a sound appreciation of his optical theories. Huygens failed to grasp Newton's theory of light. Hooke thought that Newton had revived the Aristotelian theory of light, because he interpreted Newton's rays with different refrangibilities as rays with blue, red, and green qualities. Pardies, and Hooke too, misunderstood Newton's crucial experiment.

Partly because of these bitter debates that arose from his findings, Newton stopped publishing papers on optics. Instead, he gradually accumulated materials for a grand book. The first draft had been essentially completed by 1687, but he did not publish it at that time. It was finally published in 1704, with the title *Opticks*. It was a culmination of Newton's research on optics, and at the same time the beginning of the Newtonian program of optics in the eighteenth century.

NEWTON'S REFLECTING TELESCOPE

Newton's new theory of light led to the conclusion that chromatic aberration in refracting telescopes was impossible to eliminate, because it was caused by the very property of light. One way to dispense with it was to use a reflecting telescope. Newton made his first reflecting telescope in 1668, but it was no more than a toy. His second 6-inch (15.2-centimeter) telescope was impressive, and Isaac Barrow (1630–77) demonstrated it before the Royal Society (see Figure 4). It created much sensation there, as this little telescope showed the moons of Jupiter. The day it was demonstrated before the members of the Royal Society, Newton was elected as a fellow. Even Huygens, who later criticized Newton's theory of light and color, was fascinated by Newton's new reflecting telescope.

Newton claimed that the design of this telescope had originated from his novel theory of light and colors. However, the telescope was only remotely connected to

Figure 4. Sketch of Isaac Newton's reflecting telescope.
© The Royal Society

his optical theories. Only the general idea that the reflecting telescope could avoid chromatic aberration, not its specific design, was based on Newton's theory. Today, Newton's second reflecting telescope is in the custody of the Royal Society.

HUYGENS AND THE WAVE THEORY OF LIGHT

Huygens's mathematical theory of light focused on the phenomenon of double refraction. In his *Traité de la lumière* (first communicated to the Académie des Sciences in 1679 and published in 1690), he proposed that light was a wave or pulse process in the **aethereal medium,** which permeated all substances. The notion of a wave's periodicity was lacking, but Huygens made an important contribution to the wave theory of light by proposing his principle concerning the propagation of a wavefront or pulse front. This principle considered every point on a wavefront to be a center for secondary spherical waves (wavelets) whose radius is $c \times \Delta t$, where c is the velocity of the wave and Δt is the duration of time. Huygens initially devised the idea of the wavefront to explain the **rectilinear propagation of light** by means of the wave theory. In Huygens's theory, secondary wavelets were in phase with the original wave and propagated with the same speed. These wavelets formed a new wavefront a short time later. His theory also explained the bending (diffraction) of light as it passed a narrow hole, since those unblocked wavelets were free to expand radially after passing the hole.

In his *Principia* (1687), Newton criticized Huygens, asserting that both sound waves and ripples in water diverge after passing through a hole. Newton claimed that if light was a wave, this considerable divergence must occur. In light, a slight diffraction effect did in fact occur, but it was far from similar to that of sound or water waves. Newton was reassured by this consideration that light was corpuscular.

In his reply, Huygens repeated his previous claim that those secondary waves not reinforced by the principal wave were too weak to produce any visible effect. However, why this should be so was not answered. Only in the early nineteenth century did Augustin Jean Fresnel (1788–1827) for the first time suggest a correct explanation of the rectilinear propagation of light in terms of the wave theory. According to Fresnel, the secondary waves in the shadowy region had **phase** relationships such that each canceled the other, to produce darkness.

Huygens explained refraction in terms of wavefronts. By assuming that the velocity of secondary wavefronts was diminished after entering a denser medium, he derived the sine law of refraction. He then tackled double refraction. In double refraction, Huygens elaborated, the incident wavefront was split into ordinary and extraordinary wavefronts. The ordinary wavefront, being spherical, propagated like normally refracted rays, whereas the extraordinary wavefront had the form of an **ellipsoid.** His theory roughly coincided with the experimental data available at that time.

Huygens could not, however, account for a very puzzling phenomenon, which happened when these ordinary and extraordinary refracted rays were made to enter another Iceland crystal and to refract once again. According to the position of the second crystal, three different results were obtained: The two rays were not split

* **Aethereal medium, or aether.** A transparent medium formerly supposed to fill the universe and to be responsible for the transmission of light.

* **Rectilinear propagation of light.** The phenomenon whereby light travels in straight lines in a homogeneous medium.

* **Phase.** The fraction of the entire period of an oscillation as measured at any point in time from an arbitrary time origin.

* **Ellipsoid.** A geometric surface, all of whose plane sections are either ellipses or circles.

further; each ray was split into ordinary and extraordinary rays as before; or the ordinary ray gave rise to an extraordinary ray, and vice versa. Newton criticized the wave theory because of its inability to explain these secondary refractions. On corpuscular grounds, Newton endowed a ray of light with two opposite "sides" that caused the extraordinary refraction, as well as two other "sides" for ordinary refraction. In other words, Newton believed that these two rays did not have the same properties with respect to the plane perpendicular to their direction of propagation. This was the genesis of the idea of **polarization**.

* **Polarization.** The restriction of a wave motion to certain planes or directions of vibration.

Huygens's theory was not widely accepted. Little attention was given to double refraction. From the beginning of the eighteenth century, Huygens's *Traité* was ignored almost completely, even among wave theorists. It was Newton's *Opticks* that provided a dominant paradigm for eighteenth-century optical research.

Eighteenth Century

NEWTON'S *OPTICKS* AND THE CORPUSCULAR THEORY OF LIGHT AND COLOR

Newton made the following remark about the delayed publication of *Opticks*: "To avoid being engaged in Disputes about these Matters, I have hitherto delayed the printing." Robert Hooke, Newton's lifelong enemy, became terminally ill in 1703 and soon died. Only then did Newton decide to publish *Opticks*. In contrast to the *Principia*, which was a thoroughly mathematical work, *Opticks* was largely experimental. Newton was explicit about its experimental tone by stressing his intention that "my design in this book is not to explain the properties of light by hypotheses, but to propose and prove them by reason and experiment."

Opticks was an extension of Newton's new theories of light and colors from 1672. Many of the experiments described in *Opticks* had been performed 20 years or more before its publication. However, *Opticks* also embraced fresh explanations and speculations that he had added since 1672. In particular, he gave the Newtonian account of interference (or Newton's rings; see below) and a criticism of Huygens's wave theory. In 1675–79, Newton had explained interference by using the concept of *aether waves* that were created by the action of light. But in *Opticks* he did without the term *aether*, although he retained the same idea when he remarked that "rays of light . . . excite vibrations in the refracting or reflecting medium or substance." Instead of relying on the mechanism of the aether, he used "fits of each reflection and transmission" without providing a further mechanism for fits (i.e., alternations). Fits in Newton's *Opticks* were similar to gravity in his mechanics. Each was a microscopic mechanism that belonged to the realm of hypothesis, which Newton had tried to avoid. Since Newton's fits had evolved from his aether waves, they retained an element of the wave. The Newtonian fits were therefore even appraised by a physicist in the twentieth century as "a remarkable anticipation of the twentieth-century quantum-theory explanation."

Newton had more trouble with diffraction. In the *Principia*, he had endeavored to explain diffraction in terms of short-range forces between light and matter. However, after having performed careful experiments on diffraction in the late 1680s or early 1690s, he discovered that diffraction was a highly complicated phenomenon that defied any explanation that he could possibly imagine. Diffraction remained largely unexplained in *Opticks*.

At the end of the second edition of *Opticks* (1717), Newton attached 31 Queries. Many of them were highly speculative. The optical dynamics (i.e., the mutual interaction between light and matter), which was barely mentioned in the main text of *Opticks*, was discussed in Queries. The nature of heat, light, and electricity was also conjectured upon. The effect on human vision of the transmission of ethereal vibrations, his corpuscularian attempt to solve double refraction, his criticism of Huygens and Descartes, and the possibility of various short-range forces in chemistry, electricity, magnetism, and capillarity were also addressed. Here, Newton advanced the philosophy of active principles in the universe, such as "certain powers, virtues or forces," as well as their religious implications. Although he did not explicitly discuss the aether, he mentioned the "subtle medium" and "electric and elastic spirit."

Newton's *Opticks* provided a basis for further optical studies throughout the eighteenth century. Combining *Opticks* with the *Principia*, Newtonianism became the dominant paradigm for physics. Newton's white light was a mixture of homogeneous rays of light with different refrangibilities. Light consisted essentially of corpuscles traveling at a very fast, but finite, speed. Elementary colors corresponded to the homogeneous rays of light. These, when combined, explained most phenomena of optics and color. In the eighteenth century, Leonhard Euler (1707–83) proposed an impulse or wave theory of light, but it could not compete with Newtonian corpuscular theory. In particular, Euler's theory could not easily explain the production of colors.

In the early nineteenth century, two competing developments took place. On the one hand, the Scottish Newtonian David Brewster (1781–1868) and the French Newtonians such as Étienne Malus (1775–1812) and Jean-Baptiste Biot (1774–1862) extended the Newtonian corpuscular theory one step further. In particular, they explained the phenomenon of polarization that had just been discovered. On the other hand, in 1800, the British polymath Thomas Young (1773–1829) proposed a wave theory of light, with which he beautifully explained the interference phenomenon.

ACHROMATIC COMPOUND LENS

During his first optical research in the late 1660s and early 1670s, Newton brought attention to chromatic aberration. The idea of a compound lens (i.e., a lens filled with water) had been suggested by Robert Hooke as a means to improve the sharpness of telescopic images. After having examined this idea, Newton concluded that compound lenses could be helpful in eliminating spherical aberration, but not chromatic aberration. He believed that no correction would be possible for chromatic aberration. He seemed to think that dispersion and refraction were the two inseparable properties of light, not those of matter. To him, the index of refraction alone was the property of matter. He neither knew that the **dispersive powers** of transparent substances were different, nor that the dispersive power was not identical to the refractive power of a substance.

* **Dispersive power.** A measure of the power of a medium to separate different colors of light.

Newton's discussion of dispersion in *Opticks* was based on his astonishing prism-within-prism experiment, in which a small glass prism was placed within a large water-filled prism. Newton claimed to have observed that if the outgoing beam was made parallel with the original incoming beam, it remained white, or undispersed, despite several refractions. This claim has been regarded as highly controversial. Historians have doubted whether the experiment was actually performed. The prism-within-prism device was not only highly sophisticated, but was also impossible for obtaining the result that Newton claimed to have obtained (i.e., the white outgoing beam) with the compound prism consisting of ordinary water and glass.

Newton's claim had an important impact in other directions. If the results of the experiment were accurate, it could be used as a condition for **achromatism**. But Newton had explicitly stated that achromatism was impossible. In 1754, the Swedish mathematician Samuel Klingenstierna (1698–1765) claimed that either

* **Achromatism.** The condition in which chromatic aberration is removed.

Science and Technology: Herschel's Giant Reflecting Telescope, 1789

When William Herschel (1738–1822) first saw the sky with a rudimentary telescope in 1773, he was a 35-year-old music director in the city of Bath. The telescope that he used was a long refracting telescope [15 to 30 feet (4.57 to 9.1 meters)]. Because of trouble with the long tube, he wanted to order a 5- to 6-foot (1.5- to 1.8-meter) reflecting telescope. Failing to find anyone in London who had ever constructed such a large instrument, he decided to construct one for himself. By the end of that year, he built a 5.5-foot (1.7-meter) reflecting telescope. In May 1776, he constructed a 7-foot (2.1-meter) telescope, and in 1778, he built a 10-foot (3-meter) telescope. His discovery of Uranus in March 1781 with the latter telescope made him famous worldwide.

In 1782, Herschel constructed a 20-foot (6.1-meter) telescope with a 12-inch (30.5-centimeter) mirror. Soon after this, he constructed another 20-foot (6.1-meter) telescope with an 18-inch (45.7-centimeter) mirror. In 1786, he moved to build a 40-foot (12.2-meter) telescope with a 48-inch (121.19-centimeter) mirror. King George III promised to provide £4000 along with £200 every year for its maintenance. The construction of this massive telescope provoked much public interest. After its construction in 1789, Herschel was able to see many new images of the vast universe. The king was one of the most frequent visitors. Referring to Herschel's telescope, the king once said to the archbishop of Canterbury: "Come, my Lord Bishop, I will show you the way to Heaven."

Herschel's big telescopes gave him startling images of planets, nebula, and other regions of the sky; they also brought about a totally unexpected scientific discovery. Herschel observed the Sun's spots with his large telescopes, but they frequently heated the eyepiece and his eyes to such an extent that continuous observation was not possible. He inserted glasses with different colors between the eyepiece and his eye to reduce the heating, but he found that the reduction of heating varied enormously with different colors. He became curious as to whether "certain colors should be more apt to occasion heat, others might, on the contrary, be more fit for vision, by possessing superior illuminating power." This eventually led to his discovery of infrared rays (*heat rays* or *caloric rays* in his own phrase) in 1800, the first invisible spectrum (see the discussion below on the invisible spectrum). Herschel's astronomy needed bigger and bigger telescopes, and these telescopes opened up a new scientific field of invisible radiation [see *Astronomy and Cosmology: Eighteenth Century*].

Newton's theory of the invariance of dispersion was incorrect, or his prism-within-prism experiment was faulty. Euler, a wave theorist, also criticized Newton's theory of dispersion. John Dollond (1706–61), a London optician, performed the prism-within-prism experiment in 1758 with a flint-glass prism whose dispersive power was quite different from that of ordinary glass or water. To his surprise, he secured a white outgoing ray despite several refractions. He proved in effect that achromatism was possible. In 1759, he presented his achromatic lens before the Royal Society, obtained the patent for it for 14 years, was awarded the Royal Society's Copley Medal, and became a member of the society.

However, it is suspected that Dollond stole the detailed idea of the achromatic lens from an obscure London lawyer, Chester Moor Hall (1703–71). Not much is known about Hall, but it is known that he carefully repeated Newton's prism-within-prism experiment in 1729, and within four years had produced the first achromatic compound lens. He did not grind his compound lens himself, but asked an optician to grind one piece of it and another optician to grind the other piece. Unfortunately, they both subcontracted the same optician, George Bass (dates unknown), who figured out that these two pieces were for the single achromatic lens. A contemporary witness indicates that it is highly probable that Dollond obtained the secret of Hall's achromatic lens from Bass around 1750.

Nineteenth Century

THOMAS YOUNG AND THE WAVE THEORY OF LIGHT

During the eighteenth century, the Newtonian corpuscular theory and the wave theory of light competed with each other. Wave theories viewed light as a disturbance of some sort in an all-pervading aether; the corpuscular theory considered light in terms of particles and the Newtonian forces acting upon them.

Thomas Young was a London physician by profession, widely knowledgeable about various topics in natural philosophy. As a physician, he had been interested in the physiology of the ear and eye and in the theory of acoustics. By extending the analogy between light and sound, he employed and extended the wave theory of light. During 1800–02, he published a series of papers in the *Philosophical Transactions of the Royal Society* in which he discussed diffraction and Newton's rings (for which Young gave the name *interference*) in wave-theoretic terms.

Newton's colored rings appear when a slightly convex glass is pressed upon a plain glass. To explain this, Young applied to optics the superposition principle in acoustics, which he had developed to explain acoustical beats of sound. An incident light wave is divided into a reflected wave and a refracted wave, the latter traveling farther and being reflected on the lower glass. These two waves, reflected differently, are superposed on each other. When the superposition results in the addition of two waves, the bright rings appear. When the superposition cancels the two waves, dark rings appear.

For this explanation, the essential concept was wavelength. When the crests of two waves coincide, the light is at its maximum intensity. On the other hand, when the crest of one reflected wave is superposed on the trough of the other reflected wave, the light is at a minimum. The distance between two successive crests or troughs is called the *wavelength* of the wave. On this theoretical basis, Young calculated the wavelength of yellow light. It turned out to be 5.76×10^{-5} centimeters, which is fairly close to the accepted modern value.

Young also explained diffraction (colored fringes around a sharp object) in terms of the superposition of the linearly propagated wave with the diffracted wave at its edge. Finding Young's explanation of diffraction was unsatisfactory, Fresnel devised his own by employing Huygens's concept of the wavefront. Fresnel stated that diffraction is due to the interference between the wavefronts of the direct beam and those wavefronts that are deflected around the edge of an object.

Young started with an analogy between light and sound. In his theory, light was a longitudinal wave, like sound. The theory explained reflection, refraction, interference, and diffraction. However, as a longitudinal wave, light had to be symmetrical around the axis of propagation. Hence it was not easy for Young's wave theory to explain polarization (which accompanied asymmetry), which was discovered shortly afterward (see below).

LIGHT AS A TRANSVERSE VIBRATION

At the turn of the nineteenth century, French optics was dominated by Newtonian theory. Pierre-Simon de Laplace (1749–1827) was a staunch Newtonian. Laplace had himself explained atmospheric refraction by analyzing mathematically the interaction between light particles and air. In 1802, a British natural philosopher, William Hyde Wollaston (1766–1828), confirmed Huygens's construction of double refraction. This posed a challenge to Laplace and the Laplacians. Laplace's protégé, Étienne Malus (1775–1812), successfully explained double refraction in terms of the emission theory. Malus also discovered and then explained polarization in terms of the Newtonian concept that rays of light have different "sides." By using the same idea, Jean-Baptiste Biot explained a new phenomenon, that of chromatic polarization. Young's wave theory was simply ignored in Paris.

However, the Newtonian successes in Paris were short-lived. The virtually unknown provincial engineer Augustin Fresnel, armed with mathematics and with

Dominique François Jean Arago's (1786–1853) support, revived the wave theory of light in 1815. He won the French Academy's prize for the theory of diffraction in 1819. Fresnel used Huygens's principle and developed a more general mathematical formula for diffraction than Young. His theory beautifully explained the experimental data within a margin of error of 1.5 percent. Three out of five members of the prize committee, Laplace, Biot, and Siméon-Denis Poisson (1781–1840), were Newtonians (or Laplacians), but they nevertheless awarded the prize to Fresnel.

By 1820, Newtonian theory was more successful in explaining polarization and related phenomena, while the wave theory explained various aspects of diffraction. Laplacians regarded diffraction as a less important phenomenon than polarization, since they thought diffraction was a secondary phenomenon caused by the interaction between light and material objects. Polarization mattered most to wave theorists. But with longitudinal waves, it was not possible to devise a plausible mechanism of light that could adequately explain polarization.

In the early 1820s, Fresnel introduced the idea of transverse waves to explain polarization, but to do this he had to accept that aether must be highly elastic, like a jelly, a hypothesis that even Fresnel himself found hard to accept. The elastic solid aether model was later explored by Augustin-Louis Cauchy (1789–1857), James MacCullagh (1809–47), George Green (1793–1841), and William Thomson (later Lord Kelvin, 1824–1907), but it constantly posed difficult questions for wave theorists. The problem with aether was that it had to be rigid enough to transmit a transverse vibration, but also offer no resistance to the motion of Earth.

Fresnel's wave theory was confirmed by the white-spot experiment. This experiment proved Fresnel's counterintuitive prediction that the center of the shadow of a circular object must be the brightest. Wave theory was firmly established by 1850. Its validity was again reaffirmed by the measurement of the speed of light in water by Armand Fizeau and Jean Bernard Léon Foucault. It turned out to be less than that in air, as had been predicted by the wave theory of Fermat, Huygens, and Fresnel.

INVISIBLE SPECTRUM IN THE EARLY NINETEENTH CENTURY

Throughout the eighteenth century, the spectrum referred to something visible and colored. In 1800, William Herschel discovered an invisible ray. He had accidentally noticed that glasses of different colors used in his telescope revealed different heating effects. This led him to examine the heating action of various parts of the colored spectrum. With a prism and thermometers, he detected a rise in temperature beyond the red end of the solar spectrum, where no visible light existed, but none beyond the violet end. He named the invisible rays to which he ascribed the effect *heat* or *caloric rays.*

He went on to demonstrate that these new rays could be reflected and refracted, which raised the question of the relationship between the rays and light. Herschel discovered that while an uncolored glass might be perfectly transparent to visible light, it nevertheless absorbed about 70 percent of the heat rays. He performed several different kinds of experiments. The results always pointed to a difference between light and heat rays. Herschel concluded that two independent spectra existed.

Herschel's discovery of invisible heat rays was at first widely questioned. Meanwhile, Johan Wilhelm Ritter (1776–1810) in Germany discovered in 1801 the chemical effect of invisible rays lying beyond the violet end of the solar spectrum. As a follower of German *Naturphilosophie,* Ritter had discovered what he called *deoxidizing rays* while performing his research under the strong conviction that polarity in nature should reveal the cold counterpart of heat rays, with the cold rays lying beyond the violet spectrum. Three years after Ritter's discovery, Thomas

Young produced an interference pattern for Ritter's "chemical" (ultraviolet) rays by using paper treated with silver chloride.

After Herschel and Ritter's discoveries, three different rays—heat, light, and chemical—had been identified and were available to scientists. The point of controversy remained whether the heat and chemical rays were extensions of the visible spectrum into the invisible regions, or whether they were utterly different from light. A consensus that heat, light, and chemical rays belong to one and the same spectrum, distinguished solely by wavelength, emerged gradually in the 1840s and 1850s.

EMERGENCE OF SPECTROSCOPY

Spectroscopy began with the discovery by Wollaston in 1802 of several dark lines in the solar spectrum, which he thought to be an instrumental anomaly. The German Joseph Fraunhofer (1787–1826), who was engaged in manufacturing glass for telescopes and prisms, interpreted this anomaly as a natural phenomenon and eventually exploited it in an extremely influential instrument. Fraunhofer had produced superb prisms and achromatic lenses. With them, he discovered more than 500 fine dark lines in the solar spectrum (see Figure 5). He immediately utilized these lines for the calibration of achromatic lenses and prisms, since the sharp dark lines were good benchmarks for distinguishing the hitherto rather obscure boundaries between different colors [see *Astronomy and Cosmology: Nineteenth Century*].

Although Fraunhofer noticed that two very close yellow lines obtained in the spectrum of a lamp agreed in position with two dark lines in the solar spectrum (which he named "D"), he did not take further theoretical steps to speculate about the reason for this coincidence. In the 1830s, in the context of the dispute between the wave and emission theories of light, a heated debate arose between David Brewster (1781–1868), George Airy (1801–92), and John Herschel (1792–1871) over the cause of the dark lines in the spectrum. But Fraunhofer was far removed from such matters.

The idea that line spectra might be related to the structure of the atoms or molecules of the light-emitting or light-absorbing substance was suggested by several scientists. In 1827, John Herschel interpreted dark and bright lines as indicating that the capacity of a body to absorb a particular ray was associated with the body's inability to emit the same ray when heated (this is exactly the opposite of what is now known to be the case). Foucault in 1849 and George Stokes (1819– 1903) in 1852 conjectured on the mechanism of dark and bright lines, and William Swan (dates unknown) in 1856 attributed the D lines to the presence of sodium in light-transmitting media and in light-emitting sources.

However, it was not until 1859 that Gustav Robert Kirchhoff (1824–87), at the request of Robert Bunsen (1811–99), proposed the law of the identity of emission and absorption spectra under the same physical conditions, which made spectroscopic investigations of light widespread. Nearly simultaneously, Balfour

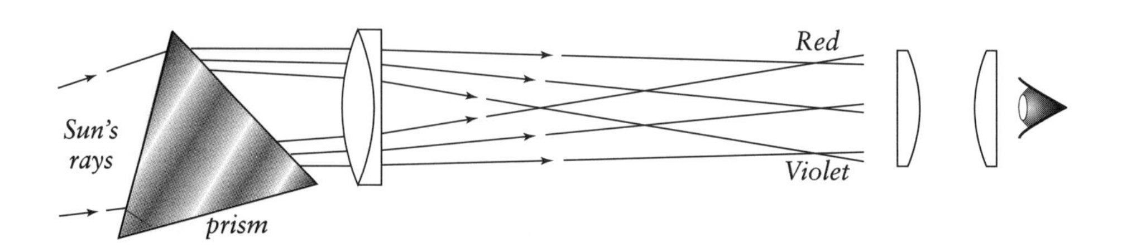

Figure 5. Schematic illustration of Fraunhofer's apparatus for viewing the solar spectrum and its absorption lines. The sun's rays were almost parallel when they met the prism, since the slit in the window was some distance from the apparatus.

Stewart (1828–87) in England suggested a similar idea. The difference between Kirchhoff and Stewart stemmed from the fact that Kirchhoff's idea was based on general principles of thermodynamics and rigorous demonstration, whereas Stewart's concept was rooted in Pierre Prévost's (1751–1839) much older, and looser, theory of exchanges. The priority dispute was fought not only by themselves, but also by their followers until the end of the nineteenth century.

Kirchhoff and Bunsen constructed the first spectroscope in 1860, and the word *spectroscopy,* or spectrum analysis, began to be widely used in the late 1860s. Various kinds of spectroscopes were constructed. In the mid-1860s, for example, William Huggins (1824–1910) combined spectroscopy with a stellar telescope for the purpose of examining the stellar spectral lines. This marked the beginning of astrospectroscopy, which made astronomy a laboratory science.

SPECTRUM ANALYSIS

How did scientists understand the origins of spectral lines? At first, it was commonly believed that the banded spectrum represented the effect of a molecule, while the line spectrum represented the effect of an atom. This belief was not unchallenged. Norman Lockyer (1836–1920), who had noticed changes in line spectra under certain conditions, proposed in 1873 a scheme involving the dissociation of an atom. Line spectra, according to Lockyer, were caused by more elementary constituents than atoms. Lockyer's hypothesis was not seriously considered by his contemporaries, mainly because atoms had long been considered indivisible.

In the 1870s and 1880s, several scientists tried to find mathematical regularities among the various line spectra of a given substance. In 1871, the Irish physicist George Johnstone Stoney (1826–1911) thought the hydrogen spectrum was due to the splitting of the original wave into several different parts. He suggested that this splitting could be analyzed by employing **Fourier analysis** and by matching the harmonics that appeared in the theorem with the spectrum lines observed. He noted three hydrogen lines, at 4102.37, 5862.11, and 6563.93 angstroms, and found their ratios to be approximately 20, 27, and 32. In 1881, Arthur Schuster (1851–1934), who claimed that Stoney's ratio could not be considered to be a mathematical regularity, cast strong doubt on the harmonics hypothesis. Schuster, however, could not suggest a plausible alternative theory.

* **Fourier analysis.** The analysis of any single-valued periodic function into its simple harmonic components.

In 1884, Johann Balmer (1825–98), a virtually unknown Swiss mathematician, examined four hydrogen lines and formulated the series now named after him: $\delta_n = [n^2/(n^2 - 2^2)]\ \delta_0$, where δ_n are the wavelengths of the hydrogen spectra, δ_0 is a constant, and $n = 3, 4, 5, \ldots$. The Balmer series was not at all similar to the simple harmonic ratio that had been proposed by Stoney. Although Balmer's formula beautifully linked the four known hydrogen spectra and turned out to be valid for the newly discovered ultraviolet and infrared spectra of hydrogen, it posed more questions than answers, for scientists failed to suggest why such regularity should exist. Later, Niels Bohr's (1885–1962) quantum model of the hydrogen atom, in which the emission of radiation was caused by the quantum jump of an electron from a higher to a lower energy level, yielded the Balmer series [see *Overview: Physics*].

LIGHT AS AN ELECTROMAGNETIC WAVE

Michael Faraday (1791–1867) had been elaborating on his concept of lines of force since the early 1830s. Magnetic and electric lines of force implied the curved action of forces as well as action through contiguous particles that permeated the medium. Faraday also believed in the conversion of various forces. As he expressed it, he held "the very strongest conviction that Light, Magnetism, and Electricity must be connected." He had previously discovered the conversion between electric and magnetic forces in 1831 in electromagnetic induction. In 1845, at Thomson's suggestion, Faraday began to look for the action of a dielectric material on polarized light. He did not find the sought-after effect of dielectric, but he did discover the effect of magnetic fields upon polarized light. Magnetic fields and light proved to be interrelated.

This discovery pointed to the hypothesis that light, electricity, magnetism, chemical forces, and perhaps gravitation were all united. Faraday's speculation was clearly articulated in his paper "Thoughts on Ray-Vibrations," published in 1846. "The view which I am so bold as to put forth considers, therefore, radiation as a high species of vibration in the lines of force which are known to connect particles and also masses of matter together. . . . The propagation of light, and therefore probably of all radiation action, occupies time; and, that a vibration of the line of force should account for the phenomena of radiation, it is necessary that such vibration should occupy time also."

James Clerk Maxwell (1831–79) was one of the few people who took Faraday's speculation seriously. Devising an ingenious "idle wheel" model for the electromagnetic aether, Maxwell suggested in 1861 that light itself was a species of electromagnetic disturbance. Maxwell's suggestion did not undermine the status of the established wave theory, since his electromagnetic disturbances had all the standard properties of a wave, and more. In 1865, Maxwell formulated his electromagnetic theory of light in a tighter mathematical form without relying on the debated mechanism of his aether. Maxwell's theory implied that the ratio of electrostatic to electromagnetic unit of electricity should be equal to the velocity of light. Although controversial in the 1860s and 1870s, Maxwell's claim became more widely accepted in the late 1870s, although Maxwell's theory did not prove persuasive to those who, like Thomson, had not accepted Maxwell's system [see *Electromagnetism*].

Maxwell himself never attempted to generate or to detect electromagnetic waves longer than those of light. He seemed far less concerned with the production and detection of electromagnetic waves than with revealing the electromagnetic properties of light. Nevertheless, Maxwell's electromagnetic theory of light did suggest that it might be possible to create such disturbances—to produce, as

Science and Technology: Electromagnetic Waves and Marconi, 1896

Heinrich Hertz had discovered what we now call the *microwave spectrum*, extending the radiation into a thoroughly new region. Following Hertz's experiments, Lodge in Britain, Augusto Righi (1850–1920) in Italy, and J. Chunder Bose (1858–1937) in India pushed in the direction of shorter wavelengths. For this, spherical oscillators replaced Hertz's linear ones. As for detectors, the coherer, invented by Édouard Branly (1844–1940) and improved by Lodge, replaced Hertz's spark-gap resonators. With these instruments, Bose successfully generated waves with centimeter wavelengths. Experiments on the diffraction, refraction, polarization, and interference of microwaves followed.

Practical applications of Hertzian waves were not at first obvious. In 1895–96, Guglielmo Marconi (1874–1937) opened a new field by applying Hertzian waves to telegraphy. What is notable in Marconi's work is his movement against the mainstream of physics: He tried to increase, rather than to decrease, the wavelength. He erected a tall vertical antenna and connected one end of the discharge circuit to it and the other end to the earth. The antenna and the earth connections increased the capacitance of the discharge circuit considerably. This lengthened the wavelength and increased the power that could be stored in the system. When he succeeded in the first transatlantic wireless telegraphy in 1901, the transmitter used 20 kilowatts of power and the estimated wavelength was of the order of 1000 meters. Long waves were the only possible means to combine power and communication.

it were, something that could truly be called an electromagnetic wave. The spectrum would then be extended far beyond the infrared, to centimeter, and even meter, wavelengths.

In the early 1880s, Maxwellians such as George F. FitzGerald (1851–1901) and J. J. Thomson (1856–1940) suggested ways to generate such waves by purely electric methods. In particular, FitzGerald specified rapid electrical oscillations in a closed circuit, such as condenser discharges, as a correct method of generating these waves, and calculated the wavelength that would thereby be generated. But he did not know how to detect such waves. In 1887–88, Oliver Lodge (1851–1940) experimented with Leyden jar discharges, but he did not produce or detect fully propagating waves.

Heinrich Hertz (1857–94), who had been exposed, via Hermann von Helmholtz (1821–94), to both the German action-at-a-distance school and the Maxwellian field-theoretic electrodynamics, observed a curious effect displayed by secondary sparks from a pair of metallic coils—called *Riess coils*—in 1887. He first tried to abolish the sparks, but failed to do so. Then he tried to control and manipulate the effect. He fabricated a spark detector, which eventually became a means to probe the propagation of electric forces. Hertz eventually concluded, after extensive investigations, that he had produced and detected Maxwell's electromagnetic waves. Hertz measured the length of his waves to be 26 inches (66 centimeters).

MICHELSON-MORLEY EXPERIMENT

Fresnel's wave theory of light transformed the aether into an indispensable medium for the transmission of light waves. The crucial question was whether or not Earth's motion would disturb the all-pervasive aether. Throughout the nineteenth century, most scientists believed that the aether was stationary because if it had been disturbed by the motion of Earth, the passage of the light from the Sun to Earth would have been affected. But if the aether was absolutely stationary, the relative velocity between Earth and the aether should be observed, because the stationary aether functioned as the absolute frame of reference. Maxwell suggested that the effect of the stationary aether could be detected by comparing the velocities of light rays traveling in perpendicular directions [see *Overview: Physics*].

Stimulated by Maxwell, in 1881, the American physicist Albert Michelson (1852–1931) tested the differences in the velocities of light rays traveling the same distance at right angles to one another. Michelson designed his sensitive interferometer to measure the interference between these two light rays resulting from their different relative velocities, but detected no such effect. He reluctantly concluded that "the hypothesis of a stationary aether is incorrect." Since Henrik Lorentz (1853–1928) criticized Michelson's calculation, Michelson performed another set of experiments with his assistant Edward Morley (1838–1923) in 1887–88, but still could not detect any effect. If their experimental results were correct, the accepted conception of the stationary aether could not be sustained.

As a theoretical physicist and a strong believer in the aether, Lorentz did not accept Michelson's experimental result indicating the nonexistence of the aether. To "save" the aether, Lorentz showed in 1892 that Michelson's experimental result could be explained by assuming that the time coordinate of a moving body was slowed down. This was a strange assumption, since, according to our daily experience, time should not be affected by the motion of a body. Lorentz thus added that this changing time was not real time but only a mathematical time introduced to explain an anomaly such as Michelson's experiment. He called it *local time*. In

1904, Lorentz realized that his new theory would lead to another conclusion, that a moving body undergoes a contraction of its length.

In this way, Lorentz was able to retain the aether. But in so doing, his theory assumed several hypotheses, such as local time, the contraction of a moving body, a complicated theory of the relationship between aether and matter, and some inconsistencies such as the violation of Newton's third law. As is now well known, Albert Einstein's (1879–1955) novel theory of special relativity (1905) explained Michelson's experimental result without such ad hoc hypotheses, but his paper was, in fact, written without any knowledge of Michelson's experiment. According to Einstein's theory of relativity, there was no absolute frame of reference, and "the aether became superfluous."

Twentieth Century

EINSTEIN'S PHOTON AND THE PHOTOELECTRIC EFFECT, 1905

Einstein's theory of special relativity rendered the aether unnecessary. But if the aether was not needed any more, how could light be sustained as a wave phenomenon? During the last quarter of the nineteenth century, the wave theory of light was strengthened by Maxwell's electromagnetic theory of light. The wave theory, combined with Maxwell's theory, explained most optical phenomena. The aether, although problematic, remained an essential medium for light as a wave.

Besides Michelson's experiment, the wave theory of light failed to explain another optical phenomenon. This was the photoelectric effect, first discovered by Hertz, in which light did not behave as a wave. In the photoelectric effect, surface electrons of a metal were stimulated and emitted by light, but their velocity was proportional not to the intensity of the light (as wave theory predicted), but to its frequency.

In his paper published in March 1905, Einstein tackled this question by proposing the revolutionary concept of the light quantum. In 1900, the German physicist Max Planck (1858–1947) had proposed that energy in blackbody radiation be quantized by a small amount $h\upsilon$, where h is a universal constant and υ is the frequency of radiation [see *Atomic and Nuclear Physics*]. On this basis, Einstein maintained that "monochromatic radiation of low density behaves in thermodynamic respect as if it consists of mutually independent energy quanta of magnitude $h\upsilon$." He later claimed that the light quantum also had the momentum $p = h\upsilon/c$. The photoelectric effect could easily be explained with this, because the electrons' energy E is equal to $h\upsilon - P$, where $h\upsilon$ is the energy of the radiation and P is the work function of the metal, that is, the energy needed for the electrons to escape the surface.

This idea of the light quantum was not received favorably. It forced one to give up the wave theory of light, which explained almost all optical phenomena (interference in particular). Max Planck's recommendation of Einstein to the Prussian Academy in 1913 shows that even Planck did not accept Einstein's light quantum: "In sum, one can say that there is hardly one among the great problems in which modern physics is so rich to which Einstein has not made a remarkable contribution. That he may sometimes have missed the target in his speculations, as, for example, in his hypothesis of light-quanta, cannot really be held much against him, for it is not possible to introduce really new ideas even in the most exact sciences without sometimes taking a risk." Einstein's light quantum was reevaluated only in the 1920s. In 1923, the American physicist Arthur Holly Compton (1892–1962) discovered that the wavelength of x-rays scattered from free electrons was greater than that of the original rays, confirming

Einstein's $E = h\upsilon$ and $p = h\upsilon/c$. After this discovery, physicists began seriously to consider Einstein's light-particle theory.

QUANTUM MECHANICS AND WAVE-PARTICLE DUALITY

The difficulty with Einstein's light-quantum theory was its inability to explain wave phenomena of light such as interference and diffraction. Louis de Broglie (1892–1987), a French physicist, was very interested in this puzzle. He thought that Einstein's light particle should be thought of as accompanying some kind of wave to explain these wave effects. But here, he turned his thinking upside down. If light was a particle that accompanies a wave, an ordinary particle might accompany a wave as well. In his Ph.D. thesis published in 1923, de Broglie proposed that an ordinary particle could be regarded as accompanying a wave with a particular frequency. Its frequency is determined by its energy, $\upsilon = E/h$, where E is the particle's **kinetic energy** and h is Planck's constant.

* **Kinetic energy.** The energy that a body has by virtue of its motion.

De Broglie's idea of the matter wave became known to Erwin Schrödinger (1887–1961). Schrödinger first formulated a time-independent equation for the matter wave and later devised the general Schrödinger equation. Schrödinger's equation solved a wave equation in terms of the general wave function P. He then proved that his equation would produce the same results as those produced by Werner Heisenberg's (1901–76) **matrix mechanics**. However, the wave function P was a complex-valued function of position and time and, as such, direct interpretation of it seemed meaningless.

* **Matrix mechanics.** The mathematical theory of quantum mechanics that represents operators by their matrix elements.

Schrödinger, however, insisted on drawing a physical meaning from it. On the other hand, Max Born (1882–1970), a mathematical physicist in Göttingen, who had also worked on the matrix reformulation of quantum mechanics, argued in 1926 that $|P(x)|^2$ represented the probability of finding a particle at point x. Einstein supported Schrödinger's realistic, but eccentric interpretation. Niels Bohr (1885–1962) adopted Max Born's probabilistic interpretation.

Synthesizing various trends, Bohr proposed a general scheme. He argued that our ordinary language of waves and particles, which worked well in classical situations, could no longer represent the wholeness of nature in the atomic domain. Such seemingly conflicting concepts as "wave versus particle," "causality versus indeterminacy," or even "classical versus quantum" were not really conflicting, but instead, complementary. According to Bohr, it is not

Science and Technology: Invention of Polaroid Film, 1947

In 1947, Edwin Herbert Land (1909–91), the founder of the Polaroid Corporation in Boston in 1937, first demonstrated the instant, or one-step, camera before the Optical Society of America. The first commercial instant camera, Model 95, came into the market in 1948, marking a great success. This was the beginning of the instant camera, which is more widely known as the Polaroid. (The company's name was originally chosen as Polaroid since it manufactured a cheap polarizer.)

Land's motivation was humble. According to his recollection, an innocent question from his daughter triggered this innovation. "I recall a sunny day in Santa Fe, N.M., when my little daughter asked why she could not see at once the picture I had just taken of her. As I walked around the charming town, I undertook the task of solving the puzzle she had set to me. Within an hour, the camera, the film, and the physical chemistry became so clear to me."

He performed the first experiment with a sheet of blotting paper wetted with a developer. He rolled the film against the paper and left it exposed in the camera for a minute. When he peeled it off, he obtained a black-and-white image on the paper. From this simple experiment, it took him three more years of intensive research and innovation to finally produce a commercial instant film.

correct to say that light is "either wave or particle." Neither is it correct to say that light is "both wave and particle."

What, then, is light in Bohr's synthesis? Light sometimes behaves like a particle and, at other times, like a wave. Wave and particle are simply complementary concepts for the fuller understanding of light. Einstein strongly objected to Bohr's complementarity principle. Werner Heisenberg was not very happy with it, but he reluctantly accepted it. The Copenhagen interpretation, built largely on Bohr's complementarity principle, was rapidly accepted as the new orthodoxy in quantum physics.

DENNIS GABOR AND HOLOGRAPHY, 1948

The chromatic and spherical aberrations that affect the sharpness of a telescopic image are also a problem in the microscope, which also uses lenses. Calculations show that because of these effects, one cannot see an object smaller than 4 angstroms (4×10^{-8} centimeter). In the 1940s, Dennis Gabor (1900–79), who had fled to England from his native Hungary to escape Hitler's presence in eastern Europe, tried hard to overcome this optical limit. Gabor had once worked in Germany on the focusing of electron beams and their application to the electron microscope.

While waiting his turn to play tennis at Easter 1947, a novel idea came to him. He imagined a two-step imaging process: first, using an electron microscope, photograph the interference caused by the superposition of a diffracted wave with a **coherent** background **wave**; and second, correct this image by optical means. This electron microscope photograph records all the information, including the amplitude and phase. He called this electron interference pattern a *hologram* after the Greek word *holos*, meaning the whole. This entire process was called *holography* (see Figures 6 and 7). To his disappointment, however, the experiment failed to image individual atoms, as he had hoped.

* **Coherent wave.** A wave that maintains a nearly-constant phase relationship, both temporally and spatially.

During the following 15 years, Gabor lost interest in holography. After the invention of the laser, however, holography became a popular scientific topic, and Gabor's dream came true, because he could take holographic pictures of large atoms by using the laser to provide a highly coherent background.

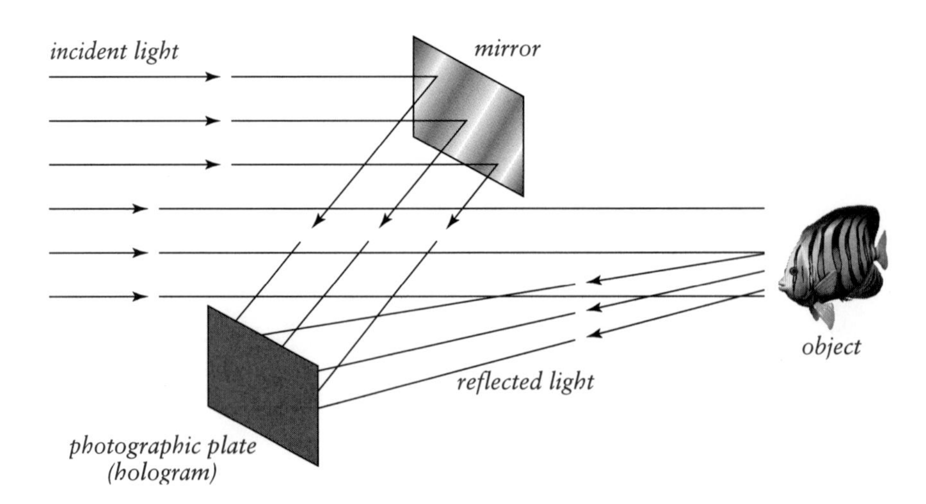

Figure 6. The way to construct a hologram.

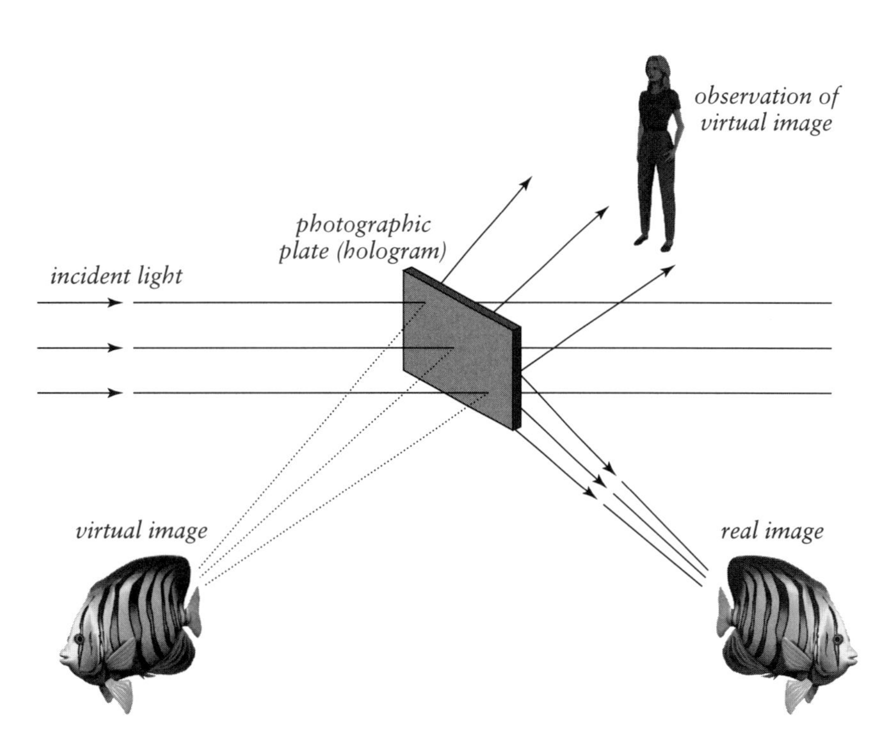

Figure 7. The way to view a hologram.

CONSTRUCTION OF LASERS, 1960–66

After World War II, the U.S. military had been working to reduce the wavelength of microwaves down to 1 millimeter (0.039 inch). Charles Townes (b. 1915) at Columbia, one of those working on microwave amplifiers and oscillators, hit upon the idea in 1951 that stimulated emission could be used to produce a millimeter wavelength by incorporating a resonant cavity to supply feedback of the wave amplified by the resonant system of atoms or molecules. In this *maser* (microwave amplification by stimulated emission of radiation), he used ammonia gas as the resonant system, but after preliminary trials and calculation revealed that the wavelength thus generated would be 1.25 centimeters (0.49 inch), instead of 1 millimeter (0.039 inch), Townes lost interest in the maser as a microwave amplifier and oscillator. Because of this, the maser was announced to the public in 1955 as the first atomic clock, with exactness surpassing that of existing time standards such as pendulum or quartz clocks for the first time.

The success of the maser stimulated several groups of scientists to turn to the problem of reducing the wavelength to that of visible light. One of the central problems was finding a material that could lase. Townes first tried potassium vapor (pumped by a potassium lamp) and signed a contract with the U.S. Air Force for financial support in 1958. Townes was working with Arthur L. Schawlow (1921–99), who had rejected the pink ruby as a lasing candidate. One of Townes's assistants, R. Gordon Gould (b. 1920), who first copyrighted his idea of the laser, joined a private company and signed a research contract with the Advanced Research Projects Agency (ARPA). Bell Laboratories also launched a research team, which chose gases—in particular, helium—instead of a solid, as a lasing medium. Another research team was working in IBM's T. J. Watson Center.

However, the scientist who first invented a practical laser did not come from one of these active centers. Theodore Maiman (b. 1927) was a researcher at Hughes Aircraft Company, who on the basis of his conviction that a proper lasing material must be a solid, discovered that a ruby could lase. Since ruby had already been disregarded as a lasing material, and since Maiman was not one of the active laser hunters, his announcement was greeted with much skepticism. However, the lasing properties of ruby were soon confirmed by other groups, and more important, within a year the IBM group, the Bell Labs group, and others discovered different lasing materials, such as helium-neon.

The laser had something for everyone. For scientists, it provided an instrument for high-resolution spectroscopy. It also generated many interesting theoretical problems to solve. For engineers, its oscillatory and amplificatory properties were enough to attract their attention. However, it was the military that took the initiative in laser research after 1960. Various research projects on the laser were supported by the Army Missile Command, the Office of Naval Research, ARPA, and other military research institutions, which saw in the laser possibilities of military communications, of guiding missiles, and of high-energy weapons for **antiballistic** missile defense. Different kinds of lasers, such as Q-switched ruby, carbon dioxide, and gas-dynamic lasers, were constructed under military support. The initial interest in high-energy laser research waned after successive failures in the mid-1970s. The interest in the laser as a weapon was revived in the form of the x-ray laser, which constituted the most important element in the proposed Strategic Defense Initiative (known as "Star Wars") of the 1980s.

* **Ballistics.** Relating to the branch of applied mechanics concerned with the motion and behavior of missiles and related phenomena.

Bibliography

Buchwald, Jed Z. *The Rise of the Wave Theory of Light: Optical Theory and Experiment in the Early Nineteenth Century.* Chicago: University of Chicago Press, 1989.

Cantor, G. N. *Optics after Newton: Theories of Light in Britain and Ireland, 1704–1840.* Manchester: Manchester University Press, 1983.

Hall, A. Rupert. *All Was Light: An Introduction to Newton's* Opticks. Oxford: Clarendon Press, 1993.

Hong, Sungook. "Theories and Experiments on Radiation from Thomas Young to X-rays." In *Modern Physical and Mathematical Sciences.* Edited by Mary Jo Nye. Cambridge: Cambridge University Press. Forthcoming in 2002.

King, Henry C. *The History of the Telescope.* London: C. Griffin, 1955.

Lindberg, David C. *Theories of Vision from Al-Kindi to Kepler.* Chicago: University of Chicago Press, 1976.

Pais, Abraham. *Inward Bound: Of Matter and Forces in the Physical World.* Oxford: Clarendon Press, 1986.

Sabra, A. I. *Theories of Light: From Descartes to Newton.* New York: Cambridge University Press, 1981.

Shapiro, Alan. *Fits, Passions, and Paroxysms: Physics, Method, and Chemistry and Newton's Theories of Colored Bodies.* New York: Cambridge University Press, 1993.

Paleontology

P. David Polly
Rebecca L. Spang

Coined in the 1820s, the word *paleontology* has three Greek components: *paleo* (ancient), *onto* (being or existence), and *logy* (study). Yet "the study of ancient existence" only poorly defines the science of paleontology, for the discipline has never included ancient civilizations (such as those of Mesopotamia) and often defers any consideration of human evolution to anthropologists. But in the 1820s and 1830s, those distinctions were not made. Educated persons might well have considered paleontology to be intimately linked with another burgeoning field, that of paleography (the study of old or ancient writing). With Jean-François Champollion's (1790–1832) deciphering of the Rosetta Stone in 1821, and William Conybeare's (1787–1857) reconstruction of the plesiosaur in 1824, the pre-Roman world was beginning to spring to life. Today, we know that 65 million years passed between the extinction of those giant marine reptiles and the building of the Pyramids, but the immensity of the geological time scale was then only dimly imagined. Georges Cuvier (1769–1832), often called the founder of paleontology (although he never used that word), described his findings as those of a "new sort of antiquary": one who collected bones and fossils rather than mosaics and coins. Paleontology first flowered because Cuvier had convinced his fellow naturalists that entire species really had become extinct. Once exploration and empire carried European science to the corners of Earth, the learned no longer imagined that plesiosaurs and giant sloths inhabited *terra incognita* but recognized that they had died out completely—ancient and modern existence had parted company and paleontology was born.

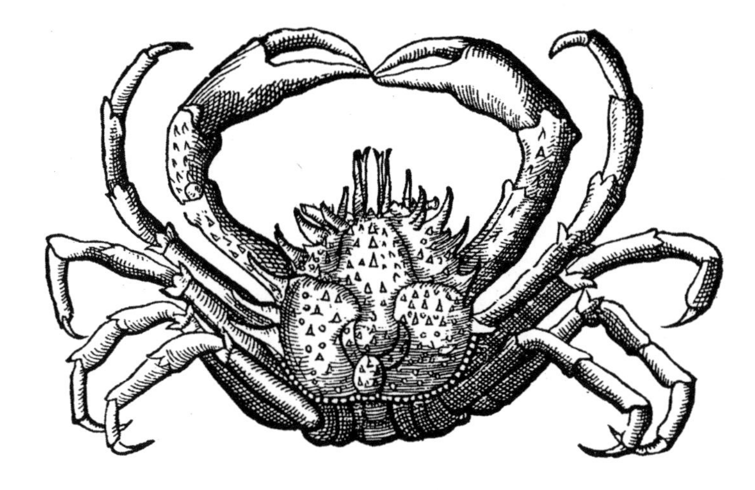

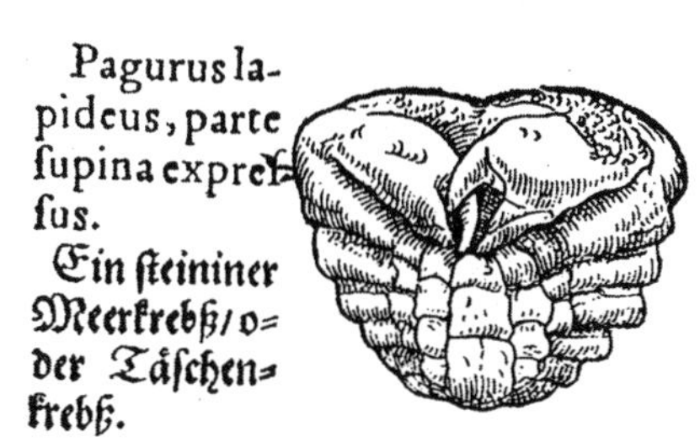

Figure 1. Conrad Gesner's 1565 illustration of a fossil and the living crab he thought resembled it. © The Natural History Museum, London

Extinction

Before the early nineteenth century, there were no disciplines of paleontology, geology, or biology. Instead, there was natural history, a broad inquiry into the objects and phenomena of nature. Some of those objects were fossils, not in the sense in which the word is used today, but in a broader sense of "things dug out of the ground." Fossils were the gems, minerals, rocks, and petrified remains that fifteenth- and sixteenth-century miners regularly brought to the surface. Some, such as gold, were highly valued. Others, such as mercury, had strange properties. Still others, such as ammonites (coiled aquatic mollusks that are distantly related to the living chambered nautilus), curiously resembled living beings. It is only the latter category that we now consider to be fossils, but in the sixteenth century, all of these objects were closely associated. As more fossils were discovered, and as more Renaissance scholars tried to explain them, many types of objects were placed in categories of their own: metals, gems, minerals, and stones. The leftover "fossils" were exactly what one now would imagine—objects from Earth that were recognizable as parts of animals and plants—and their association with living organisms was stronger (see Figure 1). For example, the Florentine anatomist Nicholaus Steno

(1638–86), demonstrated that fossil "tongue stones" were, without doubt, sharks' teeth. However, a century later, Carl Linnaeus (1707–78) still classified fossils as mineral rather than vegetable or animal.

Competing theories explained how these objects had come to be encased in stone, theories that ranged from Noah's flood to sporelike "seeds" of animals and plants that had fallen into crevices and grown in the earth. Seventeenth-century naturalists thought it was plausible that fossils were the petrified remains of once-living creatures, but they were puzzled by the fact that some of those creatures were not known among the living. Common fossils, such as the coiled ammonites, did not correspond to any known animal. Many Europeans thought it implausible that species had disappeared completely because Christian doctrine taught that God had created animals and plants. If God had designed them, why would they not have survived? The most logical conclusions were that fossils did not represent living remains, or that the creatures they represented were still living but remained undiscovered. John Ray (1627–1705), famous for his classification of animals, preferred the latter explanation, while Carl Linnaeus preferred the former. As late as 1803, it remained credible that substantial floras and faunas were unknown to science.

When Thomas Jefferson (1743–1826) organized the exploration of America's new Louisiana Territory, he asked Meriwether Lewis (1774–1809) and William Clark (1770–1838) to collect evidence of any living animals thought to be rare or extinct, including the giant elephant that had recently been exhumed at Big Bone Lick in Kentucky. A large skeleton had been found there in 1739, and it was an exciting and controversial animal. Comte de Buffon (Georges-Louis Leclerc, 1707–88) said that it was an elephant but with teeth like a hippopotamus. Controversy ensued. In 1796, Georges Cuvier presented a painstakingly detailed comparison of living and fossil elephants, emphatically concluding that the Big Bone Lick mastodon was extinct, as were many other fossil elephants. By 1814, he was convinced that periodic "revolutions," or catastrophes, had ruined vast ancient worlds, wiping their inhabitants from the face of Earth. Cuvier splendidly reanimated several different faunas, unearthed from the succession of rocks in the Paris Basin. Scientific opinion yielded to Cuvier's arguments, as the Louisiana Territory and other remote areas failed to produce living mastodons or other fossilized species. In some cases, fossil species were eventually found among the living. In the mid-nineteenth century, crinoids—stalked sea lilies commonly found as fossils—were discovered living in deep waters, and the coelacanth, the famous "living fossil" fish, was dredged from the depths in 1938. But those cases were rare, and the idea of extinction caught hold. The natural world was sufficiently well understood that it became clear that Cuvier's fossil menagerie belonged to the past. Cuvier's colleague Henri Marie Ducrotay de Blainville (1777–1850) coined the word *paléontologie* in the 1820s, not for the study of fossils per se, but specifically for the study of extinct life.

If extinction was an uncommon idea before Cuvier, it became very fashionable after him. Alcide Charles Victor d'Orbigny (1802–57), who occupied the first chair in paleontology at the Muséum d'Histoire Naturelle (Paris Museum), elaborated on Cuvier's revolutions, describing no fewer than 27 creations (which he termed *stages*). Each stage of Earth history had been wiped out by violent upheavals that destroyed contemporary species. Inspired by the floodlike nature of Cuvier's last revolution, Oxford's William Buckland (1784–1856) preached that Earth's most recent catastrophe had been Noah's deluge. Following Cuvier, Buckland showed that the bones found in a Yorkshire cave belonged to extinct species and that hyenas (no longer found outside Africa) had gnawed them. Only a flood could have swept away the hyenas but left the cave intact, he said. Even those opposed to Cuvier embraced extinction. Englishman Charles Lyell (1797–1875), a stringent opponent of Buckland, developed an encompassing system of

Mt. Etna and the Pace of Geological Time

Mt. Etna is a spectacular volcano rising almost 11,000 feet (3300 meters) from the east coast of Sicily. The tallest active volcano in Europe, Etna has served as a navigational beacon on one of the world's most traveled seaways since ancient times, and its eruptions, often accompanied by earthquakes, have long intrigued human thinkers. The Greek poet Pindar described a massive eruption that occurred in 475 B.C., historians recorded particularly destructive eruptions in A.D. 1169 and 1669, and Sicilians fled major rumblings from the mountain as recently as 1971 and 2001. Its history and visibility have made Etna the archetypal volcano for European writers, and it has played a central role in debates about geology, the age of Earth, and the history of life.

When Charles Lyell made his seminal geological trip to Italy in the 1820s, during which he reinterpreted the history of Earth's rocks in terms of gradual processes and their cumulative effects over eons, he paid particular attention to Etna. In Lyell's uniformitarian theory of Earth, geological history was to be interpreted as the accumulation of small changes rather than supernatural, catastrophic upheavals such as Cuvier's catastrophes or Buckland's floods. Furthermore, these small changes extended over an almost limitless time in Lyell's view, and Earth, by logical extension, must therefore be nearly infinitely old. Etna provided many good examples for Lyell. He combined observations on the many cones and vents, the layers upon layers of cooled lava flows, and the interbedded ash layers with the historical descriptions of major eruptions. Even Pindar's 2000-year-old lava flows could be shown to be among the most recent of Etna's layers of history. Lyell's gradual extinctions and, later, Darwin's theory of evolution by natural selection were set in the vast time stretches that Lyell measured at Etna.

geology (including paleontology) that explained Earth history in terms of gradual change by processes that can be observed today (this theory became known as *uniformitarianism*). Like the building of mountains or the cutting of valleys, extinction happened by degree [see *Earth Science and Geology*]. As part of his strategy, Lyell replaced the geological name Tertiary (which had catastrophic overtones from Cuvier) with three new ages: **Pliocene**, **Miocene**, and **Eocene**. Significantly, these ages were defined by the percentage of extinct species, implying continuous loss rather than catastrophic destruction, and, even more significantly, the rocks Lyell used to illustrate these new ages were in Cuvier's Paris Basin.

➢ **Pliocene.** See *Earth Science and Geology.*

➢ **Miocene.** See *Earth Science and Geology.*

➢ **Eocene.** See *Earth Science and Geology.*

* **Senescence.** Study of the biological changes pertaining to aging.

Other scientists had their own explanations of extinction. Charles Darwin (1809–82), who developed the theory of gradual evolution though natural selection, later compared extinction to sickness and death. Darwin considered decrepitude and extinction to be a gradual but inevitable process. Many paleontologists suggested that species have fixed amounts of "vital energy," which, when exhausted, results in extinction. Alpheus Hyatt (1838–1902) incorporated the same idea into his "racial **senescence**"; and Otto Schindewolf (1896–1971) used it in his twentieth-century "typostrophism," while Edward Drinker Cope (1840–97) asserted that evolutionary specialization caused extinction, as did Edouard Trouessart (1842–1947). This idea was later encapsulated in orthogenesis, the idea that evolution undeviatingly followed particular trends regardless of whether or not changes were beneficial. For example, some thought that overly complicated shell designs had almost literally choked some mollusk groups to death, and others claimed that the horns of the Irish elk, an extinct animal with extraordinarily huge horns, had grown too big to be viable. William Whewell (1794–1866) attributed extinction to external causes, citing the recent demise of the dodo, the odd-looking bird from the island of Mauritius whose extinction had been caused by the hunting of European sailors, as an example.

In the early nineteenth century, it was not clear that Earth was particularly ancient, nor was it apparent that humans had not existed throughout its history. John Fleming (1785–1857), a British paleontologist who was a contemporary of Lyell, claimed that extinctions occurred piecemeal, each species successively eradicated by primitive humans. The Irish elk, whose exhumed antlers graced stately homes like hunting trophies, had last lived some time after Buckland's supposed floodwater deposits. Even Lyell himself, whose gradualism implied that Earth was almost infinitely old, believed for a time that people had existed throughout its history. The origin of species, especially humans, would only later become as controversial a topic as their extinction.

Typically, twentieth-century paleontologists have ascribed extinction to any number of causes, each specific to the group that disappeared. William Diller Matthew (1871–1930), for example, said that extinction was a consequence of Darwinian survival of the fittest. The evolutionary synthesis theory implied that extinction occurs when environmental change happens too quickly for a species to adapt. George Gaylord Simpson (1902–84), vertebrate paleontologist and author of the seminal *Tempo and Mode in Evolution* (one of the classics of synthesis), said that rapid evolutionary transitions usually caused extinction. The nineteenth-century hunting theme was resurrected as conservation became an important twentieth-century issue, even though a British Museum exhibit in 1932 taught that agriculture was a more likely cause of extinction than the hunt. Although it was clear that humans had not coexisted with the majority of extinct species, they certainly had with those in the **Pleistocene** ice ages. The overkill hypothesis from the 1960s and 1970s states that hunting by prehistoric Americans accounted for the extinction of mammoths, mastodons, and many other large mammals. Catastrophes also came back into vogue in the 1970s when physicist Luis Alvarez (1911–88) proposed that a giant asteroid had struck Earth at the end of the **Cretaceous**, killing off the last of the dinosaurs.

➢ **Pleistocene.** See *Earth Science and Geology.*

➢ **Cretaceous.** See *Earth Science and Geology.*

Extinction, paleontology's oldest theme, has also been its most popular. Honoré de Balzac, the French novelist, wrote this about Cuvier in 1831: "Is not Cuvier the great poet of our era? Byron has given admirable expression to certain moral conflicts, but our immortal naturalist has reconstructed past worlds from a few bleached bones; has rebuilt cities, like Cadmus, with monster's teeth; has animated forests with all the secrets of zoology gleaned from a piece of coal; has discovered a giant population from the footprints of a mammoth. These forms stand erect, grow large, and fill regions commensurate with their giant size. He treats figures like a poet: a naught set beside a seven by him produces awe" (Honoré de Balzac, *The Wild Ass's Skin,* 1831).

Queen Victoria herself unveiled Benjamin Waterhouse Hawkins's (1807–89) mausoleum to the memory of ruined worlds to 40,000 admiring spectators in 1854. The garden diorama featuring Richard Owen's (1804–92) magnificent "pachydermal" dinosaurs was topical enough to be a prominent part of the great Crystal Palace exhibition. A smaller 1868 mount of the dinosaur *Hadrosaurus foulkii* in Philadelphia, also engineered by Hawkins, was an immediate sensation. Princeton, Chicago, and the Smithsonian begged for copies, one of which was featured at the 1876 Centennial Exhibition. An even grander Paleozoic Museum was to be built in New York's Central Park (but was destroyed by political intrigue before it opened). Like the Crystal Palace garden, the Paleozoic park

Figure 2. Waterhouse Hawkins's plans for a 'Paleozoic Museum' in New York's Central Park. Courtesy of the American Museum of Natural History.

was to feature a variety of extinct mammals (American species this time) in a sumptuous natural setting reconstructed by Hawkins (see Figure 2). Since the 1890s, towering dinosaur skeletons in terrifying lifelike poses have filled the world's museums. Casts of the giant sauropod *Diplodocus carnegiei* grace museums from Moscow to Buenos Aires, thanks to the generosity and narcissism of the creature's namesake, Andrew Carnegie. In the 1930s, Sinclair Oil Company (whose logo was Carnegie's famous *Diplodocus*) backed American museum dinosaur expeditions in return for getting Barnum Brown (1873–1963; the museum's most famous dinosaur collector) to write booklets to help them sell their gasoline. Beginning in the 1930s, several national parks were designated on the basis of their extinct fauna and flora, including Petrified Forest, Fossil Cycad National Monument, Fossil Butte, and the Ice Age National Scientific Reserve. Several early stop-motion films by Willis O'Brien (who later animated *King Kong*) featured extinct worlds, including his 1917 short *The Dinosaur and the Missing Link*. *Jurassic Park*, one of the all-time box-office draws, resurrected dinosaurs via genetic engineering gone awry. From the nineteenth century's "heat death of the Universe" (the idea that heat would slowly be lost by the Sun and Earth so that eventually they would freeze like the Moon, and life would end) to the twentieth century's global warming, the fear of extinction has also captivated the public imagination.

Evolution

In addition to explaining the nature of fossils, Cuvier saw extinction as an alternative to transmutation. Evolutionary species change was a popular idea among Enlightenment writers, who saw it as a way to explain life in mechanistic terms (i.e., without reference to divine intervention). Jean-Baptiste de Lamarck (1744–1829), a contemporary of Cuvier, lectured in the first decades of the nineteenth century that fossil species were not extinct but had transmuted into new forms. Lamarck thought that species progressed from simple microscopic organisms to complex vertebrates and that the fossil record was not composed of extinct animals but represented the various stages that life had passed through on its way to the present. As late as the early twentieth century, some paleontologists still insisted that extinction was an illusion and that all lineages persist, transformed, today. By the 1820s, Étienne Geoffroy Saint-Hilaire (1772–1844) had extended the idea of transmutation to anatomy, arguing that all animals share the same basic structure. (Cuvier disagreed vehemently, arguing that an animal had only the parts it needed to live its life and that many anatomical parts of individual species had no analogs in other species.) Geoffroy's "idealist anatomy" became popular primarily among medical students, notably British radicals who had come to Paris to escape the strictures of the medical establishment at home (which was dominated by the aristocratic religious dons of Oxford). These radicals took the French evolutionary ideas back to their medical student communities, particularly in Edinburgh and London, where the rhetoric of evolutionary progress became associated with the upward struggle of the underclasses. Revolutionary sentiment was high in Britain in the first half of the nineteenth century, particularly between the French Revolution in the 1790s and the British Chartist riots in the 1830s. Many people associated evolution with revolution. Cuvier's extinction episodes were partially designed to dispel these notions by showing that the fossil record did not represent progressive change. Some interpreted them as a warning against radicalism, since the extinction episodes represented a series of catastrophic revolutions in which most species were destroyed. Against this background of reform and revolution, it is no surprise that the Oxford-based Buckland ascribed his "Diluvian" sediments to the last of Cuvier's catastrophes, or even that the gradualist Lyell maintained adamantly that the increasing complexity of animals and plants from older to younger strata in the paleontological record was an artifact of chance preservation.

Industrial and imperial expansion brought increasing numbers of fossils into the scientific limelight throughout the nineteenth century. The British canal builder William Smith (1769–1839), the engineer who drew the first geological maps to show the geographic extension of rocks and their fossils, recovered paleontological remains from the thousands of miles of canals he helped excavate. Medical naturalists aboard naval survey ships (the British military required of its doctors five months of natural history training) sent back crates of fossils from around the globe. (Darwin's *Beagle* voyage is, in retrospect, the most famous of these voyages, but it was by no means unique.) And miners offloaded slag heaps of remains from burgeoning slate, phosphate, clay, and coal quarries. As this new material came under the study of paleontologists in the medical colleges, museums, and stately homes, the scientists vied with one another to ensure its compatibility with their own worldview. Progress and degeneration, discontinuity and association, extinction and transformation, space, and (above all) time were the components of paleontological debate in the nineteenth century.

Progress in Paleontology

Robert Grant (1793–1874), a medical anatomist at radical London University, was a disciple of Lamarck, Geoffroy, and Blainville, whom he visited regularly in Paris. Grant advocated homology (Geoffroy's idea of similar parts in different organisms), anatomical continuity, and ecologically controlled gradual transformation. In his summer course on fossil zoology for medical students, Grant taught that Earth's successive faunas had been continuously transformed through changes in climate. Accordingly, he divided geological time into the Protozoic (early life), **Mesozoic** (middle life), and **Cenozoic** (recent life) periods.

➢ **Mesozoic.** See *Earth Science and Geology.*

➢ **Cenozoic.** See *Earth Science and Geology.*

Richard Owen (1804–92), also a medical man, was a sharp contrast to Grant. Revolted by the radicalism of Edinburgh, Owen moved to St. Bartholomew's Hospital in London to finish his medical training under John Abernethy (1764–1831), the president of the conservative Royal College of Surgeons. Owen flourished in this medical environment as curator of the college's Hunterian Collections, where he even played host to the visiting Cuvier. Like Grant, Owen was fascinated by Geoffroy's idealist anatomy, but unlike Grant, he saw God's plan in the common design of animals. He made it his life's work to reconcile Geoffroy's ideas with Cuvier's functional anatomy and geological revolutions. For Owen, life's history seemed to be orchestrated by divine guidance and was not at all to be associated with progress, revolution, or reform. Owen saw humans occupying the top of the biological pinnacle but denied that progress was a natural law, since that would imply yet greater heights in the future.

In reaction to Grant's progressivism, Owen unveiled a new class of vertebrate, the Dinosauria, in 1841. The animals themselves (*Megalosaurus*, *Iguanodon*, and *Hylaeosaurus*) had been discovered as many as 20 years earlier but had been conceived of only as gigantic lizards. Owen reconstructed dinosaurs with mammal-like postures so that they would appear to be the most advanced reptiles that had ever lived. Rather than walking with a lizard's sprawling gait, dinosaurs stood erect (in fact, Owen's reconstructions of *Megalosaurus* and *Iguanodon* look remarkably like rhinoceroses). Owen even speculated that dinosaurs might have been warm-blooded. If reptiles had reached their greatest development with the Dinosauria in the Mesozoic, the later Cenozoic history of reptiles had been one of degeneration rather than progress. Life's history, by extension, was not a relentless drive by the lowly up through the ranks, but a well-choreographed play in which characters came and went following a divine script, and with humans (particularly Europeans) at center stage for the grand climax. Owen was later knighted by Queen Victoria and awarded a spacious residence in prestigious Hyde Park; Grant died a pauper. Following Owen, degeneration became an important nineteenth-century explanation for extinction. Scholars such as the American neo-Lamarckian Alpheus Hyatt incorporated it into a life-cycle model of evolution, in which species originated (or were born), they diversified (or grew up), reached their peak (their maturity), then declined and became extinct (their death).

Owen was certainly not the only one to use biological classification to further a scientific agenda. Hugh Miller (1802–56), a Scottish Protestant paleontologist, used his Devonian fish from the Scottish Old Red Sandstone against progressivism. These *ostracoderms* (the earliest fish, indeed the earliest vertebrates then known from the fossil record) were heavily armored with solid skulls, much like land vertebrates. Miller portrayed them as the most advanced fish, their successors having degenerated into the forms we know today. In support of a progressive notion of evolution, Thomas Henry Huxley (1825–95) concocted Prototheria, Metatheria, and Eutheria as stages of mammalian evolution. These stages emphasized continuity from lower grades (stages of evolution) of mammal to higher ones so as to counteract Owen's attempt to classify humans apart from other mammals. Huxley (something of a radical) hated Owen and intended to thwart him at every turn. Huxley even launched a hostile takeover of Owen's Dinosauria, deemphasizing their similarity to mammals and repositioning them as an evolutionary intermediate between early reptiles and birds. Early twentieth-century evolutionary paleontologists modified Huxley's evolutionary grades, using the idea to show that groups acquired important specializations that took them into a new, adaptive stage of evolution. In the late twentieth century, cladists (systematists who advocated a numerical approach to **phylogeny** reconstruction and who demanded that classification reflect only evolutionary branching) dismantled those **taxonomic** groups, arguing that evolution was not adaptive and that grades should not be used in classification. Instead, cladists argued that groups should be organized by smaller and smaller subdivisions that emphasized the splitting aspect of evolution rather than change over time.

* **Phylogeny.** Evolutionary history of a species.

* **Taxonomy.** Identification of living organisms within an organized system of classification.

Evolutionary Theories

If paleontological debate on transformation during the 1830s through 1850s was primarily about progress and anatomical continuity, it metamorphosed completely into explicit conflict about evolution following the publication of Darwin's *On the Origin of Species* in 1859. Owen (who was vague about his opinion on evolution but was adamantly against natural selection and survival of the fittest) prepared his counterargument in advance by reclassifying mammals. He removed humans from their familiar place in the Primates, reclassifying them in Archencephala, the "highest brained," and divorcing them not only from apes (the obvious candidates for ancestors of humans) but from all other mammals as well. Later, he emphasized the unchanging nature of species in the fossil record, pointing out that the last **ichthyosaurus** was scarcely different from the first. Similarly, Adam Sedgwick (1785–1873), paleontology professor at Cambridge, attacked Darwin with fossil **brachiopod** species that did not change at all throughout their long sojourn in Paleozoic strata. Sedgwick also reminded Darwin that Hugh Miller's ancient fish (which he claimed were both the highest and the oldest) flatly contradicted progressive evolution. Louis Agassiz (1807–73), a Swiss-born paleontologist at Harvard University, offered similar scathing criticisms of evolution. Other paleontologists, however, were just as convinced that the fossil record agreed with evolution. Karl Zittel

* **Ichthyosaurus.** A genus of extinct marine animals having an enormous head, a tapering body, four paddles, and a long tail.

* **Brachiopod.** A mollusk with a long spiral arm used in gathering food.

(1839–1904), one of the leading paleontologists in Germany, accepted that some fossil species were persistent, as Sedgwick and Agassiz insisted, but retorted that others evolved quite rapidly. Wilhelm Waagen (1841–1900) in Munich maintained that ammonites definitely evolved, although he argued that change was driven from within the organism and not caused by environmental factors as Darwin had suggested. Joseph Leidy (1823–91), a Philadelphia anatomist who had been studying the extinct faunas of North America for more than a decade, said that the *Origin* explained everything that had been puzzling him in his research. Leidy's successors in the American West, Cope and Othniel Charles Marsh (1831–99), described thousands of new fossil species, tracing through them the structural and temporal changes produced by evolution. Huxley and Owen had another showdown over *Archaeopteryx*, the Mesozoic bird from the Solnhofen deposits in Germany, Owen using it as another example of permanence of type and Huxley emphasizing its intermediate evolutionary position between dinosaurs and birds (see Figure 3).

Sedgwick attacked Darwin not only with paleontological facts but also with the accusation that Darwin's logic was not inductive. Natural selection, Sedgwick claimed, was not an unquestionable conclusion drawn from the patient accumulation of facts. (Inductive knowledge is based on repeated observation and is often distinguished from deductive knowledge, which begins with general principles and deduces conclusions from them.) The inductive ideal had been cultivated in geology by Lyell and his anticatastrophists, who portrayed Buckland and Cuvier as wild mythmakers, unfamiliar with the facts borne in the rocks. Observation and description were primary, they said; interpretation and explanation came later, if at all. Some paleontologists completely avoided entering the debate on evolution, rebuking those who did.

* **Phylogenetic trees.** Diagrams showing relationships among different groups of organisms, similar to genealogical trees. These relationships have been reconstructed using a variety of techniques, drawing primarily on comparisons of anatomical structure (or more recently, molecular structure) and dates from the fossil record.

At the end of the nineteenth century, amid controversy over ill-founded **phylogenetic trees** and evolutionary theories, many paleontologists again asserted that their field was best grounded in description. Several geneticists, such as Thomas H. Morgan (1866–1945) and Theodosius Dobzhansky (1900–75), capitalized on that attitude, which helped them to sideline paleontology as an evolutionary discipline. Many paleontologists were happy to remain on the inductivist margins. David M. S. Watson (1886–1973), professor at the University of London, stated that the paleontologist differed from the geneticist in that the former could observe only the facts of succession, not the evolutionary process itself. Simpson, on the other hand, bemoaned this attitude in the introduction to his *Tempo and Mode in Evolution*, warning the reader that he would not be parading out the standard fare of dry description. In

Figure 3. The fossil bird Archaeopteryx, *which possessed many reptilian features but was feathered like a bird, adjacent to artist's rendering of the skeleton. © The Natural History Museum, London*

1975, the journal *Paleobiology* was founded as a move away from monographic description, which was ascribed by the journal's founders particularly to invertebrate paleontology. With opposite motives, cladistic paleontologists such as Colin Patterson (1933–98) asserted that nothing about evolution, including phylogeny, could be known until the pattern of trait distributions in all organisms had been described. Only then could the question be asked whether evolution was the best explanation for that pattern. Patterson also argued that paleontology, with its inherently fragmentary specimens, would be able to contribute very little to the study of evolution and phylogeny in the face of the overwhelming data from molecular biology.

Despite what Huxley called the deplorable stream of cold water that was poured over speculative thought, many paleontologists were comfortable with the study of evolution. Vladimir Kovalevsky (1842–83), an idealistic young Russian who took up paleontology because he thought Darwinism would better the world, applied the natural-selection model to the evolution of the horse, the nineteenth century's most familiar animal. Kovalevsky traced the equine pedigree from Cuvier's *Anchitherium* through the three-toed *Hipparion* to the single-toed modern horse, explaining the structural changes as Darwinian adaptations to a world transformed from marshy forests to dry plains. Huxley and Marsh adapted Kovalevsky's story. Marsh substituted *Eohippus* for *Anchitherium*, added a host of other intermediates from the fossil fields of the American West, and passed the lineage on to Huxley, who used the horse repeatedly as an example of evolution demonstrated by paleontology.

Huxley, Marsh, and Kovalevsky were among the few paleontologists who counted themselves as Darwinians. Many others, particularly in the United States, were avid evolutionists but adamantly opposed to natural selection. Edward Drinker Cope and Alpheus Hyatt, for example, led a well-developed evolutionary movement, usually called *neo-Lamarckism* [Alpheus Spring Packard (1839–1905), an ally in the Cope-Hyatt confederacy, coined the word in a laudatory biography of Lamarck]. The neo-Lamarckians insisted that Darwin and natural selection had not explained how variation (the source of evolutionary change) arose. Opposed to the idea that only the fit survived, Cope and Hyatt substituted the idea of *use inheritance*, through which individual action produced new features that were then passed on to descendants, thus putting the individual organism in control of its own evolutionary destiny. This aspect of Lamarckism was popular with many, from church leaders to socialists. In fact, Lamarckism was officially adopted as "correct" in Stalin's Soviet Union when Trofim Lysenko (1898–1976) convinced the Politburo that Darwinian selection was capitalist propaganda. Growth and development played a leading role in nineteenth-century neo-Lamarckian evolution; bodily changes that occurred during the life of an individual (such as muscular development or bone remodeling) were permanently added to the developmental pattern and passed on to descendants. Cope, in particular, drew on experiments on cellular electricity, which appeared to build up in organs as it did in **Leyden jars**. If electrical force was responsible for growth and if individual action could redistribute it (after all, a worker's hands grew larger with lifetime use), the neo-Lamarckians could explain evolution without relying on natural selection as its mechanism. Cope offered his own version of horse evolution, which (like everything Cope did) was opposed to Marsh's. The physical act of pounding across the ancient Miocene grasslands had concentrated the electrical growth force in the central toes of the ancestral horse, causing the lateral toes to atrophy and the central toe to elongate. Cope's horse evolved without the carnage implied by Darwin's selective elimination of the unfit. Hyatt used similar logic to describe the evolution of the snail *Planorbis* through the strata of the Steinheim meteor crater in Germany. Despite their emphasis on ontogeny (growth and development), neither Cope nor Hyatt agreed with

➢ **Leyden jar.** See *Electromagnetism*.

Ernst Haeckel's (1834–1919) biogenetic law, which said that ontogeny recapitulates phylogeny, or that embryonic development repeats the sequence of ancestors from a species' history. The Americans argued that ancestors were not like early embryonic stages except, perhaps, in certain limited features.

Many paleontologists at the end of the nineteenth century were opposed to Darwinian selection for another reason—they saw linear trends in fossil sequences. Natural selection would have carried evolution in a meandering course as environments shifted back and forth. Many paleontologists thought that evolution seemed to march undeviatingly through strata, regardless of environmental changes, a process known as *orthogenesis*. Henry Fairfield Osborn (1857–1935) described multiple lineages of elephants and titanotheres (giant rhinoceros-like mammals from the Tertiary deposits of Asia and North America), each following its own nonadaptive evolutionary path in parallel to its close relatives. To Osborn, titanotheres seemed to progress through a predictable sequence of increasing size and elaboration of horns on their bony skulls (critical of Osborn, Simpson later reinterpreted Osborn's orthogenetic lines as the extremes of variation in a single evolving population). Othenio Abel (1875–1946) argued that long-term trends began as adaptive responses to particular environments that could continue indefinitely through phylogenetic inertia, resulting in unusually enlarged organs. Orthogenesis had its seed in Hyatt's evolutionary cycles, in which groups progressed through youth, maturity, and senescence, but was influenced more specifically by the idea of directed mutations, which forced lineages into unidirectional evolutionary trajectories. Beecher's hyperspiny mollusks and Loomis's enormously antlered Irish elk were used as examples of lineages that had been driven to extinction by such maladaptive trends.

Phylogeny and Missing Links

In the *Origin*, Darwin conjectured that the hidden bond of relationship that naturalists had been seeking was genealogy. Linnaeus's sexual system, Buffon's degeneration, Lamarck's ascent, and Cuvier's functional coordination had each been constructed as the essential basis of a natural classification. For many late-nineteenth-century evolutionists, the diversity of life was connected by a branching genealogical tree, or phylogeny (both word and metaphor came from Haeckel in 1866). Identifying such phylogenies became one of paleontology's primary activities. Haeckel was one of the first to see a phylogenetic program in the *Origin*. Haeckel was no paleontologist, however; he found his phylogeny in **ontogeny**. Others sought ancestors among the ruined remains of extinct life. Huxley adopted phylogenetic arguments to attack Owen's views on evolution; Owen responded by reluctantly redescribing his fossils in terms of descent, but always maintained that Huxley's specifics were wrong. Some paleontologists claimed that their fossils proved Haeckel's biogenetic law. Others, like Kovalevsky, saw the traces of selection and adaptation in their trees. To the American paleontologists Marsh, Cope, and Hyatt, phylogeny became the primary goal of evolutionary study. A minority of paleontologists, including Friedrich Quenstedt (1809–89) and Waagen, thought that phylogenetic connections could be made only between very closely related species, but most post-Darwinians presumed that life was monophyletic (stemming from a single origin) and that phylogeny could, in principle, be traced among all organisms.

* **Ontogeny**. The development of an individual living organism.

Phylogenetic paleontology was in large part the search for intermediate taxa, or "missing links," following from Darwin's confession that the gaps in the fossil record embarrassed his theory (in practice, he had purposefully emphasized the gaps to bolster his assertion that evolution was extremely slow and gradual). Huxley and Owen battled over *Archaeopteryx*, the former arguing that it bridged the evolutionary gap between dinosaurs and birds and the latter, just as emphatically, stressing that it widened the gap immeasurably. First Kovalevsky, then Marsh, filled in the steps between horse and five-toed mammals. Trilobites, extinct **arthropods** resembling many-legged beetles, were pressed into service as ancestral intermediates (or impasses) between arthropods (crustaceans, spiders, and insects) and other invertebrate groups. Amphibians, reptiles, and fish from the Carboniferous coal deposits spanned fish and land vertebrates, while fossils from the deserts of South Africa and Texas were branded mammal-like reptiles with obvious intent. Eugène Dubois's (1858–1940) fossil human, *Pithecanthropus erectus*, more popularly known as Java Man, was the most controversial "missing link" of all (Haeckel had coined the genus name *Pithecanthropus* for the then-hypothetical intermediate between human and ape).

* **Arthropod.** Animals with jointed feet; a name for some members of the Articulata, including insects and spiders.

The flamboyancy with which paleontologists proposed phylogenetic trees offended some. Alexander Agassiz, son and successor of the antievolutionist Louis, announced in 1880 that the time for genealogical trees was over. He enumerated the extraordinary number of conceivable trees connecting even a few species, concluding that it was impossible to identify the correct one (akin to the needle in a haystack). As the first generation of post-Darwinian paleontologists aged, inductivist calls for less speculation prevailed, some coming from young paleontologists and some from rival experimental physiologists and, later, geneticists. In 1893, Osborn remarked that the phylogenetic tree had virtually become extinct in paleontological monographs. In the first decades of the twentieth century, only a few paleontologists (primarily at Osborn's American Museum of Natural History) studied phylogeny. Many of the rest devoted themselves to descriptive anatomy or **stratigraphy**, often in the context of petroleum exploration.

* **Stratigraphy**. The study of the ages of rock beds, which largely involves comparisons of the fossils entombed within them. In the twentieth century, other techniques, such as radiometric dating, have contributed to stratigraphy.

Species and Paleontology

In the 1930s, the prominent British paleontologist D. M. S. Watson argued that unlike genetics, paleontology could contribute to the study of evolution only on a relatively large scale, among higher taxonomic categories (or taxa) such as genera, families, orders, and kingdoms, but could not contribute on the smaller scale of species. His argument may sound strange, because species are now considered to be the fundamental units of paleontology (and indeed, of organismal biology). Species are usually defined as interbreeding populations of organisms, and although interbreeding cannot be directly observed in the fossil record, many paleontologists accept this definition. Species are said to be fundamental because they are groups of real organisms (whether alive or dead), while other, higher categories are arbitrary because they are simply names referring to groups of related species. Paleontologists have not always shared this view (and do not always now), in part because fossils are the remains of dead organisms; groups must be recognized from their fossilized anatomy rather than from interbreeding and gene flow (this was the reasoning behind Watson's statement). The process of speciation itself (the splitting of one interbreeding population into two), however, poses deeper issues for paleontologists because they deal regularly with populations that persist and evolve for quite a long time without splitting. At the beginning of the twentieth century, many paleontologists referred to such lineages as genera, while variants within the evolving lineage were species. Speciation, for these paleontologists, meant something very different from populations splitting. Furthermore, some paleontologists did not view higher taxa as groups of species but saw them as categories defined in their own right. Thus, many of the nineteenth- and twentieth-century debates among paleontologists and biologists about evolution concerned the nature of species and their relevance to evolution.

In the nineteenth-century debates about evolution, many paleontologists were concerned about the fixity of species—whether or not species changed over time. Some (e.g., Lyell in his early career) argued straightforwardly that species could not and did not change; others (e.g., Grant) said they did. But most views did not fall into such a simple dichotomy. Whewell and Agassiz, for example, argued that species could change in superficial ways (they might become smaller, fatter, or longer-haired, for example) in response to the environment, but that organisms could not change in more substantial ways. For Agassiz, it was important to describe variation within a species (many paleontologists described single specimens only because they thought all members were fundamentally alike), but it was equally important to indicate how groups differed from one another in more fundamental ways. Agassiz had an extreme, but not necessarily atypical, system of classification. Each taxonomic rank, including species, was defined by different kinds of traits: Species traits were purely adaptive (changed with environmental conditions), family traits were structural, and class traits were related to the overall organization of the body. This conception was very different from that of paleontologists today, in which the higher categories are simply collections of species. In his later years, Agassiz used the distinction between species traits and higher traits to argue against evolution, insisting that natural selection could modify only species traits, but not generic, familial, or ordinal traits.

Cope, an evolutionist who favored mechanisms other than Darwin's natural selection, adopted Agassiz's system but in an evolutionary context. Like Agassiz, Cope limited the purview of natural selection to species, but unlike Agassiz, proposed other mechanisms whereby genus, family, and higher traits evolved. In Cope's model, organisms could evolve from species to species and from genus to genus (for example) quite independently; an organism could transform from one genus to another while remaining the same species. For Cope, selection modified species traits and explained the evolution of species, but developmental processes regulated genus traits and therefore explained the evolution of genera (Cope published his *Origin of Genera* as a response to Darwin's *On the Origin of Species*). Cope's idea was a radical (and evolutionary) variant of fixed species—the species itself did not change, but individual organisms did, transmuting from one species to another, just as individual atoms change from uranium to lead without changing the categories of uranium and lead themselves.

In contrast, Darwin conceived of species as interbreeding populations that changed over time. If a species changed enough (or split into enough different populations), it might be given a new genus name. However, that decision was completely up to the discretion of the scientist, and evolution was a process that only species went through.

Interestingly, many late-nineteenth-century paleontologists thought that genera were more important (or more "real") than species. This conception was based partially on Agassiz's idea that species traits were superficially variable (as opposed to stable genus traits) but was also based on traditions of logic that treated species as variants of a formally defined genus (the words literally mean *specific* and *general*). In twentieth-century debates about evolution, some paleontologists argued that evolution was a genus-level process rather than a species process because they viewed species as minor variants and thought of genera as the real taxa. Even today, paleontological literature often deals only with genus names (e.g., *Anchitherium* or *Triceratops*), leaving the idea of species out altogether.

There was more than semantics at stake in twentieth-century disagreements about species, however. Many biologists, such as Ernst Mayr (b. 1904), followed Darwin in defining species as populations of interbreeding individuals; consequently, speciation was the process by which one such population split into two. But in the fossil record, the splitting (or divergence) of groups was only a minor part of evolution. If speciation was splitting, a species had to be the long lineage of fossil populations that lay between splitting events. However, significant evolutionary change could happen within those species, change that paleontologists thought was important and wanted to emphasize with taxonomic names. Osborn called these minute gradations *mutations*. Depending on the trajectory of evolutionary change, he referred to intermediate populations as *mutation ascending* or *mutation descending*. Although today one usually thinks of mutations in the context of genetics, the word was coined by paleontologist Wilhelm Waagen to refer to small evolutionary jumps between one fossil group and another. Geneticists such as Thomas Hunt Morgan (1866–1945) appropriated the term to refer to saltational (jumplike) variants in fruit flies, which, they argued, were caused by genes. As paradigms changed, the word *mutation* dropped out of the paleontological vocabulary but was retained in the genetic lexicon. As a solution to the speciation dilemma, Simpson synthesized the biological and paleontological notions in his evolutionary species concept: A species was a series of populations in a lineage evolving independently of others. The biological concept of a single interbreeding population was thus stretched through geological time.

The Age of Earth

In his damning review of the *Origin*, Adam Sedgwick, the Cambridge don who had once tutored Darwin in geology, assembled a comprehensive array of arguments: There was no progression in the fossil record (Paleozoic rocks, he said, contained a diverse and complex range of life, not the predicted simple, single-celled animals); successive faunas had no continuity between them (they were true revolutions in the Cuvierian sense); and the gaps in the geological record were not as great as Darwin had claimed. Since Darwin had portrayed himself as embarrassed by those gaps, this last criticism is at first difficult to understand. But for natural selection to be plausible, Darwin needed vast stretches of time to connect the discontinuities that Sedgwick highlighted. In the same way that Lyell had postulated gaps in the fossil record of the Paris basin to argue that Cuvier's catastrophes were really gradual, species-by-species extinctions, Darwin emphasized gaps to gain the time he needed to evolve the fauna of one epoch from that of the preceding one. Sedgwick was arguing that geological time was not long enough to encompass evolution by natural selection, regardless of how plausible the idea might be otherwise.

Nineteenth-century paleontologists were concerned with geological time, both its subdivisions and its durations, not as ends in themselves but as part of a wider debate about evolution and extinction. Divisions of time in Earth's history were not new to the nineteenth century. In the 1770s, Abraham G. Werner (1749–1817), an extraordinarily popular professor at Freiburg's mining school, had formalized a neptunist system (the neptunists explained geological history in terms of receding waters rather than volcanism that categorized rocks into primary, secondary, tertiary, and quaternary). Primary rocks formed the cores of mountains, secondary rocks had eroded from the primary and been deposited in the oceans (they contained marine fossils), tertiary rocks had been deposited on top of them on dry land (Cuvier's tertiary animals were mostly land vertebrates), and quaternary deposits were left by rivers and lakes on top of the tertiary. Some of Werner's names, if not his concepts, were retained in later efforts to classify geological time. The time represented by the fossil record and the nature of relationships among strata (and their fossils) were important topics in the nineteenth century.

Cuvier, as part of his argument against Lamarck and Geoffroy's transformism, had used Egyptian mummies as evidence that species were immutable, even over vast periods of time. Since the mummified cats, crocodiles, and ibises were exactly like their living counterparts, Cuvier concluded that evolution was disproved. The roughly 4000 years that had elapsed since the mummies were embalmed was considered comparable to the time separating his ruined worlds (Buffon had estimated Earth to be about 70,000 years old based on his **nebular theory** and rates of water **subsidence**). In fact, Cuvier's description of himself as a new species of antiquary emphasized that prehistory was simply an extension of ancient history. The homophony of Blainville's *paleontology* and *paleography* was not coincidental. Even Lyell, who proposed a vast age for Earth, jokingly compared his fossil mollusks to the easily decipherable demotic script while discounting Cuvier's vertebrates as more like indecipherable hieroglyphics. Intent on obfuscating Cuvier's faunal discontinuities, Lyell asserted that the Paris basin strata were only small windows into almost limitless geological time, which meant that what appeared to be sudden mass extinctions could really have accumulated gradually in the long-missing intervals. Darwin's fossil gap argument was analogous. Since alternative theories, such as Cope's neo-Lamarckism or Waagen's mutations, moved evolution at a much faster pace than natural selection, there was a vested interest in exactly how old Earth was. Furthermore, as Sedgwick's denunciation implied, there was active interest in whether the **Silurian**, with its advanced fishes, represented the oldest chapter of Earth's history, and in whether intermediate eras might not plug the faunal discontinuities in the same way that intermediate fossils filled phylogenetic gaps.

➢ **Nebular theory.** See *Astronomy and Cosmology: Eighteenth Century.*

* **Subsidence.** The sinking of a large area of Earth's crust.

➢ **Silurian.** See *Earth Science and Geology.*

Up until the time that the *Origin* was published, Sedgwick was himself embroiled in a bitter controversy with Roderick Murchison (1792–1871), who established the Silurian age, one of the earliest geological ages then known, about a new subdivision, the Cambrian. In the 1830s, Sedgwick thought that a group of folded and twisted rocks in Wales (the region the Romans had called Cambria) was older than Murchison's Silurian. Murchison and others had denounced the idea, accusing Sedgwick of not defining the new age using a fossil fauna, which would allow the Cambrian to be recognized in other parts of the world. The Cambrian rocks did have fossils, although they were not as abundant or advanced as Murchison's Silurian fauna; however, Sedgwick refused to describe them. Two other paleontologists did just that in the 1850s, demonstrating their similarity to other faunas in Europe and America. When the Cambrian was recognized as a new period, the earliest then known, many said that its simple fauna filled the very gap identified by Sedgwick in his critique.

There were many such controversies in the nineteenth century (Buckland compared them to an enclosure act on the open commons of geological time) and into the twentieth. As late as the 1910s, William Diller Matthew (1871–1930) was a proponent for a new **Paleocene** epoch to be inserted between the Cretaceous and the Eocene. The groupings of ages were almost as emotionally charged as the grouping of humans and apes. Great tracts of time had been united as Paleozoic, Mesozoic, and Cenozoic to emphasize the progressive, evolutionary nature of the faunas (the names mean ancient animal age, middle animal age, and new animal age, respectively). These eras are still used today and are sometimes known as the Age of Fishes, the Age of Reptiles, and the Age of Mammals, respectively. A jealous twentieth-century paleobotanist later demanded that a Paleophytic, Mesophytic, and Cenophytic be established in parallel to emphasize that fossil plants also exist (*-phytic* derives from the Greek word for plant).

➢ **Paleocene.** See *Earth Science and Geology.*

The actual duration of geological time also figured heavily in nineteenth-century debates. Darwin and Huxley followed Lyell in maintaining that Earth was extremely ancient. Calculating rates of erosion in the Weald in southeastern England, Darwin estimated that Mesozoic time was more than 300 million years ago. The long Paleozoic before it gave him ample time for his slow natural selection to effect the changes he saw in the rocks. Others disagreed. The physicist William Thomson (later known as Lord

Kelvin, 1824–1907), said that the laws of thermodynamics clearly showed that Earth must be very young. Based on the temperature of Earth's interior, Kelvin's rates of cooling implied that as little as 100 million years ago, Earth's surface would have been molten. Gradual evolution was not tenable because even 1 million years ago, Earth's surface would have been too hot for life. American geologist and neo-Lamarckian Clarence King (1842–1901) allowed even less time, saying that experiments on rock-melting temperatures allowed no greater age for Earth than 24 million years. Despite Thomas Huxley's impassioned rebuttal of Kelvin's inroads against uniformitarianism and natural selection, many young paleontologists liked Kelvin's limit. Anti-Darwinians, such as Cope and Hyatt, capitalized on the young Earth in their crusade against natural selection. Even paleontologists who disagreed with the neo-Lamarckian point of view invoked Wilhelm Waagen's large-scale mutations to speed up the pace of evolution.

Radiometric dates, based on the decay of radioactive minerals such as uranium, soon turned the tide back toward an ancient Earth. Beginning in 1903, physicists working on the newly discovered **radioactivity** began saying that Kelvin's estimates were far too short. Following Pierre Curie's startling announcement that radioactive salts produced heat, physicist Ernest Rutherford (1871–1937) said that radioactivity maintained Earth's heat and that it was not cooling at all. Rutherford and his colleagues further suggested that radioactive decay itself might be used to determine the age of Earth. Robert Strutt (1875–1947) at London's Imperial College and Bertram Boltwood (1870–1927) of Yale both began measuring the radioactivity of minerals and by 1910 had concluded that Earth must be at least 2 billion years old. The new dates extended Earth's age by a factor of 20. Resurgent Darwinian paleontologists gloried in the rapidly aging Earth. Charles D. Walcott (1850–1927), famous for his work on the extremely ancient Burgess Shale animals and Marsh's successor as director of the U.S. Geological Survey, said that geology and paleontology had always assured him that Kelvin's estimates were far too short. In 1915, vertebrate paleontologist W. D. Matthew thought he had only 3 million years to explain how evolving lineages could get back and forth across oceans but was relieved to report in 1939 that he now had 60 million years. Radioactive dating could be used only on rocks of volcanic origin (fossils are usually found in water-deposited rocks), so various methods were used to extrapolate radiometric dates to fossils. Lake varves, thin layers of sediments representing the winter-summer deposition cycle, were counted to estimate the length of the Quaternary period. Matthew used the evolution of horses to estimate the duration of Tertiary periods by calculating the total time from the Dawn Horse, *Hyracotherium*, to the living *Equus* and dividing it evenly among the evolutionary stages he recognized.

➢ **Radioactivity.** See *Atomic and Nuclear Physics.*

Figure 4. Paleontologists Henry Fairfield Osborn, William Berryman Scott (1858–1947), and Francis Speir (dates unknown) dressed in adventurous western garb during one of their early field expeditions. Courtesy of the American Museum of Natural History.

The Paleontological Expedition

American expansion in the West opened vast new territory to paleontological exploration, cultivating a lasting association between fieldwork and adventure (see Figure 4). Despite heroic voyages such as Darwin's HMS *Beagle* trip, European fossil collecting in the early nineteenth century was usually associated either with gentlemanly excursions or with industrial excavation. The famous *Iguanodon* discovered at Maidstone, England, in 1834 was found in a commercial quarry. At the British Museum of Natural History, specimens were typically sent home by members of the imperial armed forces or were purchased from private collectors such as Mary Anning (1799–1847).

Albert Gaudry (1827–1908) led an expedition from the Paris Museum to Greece in the 1850s, where a remarkable mammalian fauna was found at Pikermi. However, it was fieldwork in the American West that began to make paleontology synonymous with dangerous treks, inhospitable mountains, and scorching deserts.

Armed conflict between the U.S. government and Native American tribes augmented the idea of adventure and exploration. The year after George Custer's 1876 defeat by Sitting Bull, a fossil-collecting party led by Cope was returning from the Judith River in Montana when their scout spotted a Sioux camp with some 1000 warriors. Several members of the party deserted, and the paleontologists diverted their path through rugged badlands, struggling to catch the last steamboat of the season on the Missouri River. Custer himself collected fossil clams, some of which are now housed at the University of Michigan. Despite tales of daring, violence toward paleontologists was never reported. Jacob Wortman (dates unknown), a collector for Cope, described the Bighorn Basin in the 1880s as a wild land uninhabited except for roving bands of hostile Indians. The area had been reserved in the 1870s for the Crow, Arapahoe, Cheyenne, and Shoshone, but following Custer's defeat in 1876, the U.S. Army had moved these peoples out and opened the area to miners and sheep ranchers. Although there were no real towns in the Bighorn Basin, Wortman was able to run his expeditions from Fort Washakie (near present-day Lander, Wyoming), an established outfitting post for the region.

Despite the image of the paleontologist as a rugged individual, carefully organized teams, often with substantial financial backing, carried out most of the fieldwork in the American West. Joseph Leidy (1853–91) was attached to the first Geological Survey of the Western Territories, which was organized by Ferdinand Vandiveer Hayden (1828–87). That survey and three subsequent ones were commissioned by Congress to investigate the mining and agricultural potentialities of the western territories. Leidy was to have physically accompanied the survey in 1854, but when professorial duties prevented him, the fossils were sent back to him in Philadelphia. Cope and Marsh had similar relations with the surveys, sometimes traveling with the field crews and sometimes not. Both Cope and Marsh employed teams of collectors, who were deployed throughout the West like small militias. Charles Sternberg (1850–1943) and three generations of his family worked for Cope, Marsh, and Osborn. Cope and Marsh employed special managers, such as Samuel Williston (1851–1918), to look after fossil affairs year-round or for special projects. Marsh employed two former Union Railroad men continuously for six years to manage dinosaur excavations at Como Bluff, Wyoming. The managers even considered buying the land under a homestead act to protect it from Cope's collectors. The two rivals sometimes even maintained regional coordinators, such as David Baldwin who covered New Mexico—sometimes employed by one, sometimes by the other.

Figure 5. The skeleton of Tyrannosaurus rex *mounted in the American Museum of Natural History in 1910. Courtesy of the American Museum of Natural History.*

As financial magnates such as J. P. Morgan and Andrew Carnegie amassed fortunes at the end of the nineteenth century, they reinvested much of it in social betterment programs and civic projects such as natural history museums. This allowed museum money to be devoted to high-profile paleontological field expeditions, the most notable run out of the new American Museum of Natural History in New York. These were well-funded, well-managed affairs designed to attract international glory to the museums and their backers. Barnum Brown's famous 1902 *Tyrannosaurus rex* skeleton from Hell Creek, Montana, was exactly the kind of trophy these teams were expected to bring back (see Figure 5). Earl Douglass (b. 1862) found an enormous new sauropod,

Diplodocus carnegiei. In the twentieth century, government projects also funded paleontological fieldwork. Collecting in the Bighorn Basin was funded by Theodore Roosevelt's Shoshone Reclamation Project, which aimed to convert western deserts to productive farmland. Franklin Delano Roosevelt's Depression-era work programs also supported paleontological collecting and exhibitions. After World War II, paleontology was one type of scientific project funded by the newly established National Research Council.

Transportation was a major issue in mounting field expeditions to remote areas. When crews first began working the Bighorn Basin in the 1880s, they carried their supplies and fossils on pack mules because wagon transportation was impossible. Many quarries, such as the famous dinosaur locality at Como Bluff, were near the recently completed transcontinental railroad (the Golden Spike that joined the eastern and western sections of the railroad was driven in 1869). Railroad entrepreneurs such as J. P. Morgan even provided special train cars to transport loads of dinosaur bones back to New York. Fossil sites sometimes took their names from railroad stops, and occasional railroad towns were named after paleontology. The Wasatch formation, which is famous for its fossil mammals, was named for a stop where expeditions disembarked, while Granger, Wyoming, was named for American Museum field paleontologist Walter Granger (1872–1941), who collected at nearby Bone Cabin. Other methods of transportation were also available. Cope and his crews often used overland stagecoaches and Missouri River steamboats for transportation to and from their field sites. Barnum Brown built a flatboat from which to mount American Museum excavations along the Red Deer River in the 1910s. His crew scanned the banks with binoculars for dinosaurs as they floated along in their "fossil ark." Automobiles later provided easy access to areas far removed from train lines. George Simpson learned to drive in 1924 when he and W. D. Matthew purchased a car in Amarillo for a collecting trip in the Texas panhandle. Like the railroad entrepreneurs, the automobile industry occasionally sponsored fieldwork. Dodge built the cars used by Roy Chapman Andrews's (1884–1960) 1925 Central Asiatic expeditions. In 1931, Citroën sponsored an excursion, including paleontologist Pierre Teilhard de Chardin (1881–1955), along the old Silk Road through central Asia to demonstrate its latest tires.

Some of the largest, most remarkable expeditions were organized by metropolitan museums in the first decades of the twentieth century. Perhaps the most ostentatious was the American Museum expedition, to central Asia, led by Roy Chapman Andrews beginning in 1922. Andrews enlisted paleontologists, geologists, zoologists, meteorologists, and cartographers to map the central Asian plateau and collect scientific data, including fossils, from this remote area. The expedition hoped to recover the earliest human fossils, which Osborn predicted would be found in central Asia (the only fossil that was actually known from Mongolia was a single rhinoceros tooth collected by Russian scientists in 1892). The logistics were horrendous. The area was almost completely inaccessible, with no trains, a harsh climate, and almost no food or water. Andrews proposed taking an automobile caravan, which could penetrate more than 100 miles per day across the Gobi Desert. Camel caravans, some with up to 125 animals, would be sent well in advance to predetermined staging points with tons of flour and rice, thousands of gallons of gas, and other supplies. Andrews raised more than $250,000 through subscription and large donations just to get the first season started. When the expedition found a skeleton of the giant mammal *Baluchitherium*, Osborn proclaimed the expedition a success. He said that the specimen's journey across the desert, through China, and around the world to New York in a steamer was the greatest paleontological event the world would ever know. Osborn was even more ecstatic when the expedition recovered spectacular dinosaur fossils, including actual dinosaur nests with eggs. The expedition ended after several years, in part because Andrews auctioned one of the eggs to raise more money (Chinese officials accused him of robbing them of valuable national treasures), and because he occasionally antagonized local bureaucrats by turning the expedition's guns on them. The trips made an enormous popular impact. The dramatic opening scene in the 1937 movie *Lost Horizon* included a paleontologist escaping by plane amid revolutionary gunfire and with the only fossil *Megatherium* from Mongolia clutched tightly in his hands. It is said that the Indiana Jones character from the 1981 film *Raiders of the Lost Ark* was based on Andrews.

Collapse of Evolutionary Paleontology

At the end of the nineteenth century, paleontological activity was reaching a crescendo. Paleontologists, the "high priests of the past" whose fossils revealed the patterns and processes of life's history, dominated evolutionary theory. Many considered Darwinism (although not evolution) to be dead, replaced by Lamarckian or orthogenetic models that were dominantly (although not exclusively) the work of paleontologists. By 1930, however, this had changed. Evolutionary paleontology had largely collapsed, practiced thereafter primarily by vertebrate paleontologists, paleobotanists, and a minority of invertebrate paleontologists. By World War II, the majority of paleontologists were micropaleontologists (students of fossilized shells of microscopic marine organisms of the sort that compose plankton) in search of oil. This transition was not simply economic but was driven by the burgeoning credibility of genetics, the expansion of the university system (particularly in the United States), the falling reputation of phylogenetics, and older estimates of Earth's age.

Science and Society: Growth of University Systems

Although the earliest universities were founded much earlier, the institution as it exists today is largely a nineteenth-century construction whose growth parallels that of paleontology. The earliest universities—such as those in Bologna, Paris, and Oxford—were places to study theology, medicine, law, and classics. During the nineteenth century, education (particularly in science and technology) became important for increasingly centralized nation-states. Many European universities were nationalized or reorganized in order to increase enrollments, to foster nationally important curricula (especially in relation to economic and military agendas), and to reflect changing ideas of democracy and society. In many countries, higher education was such a priority that national tax money was used in its support. The Morrill Act of 1862 in the United States granted public lands for the support of state universities. Specialized graduate training was imported to the United States from Germany, first at Johns Hopkins University in 1867.

The history of paleontology as a discipline is entwined with the evolution of the university. In the early nineteenth century, Cuvier, Lamarck, and Geoffroy were teaching natural history, comparative anatomy, and paleontology to medical students in Paris, and William Buckland at Oxford University was teaching theology students about fossils and natural history, interpreting his fossils in terms of biblical accounts. Paleontology was not a university subject in its own right in either case. The battles between Buckland and his antagonist Lyell were symptomatic of broader disputes over the position of religion in government and society (the Church of England was an integral part of the state apparatus at the time). Robert Grant, whose progressive evolution so inflamed Richard Owen, was a professor at the "godless" University of London, established in 1836 to break the religious stranglehold on higher education in England that Oxford and Cambridge epitomized. Before the 1870s, paleontologists in the United States, such as Leidy, taught medicine, whereas the next generation, such as Cope and Marsh, did not hold university appointments until the latter part of the nineteenth century, during the American expansion of university learning. Paleontology did not really become a university subject in its own right until Osborn established the Department of Vertebrate Paleontology at the American Museum as an extension to the Department of Biology while he was helping Columbia College follow the model of Johns Hopkins by transforming itself into Columbia University in the 1890s.

During the nineteenth century, the unprecedented pace of discovery of new fossil material had contributed to paleontological ascendancy in the field of evolution. Cope, Hyatt, Marsh, Huxley, and Osborn had pushed their dusty bones into the spotlight of social debate, filling phylogenetic gaps, demonstrating human ancestry, and exposing new modes of evolution. But paleontological dominance was also linked to skepticism about Darwin's particular version of the story. Natural selection was distasteful to many (largely because of its Malthusian image of the merciless weeding out of the less fit by the mighty), who were attracted to the more humanistic Lamarckian theories of paleontologists such as Cope and Hyatt. In this milieu, the scientific persuasiveness of paleontological Lamarckism (and orthogenesis) was bolstered more by technical objections to natural selection (relating to inheritance and the age of Earth) than it was by positive support for the theories themselves.

In the 1910s and 1920s, geneticists such as Thomas Hunt Morgan at Columbia launched their own offensive into evolutionary theory. They identified their own field as the only legitimate inquiry into how evolution worked; paleontologists could describe events but could not explain them. As universities expanded following educational reforms at the turn of the century, laboratory sciences (such as genetics) predominated, edging paleontologists and other natural historians into museum refuges. By the 1920s, geneticists were resurrecting Darwin's natural selection as the mode of evolution compatible with their new "hard inheritance," or genes. At roughly the same time, radiometric dating had increased the age of Earth some 20-fold, allowing a generous length of time for gradual Darwinian evolution. Popular opinion was swinging away from evolution altogether (not just Darwinism), particularly in the United States, where Christian evangelical fundamentalism was spreading (Tennessee's Butler Act banning evolution from public schools, made famous by the Scopes "monkey" trial, was passed in 1925). In this context, many paleontologists (as well as geneticists) preached against the pitfalls of grand theorizing, calling for their colleagues and students to return to the purity of descriptive empiricism. In a memorial celebrating the second centenary of Lamarck's birth, Ermine Cowles Case (1871–1953), wrote that only the geneticist could guide biologists through the labyrinth of new conceptions gathered around the gene.

Figure 6. Oil wells in California in 1932. (Anton Wagner's "Los Angeles, 1932–34.") Courtesy of the California Historical Society; FN-296180.

Petroleum and Paleontology

As evolutionary paleontology wilted, economic paleontology bloomed. Wells in the first major oilfields of Azerbaijan and Pennsylvania had been drilled in the 1850s. The wartorn decades of the early twentieth century saw oil production increase unimaginably. In 1915, the world produced 432 million tons of crude oil; by 1938, it produced 2595 million tons. Oil companies and national governments began major explorations for new oil reserves (as matters both of profit and of national security), and paleontology happened to be important for locating them (see Figure 6). Petroleum geologists search for oil largely by drilling exploratory boreholes (often in the ocean floor) into rocks of the same age and depositional environment as those of nearby reserves. Oil usually forms in marine rocks deposited as deltas at the mouths of ancient rivers (such as the Mississippi or Nile). Those rocks had to be buried and heated, metamorphosing the rich organic remains in the river **effluvia** into carbon-rich petroleum. Using microscopic fossils brought up in well cores, micropaleontologists could tell whether the bore had gone deep enough (the fossils act as a guide to the age of the rocks) and whether the rocks at the bottom of the well were of the type expected to produce oil (only certain species of microscopic organisms, usually **foraminifera**, lived in relevant environments). As the rush gained momentum, so did job opportunities for technicians trained in micropaleontology. Geology departments in postwar universities began specializing in economic geology, and their paleontological coursework was focused toward micropaleontology at the expense of more traditional topics within vertebrate and invertebrate paleontology, such as evolution and extinction.

* **Effluvia.** By-products of food and chemical processes.

* **Foraminifera.** An order of marine protozoans having a secreted shell enclosing the body.

Fossils were always associated with industry in that they were the by-products of mining. Many of Cuvier's Tertiary fossils came from mines and quarries in France: from the phosphorites of Quercy, mined for phosphates to be used in fertilizer, and from the Montmartre gypsum, which formed a main ingredient in plaster of paris. Buckland's *Megalosaurus* was discovered in slates that were quarried for roofing tiles (Buckland himself was also the chairman of the Oxford Gas and Coke Company). *Archaeopteryx* came from the limestone quarries of Solnhofen, whose rock was used for both **lithography** and building. Hugh Miller's early fish came from Scotland's Old Red Sandstone, which was used for building (central Glasgow is filled with buildings made of it), and Louis Dollo's (1857–1931) famous *Iguanodon* herd came from the extensive Bernissart coal mines of Belgium. In the eighteenth and nineteenth centuries, industry was important for paleontology, but paleontology was not necessarily important to industry. The relationship was reversed in the twentieth-century oil economy. When, in 1937, Oxford scientists showed an interest in the paleontology of Persia (which would yield some of British Petroleum's largest oil reserves), Shell Oil gave £25,000 for a new geology building. By the 1920s, the petroleum industry was making a noticeable impact on geology and paleontology. The fledgling Paleontological Society (founded in 1908 as a union of evolution-based paleontologists) was splintering over oil. A large group broke away in 1926 to found the Society of Economic Paleontologists and Mineralogists (SEPM), with the goal of promoting research in paleontology as it related to petroleum geology. Composed equally of academic and oil company paleontologists, the SEPM joined the American Association of Petroleum Geologists and the Paleontological Society in publishing the new *Journal of Paleontology*. Most of the funding for the journal came from oil company donations.

* **Lithography.** The art of making an illustration on stone so that impressions in ink can be taken from it.

The shift toward petroleum and micropaleontology alienated paleontologists working with fossil groups that were not important for oil exploration, particularly vertebrates, plants, and many invertebrate groups. A group of paleobotanists split from the Paleontological Society in 1936, allying themselves with the Botanical Society of America as its paleobotany section. Two years earlier, vertebrate workers had formed a semi-independent section of vertebrate paleontology in the Paleontological Society, but broke away completely in 1940 as the Society of Vertebrate Paleontology. The vertebrate paleontologists, notably George Gaylord Simpson and Alfred Sherwood Romer (1894–1973), complained that the more geologically oriented Paleontological Society did not properly recognize the biological emphasis of vertebrate paleontology. Simpson and Romer tried to pull vertebrate paleontology into the sphere of evolution and genetics. Toward that end, Simpson and Theodosius Dobzhansky, the population geneticist, launched the Committee on Common Problems of Genetics, Paleontology, and Systematics in 1942 as part of the new U.S. National Research Council. The paleontologists were mostly vertebrate evolution specialists [including Romer, Glenn L. Jepsen (1903–74), and Everett Olson (1910–93)], although there were a few paleobotanists and invertebrate paleontologists. In 1949, the committee published a symposium volume entitled *Genetics, Paleontology, and Evolution*, which became one of several important contributions to the "evolutionary synthesis" that attempted to unify paleontology, genetics, and population biology with Darwinian natural selection. Several members of the committee, including Mayr and Simpson, founded the Society for the Study of Evolution.

During those years, the *Journal of Paleontology* published a serialized debate polarizing the biological (specifically, evolutionary) and geological (specifically, stratigraphic) aspects of paleontology. Edwin H. Colbert (b. 1905), a vertebrate paleontologist, and Norman Newell (b. 1909), an invertebrate paleontologist at the American Museum, argued that paleontology was really a branch of biology, albeit one whose organisms were embedded in rocks. Marvin Weller (dates unknown), an invertebrate paleontologist at the University of Chicago, argued that paleontology could only be a subdiscipline of geology. Acrimonious assertions abounded about whether biological or geological aspects of the field were intellectually easier to acquire, whether paleontologists were obliged to serve the broader interests of geology or biology, and whether nonpaleontologists were qualified to practice stratigraphy. Some accused paleobotany of stagnating, its research being too botanical to be of interest to geologists.

The efforts of Simpson, Romer, Newell, and others to integrate paleontology—particularly vertebrate paleontology—into the evolutionary synthesis marked the first of two twentieth-century attempts to revitalize evolutionary paleontology. A compromise between genetics and field biology, the synthesis fused population biology, systematics, genetics, and natural selection into a framework that attempted to explain evolution in terms of **genotype** and **phenotype** (the genes of a species as distinct from its anatomical structures), selection and adaptation, and organism and environment. Simpson imported the statistical concept of species (that species do not share particular features but are variable populations whose average changes with evolution) from the synthesis into paleontology, and used it to estimate the rates of evolution using the fossil record. The question of rates was particularly important, because synthetic theory postulated relatively slow, gradual evolution rather than the rapid mutational change asserted by early geneticists. Simpson was lionized by advocates of the synthesis [e.g., Mayr and Theodosius Dobzhansky (1900–75)] for demonstrating the steady tempo of evolution in the face of counterclaims by some geneticists (and paleontologists). Romer provided paleontological examples of adaptation and radiation (the diversification of a group of organisms), notably the evolutionary transition from fish to amphibian. Newell, a colleague of Simpson's at the American Museum, debunked nineteenth-century examples of orthogenesis and evolutionary life cycles (such as Alpheus Hyatt's ammonites) using quantitative statistical arguments.

* **Genotype/phenotype.** A distinction between the actual appearance of an organism (phenotype) and the particular genetic material that the organism has inherited.

The second evolutionary revitalization in paleontology coincided with a sharp downturn in the American and European petroleum industries in the 1970s and 1980s. Spearheaded by Stephen J. Gould (1941–2002) of Harvard (a former student of Newell), Niles Eldredge (dates unknown) of the American Museum, and David Raup (dates unknown) of the University of Chicago, the paleobiology movement was conceived as a reaction to the synthesis and was designed to reassert evolution as the focus of paleontology (particularly invertebrate paleontology). These paleontologists recapitulated the arguments of their late-nineteenth-century forebears, arguing that the fossil record showed long-term patterns in evolution (macroevolution) that could not be explained by the synthesis of population genetics and Darwinian selection. Punctuated equilibrium (the idea that evolution is not gradual but happens in rapid bursts followed by long periods without change) and cyclic extinctions (the idea that mass extinctions have occurred in regular cycles throughout geological history, possibly because of regular impacts by comets or asteroids) were the most notable challenges to mainstream evolutionary theory. The group founded a new journal, *Paleobiology*, under the aegis of the Paleontological Society to promote evolutionary research. The oil crises of the 1970s and mid-1980s had a major impact on the discipline of geology (the University of Texas geology undergraduate enrollment dropped by an order of magnitude in 1986). Stratigraphic and economic paleontology waned, and many paleontologists engaged with the new paleobiology. Through roughly the same time period, cladistics also became popular, first among vertebrate paleontologists, then across the entire discipline. Also known as *phylogenetic systematics,* cladistics was the method of analyzing evolutionary relationships through the identification of characters as primitive or derived; this type of analysis relies on the assessment of the chances that similar forms are near or distant relatives, according to the pattern of how such characters may have been acquired or lost. Conceptually independent from the paleobiology movement, cladistics also set itself up in opposition to the synthesis, which cladists portrayed as lacking rigor because of its focus on adaptive scenarios for phylogeny reconstruction. Unlike paleobiology, cladistics united paleontology with other biological disciplines through a shared method and ideology. Thus, in the 1990s, paleontology was again dominated by evolution and phylogeny, as it had been a century earlier.

Paleontology and Museums

The nineteenth-century creation of public natural history museums (public both in their funding and in their admission policies) eventually did much to foster a popular audience for paleontology. Often modeled on the Paris Museum, metropolitan museums of the nineteenth century exhibited grandiose research through public lectures, galleries, and collections. As the world's cities—London, Berlin, Frankfurt,

Scientific Institutions: British Museum of Natural History

The grand structure on London's Cromwell Road, containing millions of animal and plant specimens, is popularly known as the Natural History Museum, although its official name is the British Museum (Natural History). The institution was so designated by the museum's trustees when the new building opened in 1881. At the time, the Natural History Museum was technically a part of the British Museum and had been physically located in the same Bloomsbury building that housed mummies, ancient artifacts, and the British Library. But in the nineteenth century, the British government, including Prince Albert, Queen Victoria's husband, was keen to promote British science in the service of imperial power. The Great Exhibition of 1851, housed in the glass and steel Crystal Palace, advertised the glories of British engineering and included exhibitions on paleontology. Capitalizing on these urges, Richard Owen, the superintendent of the natural history collections, convinced Parliament to move his collections away from the British Museum into a glorious new building. The museum was built on the grounds of the Great Exhibition, an area that Prince Albert imagined as a permanent exhibition of British science (today the area also contains the Science Museum, the Victoria and Albert Museum, and until recently, the Geological Museum). Working under Owen's direction, the architect Alfred Waterhouse designed a perfect "Cathedral of Science" in neo-Gothic style, with a great churchlike hall, buttresses, and gargoyles of living and extinct animals and plants.

In the decades preceding the move to South Kensington, Owen had been engaged in public battle with Thomas Huxley and Charles Darwin over evolution and the origin of species. In part, Owen viewed his "Cathedral of Science" as the triumph of his view of life history over Darwin and Huxley's. Owen's view was of a stately, ordered, divinely inspired unfolding of life, while Darwin's natural selection was random and cruel, sometimes called the law of the Higgledy-Piggledy. But Owen's triumph was brief. Despite his continued objections, a large statue of Darwin was later installed on the main staircase of the museum, the very "altar" of the "cathedral," by the Prince of Wales, Thomas Huxley, and Owen's successor as director, William Flower. Only in the twentieth century was Darwin's statue moved to a less prominent position and Owen's more meager statue put in the place of honor.

Philadelphia, New York, Mexico City, Buenos Aires—established their own institutions, museums became synonymous with natural history, literally institutionalizing the association (and differentiation) among the disciplines of zoology, botany, entomology, mineralogy, and paleontology. Museum scientists, who were usually systematists, interpreted life's seething diversity as it was shipped back to colonial capitals in bottles and boxes. Extinction, evolution, race, and culture, as public debates, have therefore been closely associated with museum research. Perhaps because this tradition firmly ensconced paleontology in museums, the subject did not become a part of twentieth-century university expansions, and where it was, universities usually created their own museums.

Museums are both a public face of paleontology and important sites of research. As physical institutions, large twenty-first-century museums typically have public galleries filled with giant dinosaur skeletons, charging elephants, breathtaking wilderness adventures, and mass extinctions. Consequently, paleontologists have long been aware of the importance of a popular audience, and public perception strongly identifies paleontology with what the public galleries exhibit. But museums also contain private research areas where scientists doggedly unravel the mysteries of the natural world. The relationship between public galleries and private research areas has changed since the early twentieth century. The research areas of the museum now designated as "no public access" are not exactly private; they are openly available to the scientific community. Scientists and students from around the world visit museum research collections, which operate much like libraries containing specimens rather than books. Although access was often restricted to "gentlemen," late eighteenth- and early nineteenth-century museum galleries were also open to the public in this limited sense. Many European public museums (in the sense of state institutions funded by tax revenue) had their origins in natural history cabinets, collections of natural objects including rocks, minerals, plants, animals, and fossils. Concerned with understanding and classifying the natural world, post-Renaissance men of letters gathered objects for study and description. The specimens were open to like-minded intellectuals from seats of learning throughout the world. During revolutions and reforms in the late eighteenth and early nineteenth centuries, many large cabinets were nationalized (or purchased). The Paris Museum was formed by Jacobin legislation from the Jardin du Roi (the King's Garden) in 1793, while the core of the British Museum of Natural History was the collection of Sir Hans Sloane (1660–1753), purchased by Parliament after Sloane's death in 1753. Although access was widened (often with the rhetoric of egalitarian liberation), admission was seldom open to the public as we understand the word today. Lectures at the Paris Museum were open to members of learned societies and students of medicine, but not to women or workmen.

Figure 7. The British Museum of Natural History (c. 1880). From The Graphic *(London, c. 1880). © The Natural History Museum, London*

Within the nineteenth-century milieu of social change that extended suffrage, emancipated slaves, and unionized the working classes, museums became more broadly public. In London, a magnificent new Natural History Museum opened in 1880 on the land that had been used for the Crystal Palace of the great International Exposition of British industrial might (see Figure 7). Like the exposition, museums displayed the supposed wonders of the world, inflecting them with nationalist and imperialist overtones. These galleries were not necessarily intended for a learned audience but were partly designed to impress citizens and subjects with the scientific might of the nation and the wonders of its colonies. Research collections and public galleries were not distinct—specimens were arranged either in display cases or in drawers that anyone, scientist or schoolchild, could pull open. The public museums of the late nineteenth and early twentieth century often embedded themes of education, improvement, and social engineering in their galleries. Andrew Carnegie's gifts of *Diplodocus* skeletons were part of his larger agenda of educating the world's laborers (Carnegie was the son of a Scottish factory weaver). Both Carnegie and financier George Peabody donated millions of dollars for museums, libraries, and theaters to improve and civilize the lives of workers in sprawling industrial cities. In New York, too, Osborn imagined the exhibits at the American Museum as an improvement for the working classes (similarly, Osborn was involved with schools, YMCA facilities, and **eugenics** societies). Viewing twentieth-century horrors such as World War I as an outgrowth of materialism and mechanized life, Osborn emphasized the natural sciences and their connection with nature. His exhibits in the Hall of Man at the American Museum placed human evolution in the context of changing environments rather than depicting progressive technologies. Like Carnegie, Osborn exported his vision to other museums in the form of books, casts, and guides to duplicating the American Museum dioramas.

* **Eugenics.** Science relating to improvements in the inherited physical and mental traits of the human race. Eugenics is largely discredited due to misuse by such groups as the Nazis in the 1930s and 1940s.

The best known and most successful museum directors took an active part in building audiences, reaching out in public lectures, and writing volumes of scientific popularization. Although social magnanimity was the stated motive, public exhibits were also important for attracting funds to museums. Public access and improvement were seductive causes, both to state treasurers and to wealthy philanthropists. With the extension of suffrage and social supports, nation-states were increasingly seen as providers of public benefits—museums were powerful symbols of that new relationship but could remain so only if they were popular. An 1860 select committee of the British House of Commons found that the natural history collections were the most popular part of the British Museum with both the middle and working classes. New galleries gave (and continue to give) pride of place to extreme and garish specimens. Reconstructed dinosaur skeletons dominated exhibits: the *Ichthyosaurus* in the 1840s, Owen's *Iguanodon* in the 1850s, and Osborn's *Tyrannosaurus rex*. Because they were large and fantastic, extinct creatures (particularly dinosaurs) became synonymous with museums in the nineteenth- and twentieth-century popular imagination. Balzac eulogized Cuvier's lost worlds in his novels, worlds reconstructed from monster's teeth; Osborn literally resurrected the past as huge skeletons mounted in life poses, caught in flagrante delicto by gawking New Yorkers.

The public museum phenomenon was paralleled in an educational sense within expanding universities. In 1859, Louis Agassiz opened the Museum of Comparative Zoology at Harvard, which was devoted to advanced education and research in natural history (including paleontology). The Paris Museum had pioneered the association between museum and education when medical students were required to attend lectures by the museum professors at the turn of the nineteenth century. Agassiz, however, specifically set out to train museum-based researchers in natural history fields. Agassiz's model of the museum within the university was replicated in other universities as his students won prominent positions. In Paris in 1880, the French government extended the museum's educational role by restricting its autonomy and putting it at the service of medical and science professors in Parisian universities. In 1891, when Osborn transformed biology at Columbia (which grew from college to university in the process), he organized vertebrate paleontology not within his biology department but at the American Museum. The University of California at Berkeley had tried to sustain a department of paleontology as early as 1912, but with only intermittent success until 1921, when Annie Alexander, sugar heiress and natural history devotee, endowed an independent Museum of Paleontology as its nucleus.

The Great Depression and World War II bankrupted museum endowments and transformed national agendas from social change to national security. The position of the museums changed in the process. Osborn resigned from the directorship of the American Museum in 1933, depressed by budgetary shortfalls. Expedition money was cut there altogether in 1932, and the museum had to cut back many staff positions. National museums, such as the British Museum of Natural History, were forced to shift their emphasis to maintain their state funding. Research rather than exhibition was seen as the breadwinner. Through the same period, many universities scaled back their paleontology programs, cutting them or changing their emphasis to petroleum exploration. Evolutionary paleontology became increasingly associated with museums, locked behind the mysterious closed doors not open to the public.

Tools of the Trade

Paleontology is not usually associated with technical or technological breakthroughs as are other scientific disciplines, such as physics, genetics, or chemistry. The paradigm shifts of paleontology, extinction, and evolution are entirely conceptual. Some episodes in paleontology are linked closely with technological or methodological developments, however.

Radiometric and magnetostratigraphic dating techniques have probably had the most dramatic effect. Many controversies about evolution and extinction have been concerned in some way with the timing of events or processes. When Strutt and Boltwood first applied the principle of radioactive decay to the age of Earth in the 1910s, they came up with dates more than an order of magnitude older than paleontologists had thought. The techniques were applied to minerals bound up in volcanic rocks, which do not normally contain fossils, but their implication for paleontology was obvious: Life was much older and evolution much slower than anyone had imagined. The new dates were reported only a few years after Vernon Kellogg (1867–1937) had reported the death of Darwinism (by which he meant evolution by gradual natural selection, which had partly been discredited in his mind because it required an extremely ancient Earth). The 1950s invention of carbon-14 dating, a technique that could be applied directly to fossil remains, allowed both paleontologists and archaeologists to determine when the last ice sheets melted and when the remarkable mammoths and saber-toothed tigers had disappeared. Magnetostratigraphy, the correlation of rocks by reversals in Earth's magnetic polarity, could be applied to rocks of all sorts, volcanic or water-deposited. The *Olduvai reversal* [named because it was found in Louis Leakey's (1903–72) famous Olduvai Gorge in East Africa] was particularly relevant for dating early fossil humans. The combination of magnetostratigraphy and radiometric dating, which could be correlated with one another using the **igneous rocks** that had been spreading out from the mid-ocean ridges for the last 160 million years, allowed paleontologists to make continuing refinements in the paleontological timescale. The Geological Society of London published an influential compendium of ages and dates in 1964, followed in 1971 by another from the Cambridge Arctic Shelf Programme, and yet another in 1982 by Walter Brian Harland (b. 1917) and colleagues at Cambridge.

* **Igneous rocks**. Rocks with a volcanic origin that were deposited in a molten state (thus prohibiting the possibility of fossil remains). Other types of rocks are *sedimentary*, those laid down by water, and *metamorphic*, sedimentary rocks that have been modified by heat and pressure.

The confirmation that Earth's continents moved through the process of continental drift also had a profound effect on paleontology. How animals got from one continent to another was a potential stumbling point for new ideas about evolutionary radiations. William Diller Matthew devoted his 1915 book *Climate and Evolution* to explaining how ancient land bridges and the accidental dispersal of plants and animals on natural rafts could explain the geographic distributions of fossil mammals. Many paleontologists who doubted these scenarios of forlorn monkeys desperately clinging to branches in the mid-Atlantic, such as the brash young Malcolm McKenna (dates unknown), who replaced Simpson at the American Museum in 1960, enthusiastically adopted continental drift as an alternative explanation. By the 1990s, vicariance (the evolutionary separation of two groups by geologic change, such as continents drifting apart) had almost universally replaced dispersal as the stock explanation of how animal groups diverged. One major research program concerned the exchange of species over the Panamanian land bridge after North and South America came together in the Pliocene. Thomas Jefferson's perplexing giant sloth, *Megalonyx*, was, upon reappraisal, part of a group that had evolved in South America for most of the Cenozoic, crossing into North America only a few million years ago before becoming extinct (today, sloths are found only in South America). Jaguars and wolves were examples of groups that migrated into South America from the north.

Paleontologists have considered many simpler techniques as important advances. In the 1870s, crews working for Marsh and Cope invented the plaster jacket to protect fossils on their bumpy journey from western outcrops to eastern museums. Samuel Williston, who set soldiers' broken bones during the American Civil War, was said to have been the first to use this technique. Invertebrate paleontologists learned to quickly etch millions of fossils from solid limestone by dissolving the latter in acid. In the 1960s, it occurred to vertebrate paleontologists that they could sieve quarry detritus to find minuscule mammal teeth. Surprisingly, that simple technique inspired research programs in paleoecology and taphonomy (the study of the process of fossilization). The automobile proved important not only for visiting remote areas, but also because government-funded highway programs cut through hills everywhere, opening clean new exposures in rocks filled with fossils. Paleontologic correlation in the 1960s and 1970s owes almost as much to road cuts as it does to radiometric dates. X-ray machines in the early twentieth century and, later, CT (computer tomography) scanners allowed paleontologists to "dissect" their stony skeletons and see structures that they could not before. Biogeochemistry and stable isotope analysis allowed them to estimate temperatures and rainfall patterns in the geologic past. Oxygen isotope curves revealed that there were not just four ice ages during the Pleistocene but that worldwide temperatures had oscillated up and down more than 20 times in the last 2 million years, leading to an acrimonious review of nearly every aspect of ice age paleontology.

Journals are often important emblems of the self-definition of academic disciplines. The first serial devoted exclusively to paleontology was *Palaeontographica*, founded in 1846 by German paleontologist Hermann von Meyer (1801–69). Although it was the first exclusively paleontological journal, *Palaeontographica* was certainly not the first journal to publish papers on fossils. The *Comptes Rendus*—or collected works—of the French Academy of Sciences had included occasional paleontology papers since 1835, and in Germany, the *Neues Jahrbuch für Mineralogie* (New Yearbook of Mineralogy) had included such papers since 1807. Notably, the latter changed its name several times [*Neues Jahrbuch für Mineralogie, Geognosie, Geologie und Petrefaktenkunde* (New Yearbook for Mineralogy, Geognosy, Geology and the Petrified)] and even later to the *Neues Jahrbuch für Mineralogie, Geologie und Paläontologie*, reflecting the differentiation of eighteenth-century mineralogy into nineteenth-century disciplines. In the United States, several former students of Louis

Agassiz, including paleontologist Alpheus Hyatt, founded the *American Naturalist* in 1867 as an evolutionary journal devoted to neo-Lamarckism (Agassiz himself was a committed antievolutionist). The name of the new journal may have recalled the mock secret society that Agassiz's students had organized in their days under the naturalist from Neuchâtel: The Society for the Protection of American Naturalists against the Oppression of Foreign Professors. Cope later purchased the journal in order to rush his descriptions of dinosaurs and tertiary mammals into print before Marsh did. The content of *American Naturalist* was largely paleontological until Cope's death in 1897. Museums also published their own journal series; the American Museum of Natural History appears to have started this trend in 1881. Nationalist journals abounded in the early twentieth century, many of them adapting von Meyer's German title: *Palaeontologia Hungarica* (1921), *Palaeontologia Sinica* (1922), *Palaeontologia Polonica* (1926), *Palaeontologia Africana* (1953), and *Palaeontologia Jugoslavica* (1958). Interestingly, the Paris Museum published *Palaeontologia Universalis* between 1903 and 1958. During the mid-twentieth century, new journals were often founded to accommodate new subdisciplines; *palaeogeography*, *palaeoclimatology*, and *palaeoecology*, for example, were devoted to environmental paleontology, and *palaeovertebrata* was devoted exclusively to vertebrates. Others were interdisciplinary journals emphasizing specific time periods, such as *quaternary research* (devoted to ice age paleontology and geology).

Science and Spectacle: The Popularization of Paleontology

Paleontology is arguably one of the most popular of the sciences (in the sense that those with no formal training in the discipline can be both interested in and technically knowledgeable about it). Whether watching a television cartoon or viewing a big-screen movie, children and adults alike can rattle off the Latinate genus names of a stunning variety of dinosaurs: *Triceratops, Stegosaurus, Brontosaurus* (many will even exclaim at this point that the name *Brontosaurus* is not technically correct but is more properly *Apatosaurus*), and, of course, *Tyrannosaurus rex*. (In fact, the spelling-check feature of the word processor used to write this article recognizes all of these names but does not know those of the cat, the horse, or the mouse: *Felis*, *Equus*, and *Mus*.) Paleontology's broad public audience is an integral part of the discipline and has arguably been responsible for its continued existence. In contrast to many other sciences, paleontology has social structures that link it with a popular audience (movies, television, books, novels, exhibitions, and even toys). Although many adults read popularized accounts of physics, children seldom play with plastic atoms. Astronomy may be one of the few other sciences that attract a similar public following.

The connection between paleontology and its popular audience is not a result of twentieth-century mass media and tourism; during the nineteenth century, paleontology was, if anything, more popular than in the twentieth century. In the early nineteenth century, fossils were the leisure activity of the affluent, the pastime of amateur naturalists, physicians, and clergy. Medical doctors and Protestant clergy, particularly in Britain and Germany, were avid naturalists and paleontologists [the Reverend William Paley (1743–1805) argued in his 1802 *Natural Theology* that nature's perfection was evidence of God's design]. William Conybeare, who reconstructed the first plesiosaur in 1824, and Gideon Mantell (1790–1852), who discovered the first dinosaur fossils, were clergyman and physician, respectively. The first **Neanderthal** fossils were presented not at a meeting of paleontologists, but at the Lower Rhine Medical and Natural History Society. On the one hand, ministers and doctors were both leisured and educated, allowing them to pursue esoteric activities such as fossil

* **Neanderthal.** An extinct hominid distinguished by a broad braincase, short limbs, and large joints.

collection that most people could not; on the other hand, the fact that such prominent members of local communities were familiar with the details of paleontology meant that fossils, evolution, and extinction figured heavily in many nineteenth-century social and political debates. In 1830s Britain, for example, sermons preached from pulpits in the established Church of England were embellished with fossils from Buckland's Noachian flood and the catastrophic remains of Cuvier's former creations (and past Armageddon-like destructions), and Chartist insurgents, radical reformers who were demanding universal male suffrage and better pay for workers, drew parallels between the inexorable progress of the fossil record and the inevitable rise of the lower classes. It was to refute the latter connection that Owen designed his Dinosauria in the 1840s, and Victoria and Albert broadcast Owen's message to hundreds of thousands in the dinosaur garden at the Crystal Palace Exhibition, while Huxley tore it apart in his *Lay Sermons for the People.*

Paleontological images were evocative because they were at once poetic, freakish, and fantastic. Paleontology exhibitions were about the monstrous and the minuscule, about the vicious and the dead. Albert Koch (1804–67), a German collector, posed a skeletal herd of mastodons like a carnival sideshow, billing them as "Missourium, the Leviathan of the Bible." Ichthyosaurs and plesiosaurs became the monsters of the deep in a nineteenth-century world where the ocean floors were offering up more plankton than passion. When the British ship *Daedalus* reported a "sea serpent" in the East Indies in 1848, a flurry of letters were mailed to the *Times*, several suggesting that the monster had been a plesiosaur (Owen dampened the enthusiasm by assuring the public that it could have been no such thing). The cover of "Wonders of the Primitive World" in *The People's Library* magazine series of 1869 featured a lurid scene with an ichthyosaur and plesiosaur engaged in horrible battle with pterodactyls perched like vultures above and a volcano erupting in the background. Although this series mentioned science, its emphasis was mainly on wonder (like the "giant carnivorous frog," *Chirotherium*). Neanderthals became freaks of nature, described by Rudolf Virchow (1821–1902) as twisted by pathological agonies. Camille Flammarion's (1842–1945) *The World before the Creation of Man* (1886) illustrated iguanodons looking through the windows of Paris apartments (see Figure 8), rising from the grave to say, "Here we are, your elders and your ancestors. Without us, you would not exist." The mounted dinosaurs in the worlds' museums and the lucrative terror of *Jurassic Park* are not twentieth-century exceptions but are parts of a longtime association between paleontology and the public. As experimental sciences gained ground in the late nineteenth century, a British Museum of Natural History guidebook apologized that "of late" there had been lots of excitement about microscopy but that it was just not as gripping as paleontology.

➢ **Oligocene.** See *Earth Science and Geology.*

Figure 8. An Iguanodon *looking into the windows of a Paris apartment. From Camille Flammarion,* The World before the Creation of Man *(Paris, 1886).*

In the twentieth century, the public interface continued to play a major role in paleontology, not just in its popular conception but also in the development and agenda of the discipline itself. The discovery of a "living fossil," *Metasequoia* (the Dawn Redwood), is a good example. The story began in 1828 when Adolphe-Théodore Brongniart (1801–76) described some unusual plant fossils from **Oligocene** deposits in France that were similar to the living sequoia of North America, but otherwise enigmatic. In 1941, a Japanese paleobotanist distinguished Brongniart's tree as a new genus called *Metasequoia* based on new fossil material from Japan. Meanwhile, in the same year two Chinese botanists discovered a tree in Szechuan (known to the locals as the water fir) that was similar to sequoia. Several years later, they realized that the tree was a living version of Brongniart's fossil *Metasequoia*. Ralph Chaney (dates unknown), a paleobotanist at the University of California Museum of Paleontology who was interested in the fossil *Metasequoia*, capitalized on the similarity between the living water fir and his native California redwoods to generate local interest in the story, renaming the tree Dawn Redwood. The living fossil was so intriguing that the Save-the-Redwoods League and the *San Francisco Chronicle* sponsored an expedition that would take Chaney to China and Japan. Both the scientific agenda and the funding for the expedition (not to mention the tree's name) were tied to popular response, rather than to the university, other scientists, or state funding bodies.

Bibliography

Appel, Toby. *The Cuvier-Geoffroy Debate: French Biology in the Decades before Darwin.* Oxford: Oxford University Press, 1987.

Bowler, Peter J. *Fossils and Progress: Paleontology and the Idea of Progressive Evolution in the Nineteenth Century.* New York: Science History Publications, 1976.

———. *Life's Splendid Drama.* Chicago: University of Chicago Press, 1996.

Buffetaut, Eric. *A Short History of Vertebrate Paleontology.* London: Croom Helm, 1987.

Desmond, Adrian. *Archetypes and Ancestors: Palaeontology in Victorian London, 1850–1875.* Chicago: University of Chicago Press, 1982.

Rainger, Ronald. *An Agenda for Antiquity: Henry Fairfield Osborn and Vertebrate Paleontology at the American Museum of Natural History, 1890–1935.* Tuscaloosa: University of Alabama Press, 1991.

Rudwick, Martin J. S. *The Meaning of Fossils: Episodes in the History of Palaeontology.* London: Macdonald, 1972.

Trinkaus, Erik, and Pat Shipman. *The Neanderthals: Of Skeletons, Scientists, and Scandal.* New York: Alfred A. Knopf, 1992.

Psychology

Kenton Kroker

Unlike the natural sciences, psychology has, more often than not, treated human beings as knowing subjects capable of expressing the nature of their own self-consciousness. In what sense, then, can we treat psychology as a science? This question, although not easily resolved, represents a unifying thread that runs throughout the history of psychology as an investigative enterprise. This article begins not with psychologists, who did not appear as an identifiable group until the late nineteenth century, but with the psychological problems raised by the new investigative practices of the early seventeenth century.

Natural Philosophy and Psyche, 1543–1800

OBSERVATION, NATURAL HISTORY, AND MECHANISM

Are the senses a reliable guide to knowledge? This is an ancient question, but its modern revision began in 1543 with the publication of Andreas Vesalius's (1514–64) *On the Achitecture of the Human Body* and Nicholas Copernicus's (1473–1543) *On the Revolutions of the Heavenly Spheres*. While dealing with radically different subjects, both texts took observation to be a practical problem. Vesalius appropriated the dominant artistic traditions of his day to argue that anatomists should be trained in realism. To see the body properly, anatomical illustrations should be realistic, not iconic. Copernicus's revival of **heliocentrism,** on the other hand, implied that a mind trained in mathematical calculations could overcome the embodied experience of observing the heavenly spheres revolve around Earth.

* **Heliocentrism.** A theory of planetary motion that places the Sun at the center of the universe.

The idea that the senses could receive knowledge directly was challenged further in the early seventeenth century. Following Galileo Galilei's (1564–1642) 1610 description of his observations with a new mechanical device, the telescope, in *The Starry Messenger*, Johannes Kepler (1571–1630) compared the eye to a simple mechanism, the camera obscura, in his *Dioptrice* (1611). According to Kepler, the camera obscura (literally, a "darkened room" in which an inverted image appeared on a screen across from a pinhole on the other side of the room), the telescope, and the eye all worked according to the same geometric principles. Kepler avoided the question of how the human soul perceived this image (which, following the camera obscura analogy, was inverted). Galileo went further, arguing in *The Assayer* (1619) that only the measurable qualities of objects—their form and motion—existed independent of the observer's mind. Other qualities that objects seemed to possess—tastes, odors, or colors, for example—were merely secondary, or accidental.

The senses were indispensable to the acquisition and improvement of knowledge, but they needed to be tempered by reason, which, Galileo argued, manifested itself in mathematics. The use of instruments, such as the telescope or the pendulum, seemed equally imperative to the development of natural philosophy. Thus Kepler and Galileo both developed analogies between these mechanisms and those more perfect instruments, the sensory organs.

But if the senses were perfect in conception, they were flawed in execution. According to the English lawyer and courtier Francis Bacon (1561–1626), this was a problem of education. In his *Advancement of Learning* (1605), Bacon called for a "natural history of the soul" that would treat the soul as an object like any other, enabling general conclusions about its nature to be drawn from an inventory of facts. The absence of anything like "psychological practice" in this period is borne out by

Bacon's refusal (or inability) to come up with such an inventory. The closest he came was in his *Novum organum* (1620). Here, Bacon described four "Idols of the Mind," all of which placed the individual human mind at the center of an epic conflict between socially maintained prejudice and the new natural philosophy, which relied on experimental demonstration and what he called the *method of induction*.

The series of civil and religious wars fought over the course of the seventeenth century in England and on the continent encouraged intellectuals to turn toward natural philosophy in search of a solution. If the nature of the individual mind could be understood, then social order could be realized. In *Leviathan* (1651), Thomas Hobbes (1588–1679) rejected the idea that kings had a divine right to rule. Political absolutism (which Hobbes supported) had evolved out of the primal anarchy of a "war of all against all." But once absolute monarchs ceased offering protection to the individual, their rule was no longer justified. How had such a political order appeared in the first place? Hobbes argued that language was the key. Criticizing Bacon's claim that the invention of printing, gunpowder, and the compass were largely responsible for man's improved lot on Earth, Hobbes argued that "the invention of printing, though ingenious, compared with the invention of letters, is no great matter." Language enabled thought. As a staunch **materialist**, Hobbes insisted that all thought began with sensation, which was caused by an external object pressing (either directly or through a medium) onto the appropriate sensory organ. This motion was in turn transmitted to the brain, where a "resistance" was formed, causing an image to evolve. Understanding, argued Hobbes, was nothing more than those images "raised in man by words." Language was a human creation that made knowledge possible. Like the rule of the sovereign, it was an artificial tool for the maintenance of social order.

* **Materialism**. A philosophy claiming that all facts, including those concerning the human mind, can be explained completely in terms of physical processes.

Hobbes was reputed to be an atheist, and his ideas were widely shunned. The philosophy of René Descartes (1596–1650), on the other hand, proved to be enormously influential. Descartes felt similarly that many religious disputes originated with incoherent language, such as the Aristotelian doctrine that described the soul as "substantial form." But Descartes's solution was not in a reform of language but in the introduction of a new philosophical method, one that definitively separated the immaterial soul from the body (*Discourse on Method*, 1637). The advantage of the Cartesian method was that anyone could discover it for themselves through self-reflection and radical doubt. Everything could be doubted, Descartes argued, except the fact that one is doubting. Thus an "I" must exist, and its essence is thought. The body, on the other hand, was comparable to a machine, whose operations could be explained strictly in terms of matter in motion. Such an analogy was unpalatable to Hobbes. But for Descartes it made sensation, perception, and imagination comprehensible to an experimental natural philosophy, even as it sealed off the nature of the soul as reason's own preserve.

EMPIRICISM AND RATIONALISM

Descartes had argued that God must exist because everyone had an innate idea of what he must be like. Unorthodox as this method was, it at least maintained a modicum of faith and was therefore tolerable, particularly to Protestants in northern Europe. But the mechanical analogy's capacity to explain voluntary motion or thinking was severely limited. During the eighteenth century, the two philosophical schools of empiricism and rationalism attempted to address the problems of mind created by Cartesian dualism.

Empiricism argued that sensations were the raw materials of thought. John Locke (1632–1704), a political philosopher and an early member of the Royal Society, was the first to set out the empirical position in any detail. In his *Essay Concerning Human Understanding* (1690), Locke described the human mind as a *tabula rasa*, or "blank slate," upon which sensory experience was impressed to form "ideas." Unlike Plato's (427–348/7 B.C.) "ideal forms," Locke's ideas were not opposed to appearances. Simple ideas, such as that of hardness, came from sensory experience. The mind then combined these simple ideas to form the complex ideas of substance, modality, and relation. "Internal sensation" or "reflection," which emerged out of activities such as willing and deliberating, provided ideas about the self.

* **Epistemology.** A branch of philosophy that deals with the nature of knowledge.

Locke's **epistemology** dovetailed with his plea for religious and political tolerance. Descartes's universal, intuitive, and innate ideas of the self and of God were unacceptable to Locke. They revealed little that was "clear and distinct," as Descartes had argued. Nor did they authorize further deduction. It was the activities or faculties of the mind, not its objects, that were universal. An understanding of these activities provided the ground upon which natural philosophy could blossom into a body of knowledge capable of attaining common assent and lasting peace.

Eighteenth-century philosophers began to take Locke's doctrines in a more radical direction. In his *Essay Towards a New Theory of Vision* (1709), George Berkeley (1685–1753) argued that many concepts that Locke had taken for granted were created out of *associations* (Locke had used the term only to describe "unreasonable" mental activity). The concept of extension, for example, was nothing more than the repeated *association* of visual images with the perception of minuscule eye movements. After applying a similar analysis to other primary qualities, such as motion, number, and form, Berkeley eventually concluded that these qualities did not inhere in objects at all. Ideas did not correspond to objects. They were the only objects that existed.

In his *A Treatise of Human Nature, Being an Attempt to Introduce the Experimental Method of Reasoning into Moral Subjects* (1739–40), David Hume (1711–76) used Berkeley's arguments to turn induction into a problem rather than a method. Comparing the "law of association" to Isaac Newton's (1642–1727) gravitational law, Hume argued that every category the mind used to connect one sensation to another was nothing more than a habit of association. Causation, for example, did not describe the relationship of one event to another. Rather, it expressed a feeling experienced when two events, which had regularly appeared in succession in the past, followed each other in the present. Even the idea of self-identity, so readily apparent to Descartes and Locke, was the product of association. Reason, argued Hume, was the "slave of the passions." But Hume did not thereby repudiate reason and knowledge. Practical knowledge of the world, which was entirely dependent on the association of ideas, was usually helpful to human existence. Historical research, which Hume took up after his philosophical inquiries prompted a mental breakdown, revealed that habit and custom, not reason, had produced civilized society and all its benefits. Custom was a natural guide to moral and social progress; an excessive reliance on reason, on the other hand, could lead into obscurities and error.

Where empiricism modeled the mind upon objects in the world, rationalism treated the world as a form of mind. Baruch Spinoza (1632–77) introduced the notion of *psychophysical parallelism*, which suggested that mind and body were not causally related, but rather, were two different expressions of the same underlying event. Gottfried Wilhelm Leibniz (1646–1716) argued that sensationalism (a variant of empiricism) could never explain how an endless stream of infinitely small sensations could ever become the sensation of, for instance, hearing a wave crash upon the shore. The notion that mind experienced the world of objects through sensations was an error. In its place, Leibniz's monadology described the world as a dynamic system of mathematical points (monads) that possessed a rudimentary capacity (entelechy) for perception. This ability culminated in the ability of mind to become aware of itself, which Leibniz dubbed *apperception*.

Rationalism was intimately associated with the mathematical practices of manipulating signs and symbols rather than the natural philosophy that demonstrated through experiment. In fact, it was a professor of mathematics at Halle, Christian Wolff (1679–1754), who first used the term *psychology* to describe the specialized study of the human mind. In his *Psychologia empirica* (1732) and *Psychologia rationalis* (1734), Wolff divided psychology into two major components: a body of empirical facts derived from introspection, and a series of concepts, such as unity, immortality, and free will, that could be deduced from first principles.

Wolff's attempt to accommodate both the empiricist and the rationalist position was short-lived, however. His student Immanuel Kant (1724–1804) rejected the assumption that philosophical reflection was the psychological equivalent of scientific observation. The fundamental unity of mind meant that the mind could not take itself as an object without transforming, rather than representing, psychological phenomena. Because mind could not be known directly, the critical philosopher had to intuit its principles by observing how scientific knowledge was structured. The hope that psychology could develop as a science, however, was illusory. Epistemology was a logical, philosophical investigation into the nature of knowledge; it did not refer back in any way to the workings (mechanical or otherwise) of the individual mind.

Kant's critical, rationalist philosophy ultimately forced self-reflection into retirement as a method of psychological research. A similar movement was afoot among British empiricists. Thomas Reid (1710–96) advanced the notion of *common sense* against what he felt to be Hume's absurd reduction of the doctrine of ideas. In his *Inquiry into the Human Mind* (1764), Reid argued that perceptions conveyed the true nature of objects, and this fact could be demonstrated by noting which truths were recognizable by all. On the basis of such self-evident facts, Reid offered an "anatomy" that classified the mind's various powers, or faculties, as either active or passive, natural or acquired, which, like natural history in general, testified to the rational order of God's creation. Reid pitted his "psychology of faculties" against the notion that all elements of mind could be explained in terms of the association of ideas. But Reid's student Dugald Stewart (1753–1828) eventually merged the two in his *Philosophy of the Human Mind* (1792). Stewart's writings were quite popular, and his work became a vehicle through which the Scottish school of common sense came to the United States.

Common sense and critical philosophy might seem to be worlds apart, but they had a similar impact on the development of psychology. They both overturned the epistemic authority of individualistic self-reflection. From now on, psychological research would always require two participants rather than one.

Physiognomy and Phrenology, 1775–1850

Although Kant rejected self-reflection as the foundation of scientific psychology, he acknowledged the practical moral value of carefully observing oneself and others. By the beginning of the nineteenth century, physiognomy and phrenology were the two most popular forms of such a practice.

THE BODY AS VISIBLE SOUL

Physiognomy's central premise was that a person's physical appearance indicated the true nature of his or her temperament. It had long been a cornerstone of Hippocratic medical practice, which took physical appearance as indicating susceptibility to certain diseases [see *Anatomy and Physiology: Pre-Seventeenth Century*]. During the last quarter of the eighteenth century, however, the art of interpreting countenance as the outward expression of the soul's inner activity achieved an unprecedented popularity in Europe, largely through the writings of Johann Kaspar Lavater (1741–1801), a pastor of the Zurich Reformed Church.

Lavater's philosophy, which culminated in his four-volume *Physiognomic Fragments for Furthering the Knowledge and Love of Man* (1775–78), proved enormously popular among painters, poets, and writers, all of whom encouraged their audiences to consider the link between visible form and hidden essence in terms of individual character. Like the self-observation cultivated by literate society, physiognomy taught a code of self-presentation and emotional expression in bourgeois society. Physiognomy was the domesticated side of the colonialist trend to classify "exotic" (i.e., non-European) cultures. It was part of the culture of classification that permeated the sciences of life during the Enlightenment.

Physicians used physiognomy as a professional method of portraying madness. Enlightenment writers had tended to depict mad people as "beasts," utterly deprived of reason. But in the midst of the revolutionary political changes of the late eighteenth century, many reformers began to depict madness as a moral defect that could be studied and cured systematically. The French alienist (an early term for psychiatrist) Philippe Pinel (1745–1826) and his student J.-E.-D. Esquirol (1772–1840) used physiognomy to illustrate how the mad suffered from an exaggeration or defect of a particular character, which thus pointed toward a classification and a cure. Anatomists were equally interested in visualizing the soul, and physiognomy found its way into important new treatises such as *The Anatomy and Philosophy of Expression as Connected with the Fine Arts*, published by the English anatomist (and talented painter) Charles Bell (1774–1842) in 1806.

PHRENOLOGY: THE MIND AS BRAIN

Lavater's physiognomy, which endured well into the nineteenth century, popularized the idea that the soul left its imprint on the body. But it was through the materialistic doctrines of phrenology that psychology began to shed its affinities with the soul, and began to speak of the brain as the organ of mind. By the early nineteenth century, anatomists and physiologists throughout much of Western Europe were struggling to fit their knowledge into a comprehensive "Science of Man" that could provide new foundations for knowledge that harmonized with the democratic revolutions in the United States and France. There was little room to maneuver in the Germanic philosophical tradition of identifying nature with spirit. But the political and social transformation of France after the French Revolution in 1789 brought with it an all-embracing, materialist, and anticlerical system of knowledge called *idéologie* (from which our modern concept of ideology as a revolutionary doctrine is derived). In such a context, the physician Jean-Georges Cabanis (1757–1808), for example, could confidently announce that the brain secreted thought just as the liver secreted bile.

Phrenology, which identified mind with brain, was part of this "materialist revolt" of the early nineteenth century. The founder of phrenology, the Viennese anatomist Franz Josef Gall (1758–1828), first set down some of his theories about the brain in a letter to the *Deutscher Merkur* in 1798. By 1800, he was known throughout Europe. Within the borders of the conservative Habsburg Empire, many saw Gall's claim that the shape of the skull was a reliable indicator of a person's character as the thin end of a materialist wedge that could end only in revolution. Gall was forced to quit lecturing in 1802, in the midst of Napoleonic efforts to extend the French Revolution throughout the whole of Europe. Accompanied by his student Johann Christoph Spurzheim (1776–1832) Gall embarked on a lecture tour that eventually led him to Paris in 1807.

Parisians welcomed the intellectual controversy surrounding Gall's theories, even though many prominent scientists disagreed with his ideas. The anatomist Marie-Jean-

Pierre Flourens (1794–1867), for example, published a highly critical review of phrenology in 1842. Flourens was particularly disturbed by Gall's assertion that the soul or mind was somehow divisible into "parts" or "faculties" that operated independently of each other. Flourens surgically modified pigeons' brains to demonstrate that although some brain regions might control sensation and motor activity, there were no distinct organs responsible for any of the dozens of faculties described in the phrenological literature. (Flourens dedicated his book to Descartes, who claimed to have found the soul's material vehicle in the brain's only singular structure, the pineal gland.)

Like the method of pathological anatomy that was taking shape in the medical schools of Vienna and Paris around 1800, phrenology (Gall called it *craniology*) relied on an epistemology of localization. Just as disease could be localized in an organ, so could a person's moral nature be understood as an excess or deficit of any number of characters that were located in specific areas of the brain. But whereas the pathologist relied on the internal autopsy to confirm a diagnosis, the phrenologist insisted that the external shape of the skull, alive or dead, reliably indicated the over- or underdeveloped state of those faculties in the brain responsible for character.

Gall and Spurzheim defended their arguments through case studies and comparative examples. The organ of destructiveness, for example, which was located just above the ear meatus, was the widest part of the skull in carnivorous animals. Gall and Spurzheim found this organ to be unusually large in both an apothecary and a university student. The apothecary eventually became an executioner, and the student, who had a fondness for torturing animals, became a surgeon. The organ of causality (called the *esprit métaphysique* by Gall) was located on either side of the forehead (all organs were doubled, to correspond to the balanced role of the two hemispheres). Perhaps not surprisingly, a bust of Kant showed this organ to be extremely well developed.

Many contemporary anatomists criticized phrenology, but its practices were well within the boundaries of the scientific research of the day. Gall's critics were typically more worried about the materialistic inclinations of his doctrine than they were concerned with his supposed lack of scientific rigor. In fact, Gall made several significant contributions to anatomy. After the introduction of phrenology, it was no longer acceptable to argue that the shape of the convolutions of the brain was merely accidental. Phrenology also encouraged anatomists of the nervous system to begin to investigate structure and function together, through vivisection experiments, rather than merely studying the brain's structure through anatomical dissection.

THE FIRST POPULAR PSYCHOLOGY

Phrenology was largely an intellectual concern in France. After breaking with Gall in 1813, however, Spurzheim brought phrenology to Britain and the United States, where it rapidly became a popular pastime. By 1832, there were at least 29 societies and numerous journals dedicated to phrenology in the Anglo-American world. The undisputed center of phrenology in Britain was Edinburgh, where the lawyer George Combe (1788–1858) edited some 20 volumes of the *Phrenological Journal* (1823–47). Combe's phrenological tract, *Of the Constitution of Man and Its Relation to External Objects* (1828), was one of the best-selling books of the nineteenth century: By 1900, it had sold more than twice as many copies (100,000) in Britain as had Charles Darwin's (1809–82) *On the Origin of Species* (published in 1859).

Why was phrenology so popular? In Britain, phrenology seemed to play a dual role. On the one hand, it offered a means of self-classification that seemed free of authoritarian clericalism and the biases of class structure. Those who had been somewhat displaced by the first stages of industrialization (e.g., artisans, shopkeepers, laborers) readily associated phrenology with social reform and even revolution. But at the same time, phrenology reinforced the new industrial order by encouraging intense competition among individuals, encouraging them either to overcome their character defects through self-improvement (Combe advocated educational reform to this end) or simply to resign themselves to their biologically determined fate.

Phrenology was thus many things to many people. It provided a comprehensible and self-referential order in a society that was increasingly structuring itself around the classification, organization, maintenance, and deployment of populations in the modern nation-state.

Psychophysics and a New Experimental Tradition, 1840–79

One of the reasons that phrenology never developed into a full-fledged psychological science was because it lacked an institutional basis. Phrenologists were usually itinerant, self-taught practitioners, not the highly trained products of a specialized academic discipline. This latter aspect of psychology's history is captured in the institutional development of psychological research.

CREATION OF DISCIPLINARY SCIENCE

In the wake of the Napoleonic Wars, the Prussian minister Wilhelm von Humboldt radically reformed the German university system. Beginning with the universities at Berlin and Bonn in 1810, Humboldt advocated the creation of specialized disciplines with a research imperative in order to bring the power of organized knowledge into the service of the modern nation-state. By the 1840s, industrialization and foreign intervention had pushed the loosely organized Germanic states of the old Holy Roman Empire away from the disinterested reign of the Habsburg monarchs of Austria and toward a federal system dominated by Prussian bureaucracy. This period witnessed the formation of the independent disciplines of physics, chemistry, and physiology out of natural philosophy.

Psychological questions, however, remained in the hands of philosophers. Kant's successor at Königsberg, Johann Friedrich Herbart (1776–1841), hoped to establish a psychological discipline by modeling the relative "strengths" of ideas mathematically. Philosophers largely rejected Herbart's call and turned instead to a discussion of the development of "conscious spirit," as typified in the works of Georg Hegel (1770–1831) and Friedrich W. J. von Schelling (1775–1854). Physiologists, on the other hand, wanted to extend their field of knowledge to include problems of mind via an experimental analysis of sensation. Around 1830, "psychophysics" emerged when Ernst Weber (1795–1878), a physiologist at Leipzig, started to integrate subjective responses into physiological investigation.

WEBER'S PSYCHOPHYSICS AND THE PHYSIOLOGY OF SENSATION

In his enormously popular *Handbook of Human Physiology* (1833–40), the Berlin physiologist Johannes Müller (1801–58) proposed a doctrine of specific nerve energies, which stated that the same stimulus caused sensations appropriate to each sense organ. A blow to the head (to borrow a crude example from Müller) would cause pain as well as a ringing in the ears and spots of light appearing before the eyes. The physiology of the nervous system, then, could advance only if structural differences in the nerves and sense organs were associated with functional differences.

Weber studied the problem of touch along just these lines. To this point, the study of the nervous system had followed two lines of research: the anatomical description of the brain and nerves, and the study of reflex action, which had recently been extended to the analysis of nervous disease by the British physician Marshall Hall (1790–

1857). Each relied on the visible phenomena of body parts on the one hand, and twitching frog legs or diagnostic signs (e.g., hysterical paralysis) on the other. Weber adopted a different approach. He created an experimental situation in which human subjects responded to questions about the nature of a measurable stimulus.

Weber demanded that his experimental subjects compare one magnitude of sensation with another of lesser or greater magnitude. Applying Herbart's concept of *thresholds* or *limen*, Weber asked his subjects to state whether they felt one or two points of pressure as he poked them with compass needles separated by various lengths. In *De Tactu: Annotationes Anatomicæ et Physiologicæ* (1834), Weber quantitatively mapped the level of sensitivity across the body's surface and argued that wherever only one sensation could be derived from two points of pressure, that area was serviced by a single sensory nerve. The tip of the tongue, for example, was found to be 60 times more sensitive than the middle of the back. By the 1840s, he had conducted similar experiments in which his subjects compared varying sensations of light, sound, and weight.

Weber's method of "just noticeable differences" implied that the experience of sensation could be expressed as a mathematical function of the change in stimuli. But while Weber related structure and function, he offered no causal analysis. His psychophysical parallelism seemed like a relic of the old-fashioned **vitalism**, which was definitively rejected by the brash materialist and reductionist physiologists who came out of Müller's lab in Berlin after 1848. Weber's work was thus largely ignored by physiologists for several decades.

* **Vitalism.** A doctrine according to which a vital principle, not physical or chemical forces, generates the phenomena of life.

FECHNER AND THE GROWTH OF PSYCHOPHYSICAL EXPERIMENT

Gustav Theodor Fechner (1801–87) supplied a somewhat idiosyncratic response to such developments. While a professor of physics at Leipzig, Fechner suffered a "crisis" during which he experienced psychotic episodes, became partially blinded (perhaps because he stared at the Sun through colored lenses in order to study the phenomenon of afterimages), and was unable to eat. His recovery began in 1843, when he removed the bandages that had been covering his eyes for the last several years. His sight had been mysteriously restored, and even enhanced to the point that he claimed to be able to see the souls of plants.

Fechner became convinced that the force that generated mental phenomena was equivalent to a physical force and was therefore measurable. Inspired by Hermann von Helmholtz's (1821–94) argument, first published in 1847, that matter and energy could be neither created nor destroyed, Fechner attempted to describe the way in which stimulation, a measurable quantity, was converted into sensation, an experienced quality. Like Helmholtz, who reported his attempts to calculate the speed of nervous transmission in 1850, Fechner considered the human body to be a kind of motor that produced and consumed energy. Rather than making analogies with simple mechanisms to suggest the identity of causes in the organic and inorganic worlds, these investigators tried to describe systems in terms of calculable amounts of energy.

In his *Elements of Psychophysics* (1860), Fechner took Weber's "just noticeable difference" as a unit by which sensation itself could be measured. Fechner argued that as the physical intensity of any given stimulus increased or decreased by a constant ratio, the magnitudes of sensations would change by equal increments. By enforcing a particular relationship between experimenter and experimental subject (the subject's responses, for instance, had in all cases to be either "yes" or "no"), Fechner arrived at a description of sensation that is now known as the *Weber-Fechner law*.

Fechner's experimental program aroused considerable scientific and philosophical interest, but experiment alone was not enough to create a psychological discipline in the 1860s, in Germany or anywhere else. Physiologists continued to pursue their reductionist ends and tended to restrict their study of sensation to the examination of the relationship between the objects of sensation and the sense organs, without using introspection to study the subjective side of the process. Perhaps in reaction to the materialist orthodoxy of midcentury science, Fechner published numerous tracts under a pseudonym, "Dr. Mises," attacking hapless physicians and satirizing natural philosophy. His tracts proposed a form of panpsychism in which all beings possessed some form of mind. This tendency of investigators to use false names or fiction to propagate their theories continued well into the twentieth century.

Some philosophers, however, began to take up Fechner's new form of experimental reasoning about the mind. Théodule Ribot (1839–1916), who held the first chair in experimental psychology at the Collège de France (but conducted few experiments himself), introduced Fechner's work to a French audience in 1879. Joseph Delboeuf (1831–96), a philosopher and mathematician at Lièges, began his own psychophysical research in 1873, culminating in a critical examination of Fechner's work 10 years later. By 1889, the very idea that sensory experience could be measured was prominent enough to warrant an attack by a young French philosopher, Henri Bergson (1859–1941). Bergson's argument that sensations were qualities, not quantities, and were thus impossible to measure proved enormously influential in France, where such claims were invoked against the introduction of psychophysical experiment for the next 30 years. Bergson's work also received great publicity in Britain and the United States, but the less centralized control of education in those countries prevented Bergsonism from becoming an institutionalized philosophical form.

Figure 1. Photograph of German physiologist and psychologist Wilhelm Wundt. © Bettmann/ CORBIS

Ultimately, psychophysics found a home where it had begun. Leipzig, located on the border between Saxony and Prussia, had become one of the most important and vibrant centers of research in industrialized Germany, particularly after the 1830s. It was here that Wilhelm Wundt (1832–1920) began to incorporate Weber and Fechner's experimental techniques into a systematic program of psychological research, a program that would be exported to the United States at the dawn of the twentieth century.

From Adaptation to Adjustment: Population and Social Psychology, 1860–1990

German psychophysics adopted the investigational style of physiology and used it to study the relationship between mind and body. The psychology of populations, on the other hand, originated in the problems raised in British evolutionary debates during the middle of the nineteenth century.

SPENCER AND EVOLUTIONARY PSYCHOLOGY

In *Principles of Psychology* (1855), Herbert Spencer (1820–1903), who had worked as both a civil engineer and a journalist before settling on writing popular scientific treatises, argued that the debate over innate ideas was fundamentally misguided. Phrenological "faculties" and "innate ideas" were simply the products of the inheritance of acquired characteristics. In other words, the mind itself had evolved through adaptation.

Spencer aimed at overturning the notion that mind was in any way immaterial. Spencer was inspired by arguments put forward by writers such as Thomas Malthus (1766–1834) and Jeremy Bentham (1748–1832), which suggested that the ideal political system would provide the greatest amount of happiness to the greatest number of people. Framing this utilitarianism in terms of social and biological progress, Spencer attempted to create an ethical system in which mind became both the vehicle and the result of the improvement of the species. His system suggested that mind should be analyzed in terms of how its phenomena benefited the group rather than the individual.

FRANCIS GALTON AND INDIVIDUAL DIFFERENCES

By the late 1860s, the success of Darwin's *On the Origin of Species* had brought retrospective fame to Spencer's ideas. But Spencer's system building contained little empirical research, and Darwin's forays into psychology were largely anecdotal. It was Darwin's cousin Francis Galton (1822–1911) who, by applying statistical analysis to psychological tests, turned the analysis of psychological differences between individuals into an important facet of psychological knowledge. Galton had studied to be a physician, but, after inheriting a small fortune (and consulting a phrenologist), he decided to travel through Europe and Africa conducting geological and meteorological investigations. He eventually settled in London and took up the life of a Victorian "gentleman scientist," conducting research without ever holding an academic position.

Statistics had become an integral tool for the organization of industrial society over the course of the nineteenth century. In 1845, the Belgian astronomer Adolphe Quetelet (1796–1874) had demonstrated that height followed the patterns of a **normal distribution** within a population. In *Hereditary Genius: An Inquiry into Its Laws and Consequences* (1869), Galton argued that intelligence followed similar patterns. By equating the "ability" of "eminent" men with "intelligence," Galton used English biographical directories to demonstrate that intelligence was inherited.

* **Normal distribution.** A formulation for calculating the expected value of sampling a random variable, expressed graphically as a bell-shaped curve.

To increase the amount of "eminence" in a population, Galton argued that one simply had to encourage eminent individuals to reproduce more frequently than others. This became the central plank of his science of "racial improvement," or eugenics. In 1874, Galton introduced a questionnaire to study those most eminent men of Victorian society, scientists. In the 1880s, he extended his analysis to the British population at large. His *Inquiries into Human Faculty and Its Development* (1883) and the Anthropometric Laboratory that he set up at the International Health Exhibition in London in 1884 featured a series of innovative new tests. Some of these were psychophysical (a whistle with an adjustable pitch was blown by an experimenter until it was out of the subject's range of hearing), some were physiological (subjects were invited to use a sledgehammer to drive a marker up a pole), and others were purely introspective (subjects were asked to visualize series of numbers and describe what they saw). But Galton quantified his results in every case and used them to further refine his classifications and to consolidate his argument that "nature" counted for more than "nurture" in the quest for social improvement.

Galton's project reflected many typical Victorian prejudices regarding gender, class, and race. When his studies revealed, for instance, that most of his gifted scientific colleagues (all white men) had great difficulty in visualization, he reminded his readers that it was actually women and children who typically took delight in proficient introspection, and decided, against his original inclination, that such a skill was actually detrimental to the scientific enterprise. Yet, as was the case with phrenology, his studies proved to be popular, and not just among the ruling class. The 9000 spectators who lined up to pay three pence each to become subjects in Galton's Anthropometric Laboratory in 1884 must have shared, to some extent, his conviction that "mental energy" and "sensitivity" provided the bedrock upon which human ability was constructed, and that such measures could tell them something significant about themselves.

EUGENICS AND DEGENERATION

Making this "aggregate psychological subject" fit into a new social order was precisely Galton's intention. In his *Inquiries*, Galton coined the term *eugenics* to describe

Science and Society: The IQ Test

The advent of compulsory education provided psychologists with both a large group of experimental subjects and a social rationale for conducting intelligence testing.

In 1898, the French psychologist Alfred Binet (1857–1911) and his Rumanian assistant Nicholas Vaschide (1873–1907) were having trouble finding a population large enough to study individual differences, so they used schoolchildren as subjects. In 1899, Theodore Simon (1873–1961), an intern at a colony for retarded children, who had an interest in educational reform, replaced Vaschide as Binet's assistant. In 1905, a year after the French government had appointed Binet to a commission to investigate the situation of abnormal children, he and Simon published the first version of their intelligence test. It tested abstract reasoning as well as sensory-motor performance to identify "abnormal" children who were deemed incapable of benefiting from the existing educational system. In 1908, Binet and Simon introduced the idea of *mental age*, which assumed that intelligence developed in a consistent and predictable manner. In 1912, William Stern (1871–1938), a student of Hermann Ebbinghaus (1850–1909) in Berlin, proposed dividing mental age by chronological age, thus arriving at an intelligence quotient, or IQ.

Henry Herbert Goddard (1866–1957), a research psychologist at the New Jersey Training School for Feebleminded Boys and Girls in Vineland, supplanted the medical diagnosis of feeblemindedness with psychological tests around 1908. Lewis Terman (1877–1956) created the first test for grading normal schoolchildren. In 1917, Terman and Robert Mearns Yerkes (1876–1956) created multiple-choice tests to speed up the process of mass testing of American recruits and officers. With the passing of a more restrictive immigration law in 1924, all newcomers to America were subjected to similar examinations. In France, Henri Piéron (1881–1964) used tests to screen train and tram drivers in the 1920s.

The twentieth century saw a dramatic increase in the application of such tests, without any resolution to the controversy over what exactly was being tested. Intelligence tests rapidly became the most visible sign of the scientific authority of psychology. Through the IQ test, intelligence dethroned reason as the focal point of twentieth-century psychological debate.

the cultivation of human ability by selective breeding, just as had been done for centuries with plants or animals. He proposed a state-run program in which a series of examinations would discover the 10 most talented men and women in the nation, who would then be offered £5000 if they chose to marry.

Galton's plan made little headway until Britain nearly lost the Boer War (1899–1902), which served to demonstrate the frailty of the British male population and to drive home the idea that the empire was in decline. By 1900, fear of the "degeneration of the race" had spread throughout Europe. Eugenics societies soon emerged in Germany (1905), England (1907), the United States (1910), and France (1912), and the first international conference took place in London in 1912, just a year after Galton's death. Europe was little more than an armed camp in the years leading up to 1914, and the dramatic destruction of Russia's Baltic Fleet in the Russo-Japanese War in 1905 served only to fuel racial tensions and the specter of degeneration among Europeans and Americans. The enormous growth of psychological testing in the early twentieth century was one response to such developments.

PERSONALITY AND THE SOCIAL PSYCHOLOGY OF ADJUSTMENT

Intelligence was an applied concept. It referred to performances in certain situations and offered people only limited insight into the nature of self in an increasingly secular, industrial, and urban society. Personality and character moved in to fill this vacuum of identity, the study of which also had practical applications in the management of a modern nation-state.

In French and in English, personality had long referred to the character or identity of the human soul. But by the 1880s, its meaning had begun to change. Ribot and the French psychiatrist-philosopher Pierre Janet (1859–1947) argued that personality could become "diseased" and "split." In the United States, the Boston neurologist Morton Prince (1854–1929) described, in 1906, how personalities could "dissociate." Multiple personality began to take shape as a psychological illness. As personality began to assume a medical and even material nature, religious sentiment was also taken up as a psychological topic by the American philosopher William James (1842–1910) and the Swiss psychologist Théodore Flournoy (1854–1920). Religion and personality had been incorporated into secular psychological discourse.

Personality was equally a part of social management. By World War I, officers and salesmen alike began to be selected on the basis of a psychological analysis of their character. During the 1920s and 1930s, psychology gave direction to social research in the United States, as philanthropic organizations such as the Rockefeller Foundation and the Josiah Macy, Jr., Foundation funded the study of emotions, attitudes, and motivations as well as intelligence in an effort to promote adjustment, prevent social breakdown, increase business efficiency, and improve interracial relations. In 1937, William Stern's American student Gordon W. Allport (1897–1967) reviewed such research in *Personality: A Psychological Interpretation*. The study of personality, argued Allport, had outpaced that of intelligence. It was more comprehensive and, as he claimed in 1954, integral to the stability of future societies, because it could help individuals adjust to the rapidly increasing complexity of modern life, just as psychologists had helped soldiers adjust to the emotional and mental challenges unique to modern warfare.

One practical application of such knowledge was the study and promotion of child-rearing practices. The Laura Spelman Memorial Fund (consolidated with the Rockefeller Foundation in 1928) financed child research in various centers throughout North America. Arnold Gesell (1880–1961) at the Yale Psycho-Clinic spearheaded a parent education movement that by the 1920s blossomed into a veritable industry in child-rearing manuals, which continued throughout the century. As eugenist arguments declined after World War II, psychologists claimed increasingly that their expert knowledge was necessary if parents were to raise their children to fulfill particular developmental goals.

Personality psychology also entered the clinic in the postwar era. After leaving Nazi Germany, Hans Eysenck (1916–97) studied mentally ill patients in a London hospital, replacing psychiatric diagnosis with tests. His *Dimensions of Personality* (1947) classified people according to a two-axis system of neuroticism and introversion-extroversion. At the University of Illinois in 1949, R. B. Cattell (1905–98) began to develop complex personality tests with the help of ILLIAC I, the first computer owned by an American academic department. As the use of such computers spread, tests involving as many as 16 personality factors (Cattell's 16PF test) became more commonplace, even though their empirical foundations were often sharply criticized.

The enormous postwar growth of academic departments in the United States also helped make psychological interrogation and classification more routine. Undergraduate psychology students provided a massive pool of eager subjects that was simply not to be had in the tiny psychological laboratories that sprang up during the 1890s. Between around 1900 and 1950, for example, the study of dreams was completely transformed by the questionnaire. Sigmund Freud (1856–1939), the Austrian neurologist and founding father of psychoanalysis, had developed his dream theory from self-analysis and a handful of clinical examples. The British physician and sexologist Havelock Ellis (1859–1939) collected 100 or 200 dream reports for his *The World of Dreams* (1911). By 1953, the American psychologist Calvin S. Hall (1909–85) created a picture of a "normal dream" by analyzing more than 10,000 undergraduate dream reports in terms of plot, character, setting, and the like. Hall's studies held out the possibility that his readers could assess their own personalities by comparing their own dreams to a statistically generated model dream.

The measurement of individual differences began as an attempt to shape social policy to fit the evolutionary theories of the mid-nineteenth century, but by the late twentieth century, it had become a ubiquitous and mundane way of gaining knowledge about oneself and others. Galtonian eugenics had been jettisoned by most psychologists in favor of the widespread application of testing as a tool that could help regulate the vast apparatus of the modern nation-state.

Psychology as Experimental Practice and Discipline, 1879–1917

Historians traditionally cite Wundt's creation of a psychological laboratory at Leipzig in 1879 as the beginning of modern scientific psychology, but this is an origin myth propagated primarily by American psychologists who had trained with Wundt and wanted to shore up their own scientific authority. Experiment alone was not enough for psychology to become a scientific discipline—Wundt, an academic philosopher, never restricted his own psychological investigations to laboratory experiment. The break between philosophy and psychology occurred not in nineteenth-century Germany, but in early-twentieth-century America, where psychological knowledge came to be respected and valued as a form of professional expertise.

WUNDT'S LABORATORY

After studying at Tübingen and Heidelberg, Wundt abandoned his medical studies for physiological research, which had only recently become part of the German medical curriculum. After studying under Müller and Emil Du Bois-Reymond (1818–96) in Berlin, Wundt worked for several years (1858–64) as Helmholtz's assistant at Heidelberg. During this time, he became convinced that philosophy needed to cultivate investigative methods appropriate to its objects. Sensation should be studied experimentally, like any physiological function, while thought and language should be addressed by humanistic and historical research, or *Völkerpsychologie* (folk psychology). The experimental side of Wundt's approach generated his *Principles of Physiological Psychology* (1873), which soon became a standard textbook. Wundt's recommendation of a closer connection between philosophy and the life sciences earned him the prestigious chair of philosophy at Leipzig in 1875, where he remained until 1917.

Wundt rejected philosophical introspection (*Selbstbeobachtung*, or self-observation) in favor of inner perception (*innere Wahrnehmung*). Like Kant, Wundt argued that self-reflection changed the nature of consciousness. The inner perception required by psychophysical experiments, however, could produce reliable data because it was immediate and uncorrupted by memory. Unlike Fechner, Wundt particularly favored reaction-time experiments (made possible through invention of the chronoscope by a watchmaker, Mathias Hipp (dates unknown), in the 1840s), which forced a subject to perform practically without reflection. Through such experiments, Wundt assembled what he called the *elementary units* of consciousness (such as sensations and feelings).

By 1900, much of Wundt's work had either slipped into obscurity or was greatly transformed. His *Völkerpsychologie* was completely ignored, and his American and German students had completely changed both his methods and their object. Wundt's unique contribution was his creation of the laboratory as the preeminent space for both psychological research and training. Psychological research had existed long before Wundt's laboratory, but it had been conducted by individuals in relative isolation. After Wundt, the laboratory became a way for psychology to perpetuate itself in an institutional setting that paralleled those of physiology or physics.

CRITICISMS OF WUNDT

By the end of the century, however, several German philosophers began to react against what they felt to be the reductionist tenor of the new experimentalism. Act psychology, developed by Franz Brentano (1838–1917) in his *Psychology from an Empirical Standpoint* (1874), attempted to reconstruct perception from the subject's active, experiential perspective through intensive and detailed descriptions. In 1893, Brentano's student Carl Stumpf (1848–1936) beat out Hermann Ebbinghaus (1850–1909) and G. E. Müller (1850–1934) for a position in the philosophy department of Berlin. The latter two were noted experimentalists who had founded laboratories at Göttingen and Berlin, respectively. Stumpf's critical approach to experimentalism manifested itself in the phenomenology of Edmund Husserl (1859–1938) and in the Gestalt psychology of Max Wertheimer (1880–1943) and Kurt Koffka (1886–1941).

Psychologists had to defend their ideas in front of philosophers because before World War I, practically all senior experimental psychologists in Germany were housed in philosophy departments. By 1913, German philosophers were so hostile to experimentalism that Wundt was obliged to publish a defense of the field. Leipzig had produced research techniques, instruments, journals, and concepts, but not an independent discipline. The situation was similar in France, where Bergson's subjectivist philosophy denigrated most psychological experiment. In 1908, Binet, frustrated with his failure to gain an academic position, published his survey of philosophy instruction at *lycées* (upper-level secondary schools) and colleges in France. Binet argued that Bergsonism was embraced by almost all philosophy professors, even while the students clamored to be taught experimental techniques. At this time, psychological experiment in France referred to Janet's clinical studies of hypnotism and suggestion. Supported by Bergson, Janet had beaten out Binet to replace Ribot as chair of experimental psychology at the Collège de France in 1902.

THE NEW PSYCHOLOGY IN THE UNITED STATES

Things were quite different in the United States. Ready access to raw materials and rapid industrialization and urbanization, coupled with massive waves of immigration between 1890 and 1917, quickly transformed that nation into one of the world's most populous and wealthiest powers. Progressivism responded to the social and political challenges of modernization by creating new institutions (such as Johns Hopkins University and the University of Chicago) dedicated to turning out social scientists with the expert knowledge appropriate to reforming the American educational system.

Progressivism's institutional counterpart was the *New Psychology*, a term coined in 1897 by one of Wundt's American students, Edward Wheeler Scripture (1864–1945). Unlike the nineteenth-century tradition of mental and moral philosophy, which inculcated a sense of piety and responsibility in the well-informed U.S. citizen, the New Psychology depicted the mind as an object that served the ends of social progress and that developed (or could be developed) according to discoverable laws. Where phrenology had blossomed at the level of popular culture in Britain in the mid-nineteenth century, the New Psychology became institutionalized as the practical, expert knowledge of an educated elite in the United States.

WILLIAM JAMES AND FUNCTIONALISM

This transition was overseen by the American philosopher and psychologist William James (1842–1910). James's educational background was broad and eclectic, but much of it centered around the new physiology he was taught as a medical student in the United States and Europe. Using Herbert Spencer's *Principles of Psychology* (1855) as a text, James taught the first psychology course in the United States while teaching anatomy and physiology at Harvard in 1875. He entered the philosophy department in 1880, and by 1889, his title was changed to professor of psychology, a field he created almost singlehandedly through publication of his enormously popular *Principles of Psychology* (1890).

James was a gifted theorist and a brilliant writer, blending British evolutionary theory and German physiological psychology with the American traditions of self-help and self-determination. He was intrigued by the experimental approach to the study of mind, but not seduced by it. Introspection, he argued, be it philosophical or experimental, led directly into philosophical error, because it assumed that the "flow of consciousness" could be interrupted and inspected. In its place, James advocated the analysis of consciousness in terms of its function. In an 1884 article, for example, James argued that the perception of a fearful object did not bring about the classic physiological signs of fear, such as trembling or perspiring. This was an illusion of introspection. Instead, the body reacted immediately to the appearance of the object, which then raised the emotion in consciousness. "We do not tremble because we fear," argued James, "we fear because we tremble." Mind was not a mere "epi-phenomenon," however. Its "cash value," according to James, was in how it and the body were united in a common function or purpose. Functionalism became the rallying cry for most American psychologists by around 1910, just as their increasing participation in educational reform provided them with a poignant example of how consciousness could be made to be useful.

PSYCHOLOGICAL ENTERPRISE

James's enormous popularity gave a certain continuity to American psychology. American psychologists created their professional and scientific authority by training in Europe and then returning to found their own laboratories in North America. This migratory pattern began around 1880 and continued until around 1910, by which point U.S. psychological research had finally come into its own.

The careers of Granville Stanley Hall (1844–1924) and James McKeen Cattell (1860–1944) provide two cases in point. Hall did his Ph.D. under James, then visited Wundt's laboratory in 1879. In 1884, he became professor of psychology and pedagogy at the recently founded Johns Hopkins University in Baltimore. Hall's laboratory turned out a number of notable students before he left for Clark University (Worcester, Massachusetts) in 1888. Child psychology, religion, and evolutionary theory provided much of the content of the major journals that he established, which included the *American Journal of Psychology* (1887), *Pedagogical Seminary* (1891), the *Journal of Religious Psychology* (1904), and the *Journal of Applied Psychology* (1915).

Cattell was Wundt's first official assistant. He founded laboratories at the University of Pennsylvania in 1888 and at Columbia University in 1891, where he remained until he was expelled for his pacifist stance against World War I in 1917. Along with James Mark Baldwin (1861–1934), who founded laboratories in Toronto (1889) and Princeton (1893), Cattell developed a number of mental testing methods, out of which he formed a business, the Psychological Corporation, after leaving Columbia. Cattell, whose thesis on psychometry Wundt described derisively as "*Ganz Amerikanisch*" ("typically American"), excelled in the mundane organizational work integral to establishing psychology as a scientific discipline. He founded a number of important journals, including *Psychological Review*, *Psychological Index/Abstracts*, and *Psychological Monographs*. In 1895, he purchased

Science from Alexander Graham Bell (1847–1922), and used it and the *Scientific Monthly* to propagate the New Psychology. The success of his endeavor is perhaps indicated by the extensive use of psychological testing to determine the mental "fitness" of some 1.75 million U.S. conscripts in 1917.

Women participated in American psychology from the beginning, despite enormous obstacles. Women were active members of the American Psychological Association (APA) from its inception in 1892. A few of them, such as Mary Whiton Calkins (1863–1930) and Margaret Floy Washburn (1871–1939), gained considerable prominence in the years leading up to World War I. Calkins studied at Harvard under James, but the university refused to grant her a Ph.D., even though she had completed all necessary requirements. In 1893, she produced a study of dreams in which she collected more than 600 dream reports from her students at Wellesley College. Her use of undergraduates as subjects was an early example of an important trend. She published numerous introductory texts and became the first woman president of the APA in 1905.

Working out of Vassar College in New York State, Washburn published several important books, including *The Animal Mind* (1908) and *Movement and Mental Imagery* (1916). The latter was a key text for those who held to the "motor theory" of consciousness, which argued that thought was internalized movement. Her ideas, which included elements of both behaviorism and structuralism, were very influential in the period between the world wars.

STRUCTURALISM, THE WÜRZBURG SCHOOL, AND THE "IMAGELESS THOUGHT" CONTROVERSY

By 1903, there were at least 40 psychological laboratories across the United States and Canada, and they awarded more doctorates than any other discipline except chemistry, physics, and zoology. France had but three psychological laboratories at this time. By 1910, most psychologists described themselves as "functionalists," and many were engaged in "applied research."

Edward Bradford Titchener (1867–1927) was an important exception to the functionalist trend. He was Wundt's most famous student in the United States, although he was not always the most accurate guide to his master's thought. It is due to Titchener's influence, for example, that Wundt is commonly mistaken for an introspectionist. Titchener had become acquainted with Wundt's work while studying philosophy at Oxford. Unable to find a position in his native England after taking his doctoral degree at Leipzig, Titchener was called to take over the psychology laboratory at Cornell in 1892, where he remained for the rest of his career. Margaret Washburn was his first Ph.D. student there.

Where Wundt had emphasized the role of sensations, feelings, and images, Titchener argued that these were the sole elements that made up the "structure" of consciousness. His method of experimental introspection rejected the strict limitations that Wundt (following Kant) had placed on "inner perception." Titchener's subjects, who were also his students, were trained to "directly scrutinize" the changes that took place in consciousness immediately after they performed psychophysical tests. These experimental introspectionists underwent lengthy and rigorous training to avoid the stimulus error, the habit of confusing the interpretation of a stimulus with the experience of the stimulus itself.

In a sense, Titchener created the category of functionalism by contrasting it with his own program of structuralism. Against virtually all his American colleagues, he argued that psychology was too underdeveloped to generate reliable technology or practical knowledge. Just as anatomy had to precede physiology, the investigation of mental structure must be completed by competent introspectionists before an analysis of function could take place.

Oswald Külpe (1862–1915), a former student of Wundt and G. E. Müller, and a professor of philosophy at Würzburg, attempted to use introspection to examine thought itself. By 1907, the Würzburg school (whose ideas were welcomed in Bonn, Munich, and Königsberg) had become embroiled in a dispute with Wundt and Müller, who argued that thought was derived from sensory experience and that the psychologist's job was to demonstrate this through experiment. The Würzburgers, however, claimed to have discovered the existence of imageless thoughts, in which trained observers, when asked to introspect on the meaning of certain statements (a Nietzschean proverb, for example), found that these meanings often appeared immediately, without any preceding imagery or sensations. Thus they argued that the primitive elements of consciousness were not restricted to sensations, feelings, and images; they included situations of consciousness (*Bewußtseinslagen*), awareness (*Bewußtheiten*), and thoughts (*Gedanken*).

Wundt and Müller attacked the Würzburgers' methodology. Titchener, on the other hand, was obliged to attempt to replicate their results, because he, too, had approved of a more extensive program for introspection. Titchener, who had ruled any study of the unconscious (if such a thing even existed) out of psychological research, argued that the Würzburgers' experiments had been corrupted by suggestion. Between 1909 and 1912, Titchener's laboratory published a series of studies in which he and his students demonstrated that sensory elements could be found in every case of imageless thought, even if these elements amounted only to vague impressions of color or dull pressures felt in the neck or legs. A methodological crisis ensued, as critics pointed out that introspectors inevitably found exactly those results they had been trained to discover. The arrival of behaviorism around 1913 (see below) marked the beginning of a rapid decline in introspection's fortunes. Academic psychologists buried it with Titchener in 1927.

Psychoanalysis and the Unconscious, 1895 to the Present

ANIMAL MAGNETISM AND ITS DERIVATIVES

Despite its antirational reputation, the origins of psychoanalytic theory can be traced back to the late-eighteenth-century concept of *animal magnetism*, the precursor of hypnotism propagated by the Viennese physician Franz Anton Mesmer (1734–1815). Mesmer thought that he had discovered a "magnetic fluid" analogous to the "electrical fluid" captured in Leyden jars (an electrical capacitor invented in 1746) and used to animate the bodies of decapitated frogs by Mesmer's contemporaries, Alessandro Volta (1745–1827) and Luigi Galvani (1737–98) [see *Electromagnetism*].

Mesmer treated his first patient, an Austrian noblewoman, for a nervous complaint in 1774 by passing a magnet over her body. In the years that followed, Mesmer developed a number of therapeutic techniques, many of which involved a *baquet*, a device that Mesmer claimed could store and redistribute the magnetic fluid among as many as 20 patients at a time. Mesmer's cures were equally public performances, and his work found fertile ground among anticlerical members of the Parisian bourgeoisie. In 1784, however, Louis XVI appointed a commission to examine Mesmer's work. The commission, which included a number of notable French physicians and scientists, concluded that the effects of animal magnetism were entirely the product of "imagination." Mesmer was publicly ridiculed, but variants of animal magnetism persisted throughout the nineteenth century as artificial somnambulism and hypnotism. All these techniques seemed to alter the nature of the subject's mind, allowing him or her to self-diagnose disease, endure surgery without anesthesia, or even communicate with spirits.

What was the nature of this mysterious state? In the early 1880s, Jean-Martin Charcot equated it with disease, arguing that only hysterics could be hypnotized. His detractors at Nancy, A.-A. Liébeault (1823–1904) and Hippolyte Bernheim (1840–1919), disagreed, claiming that anyone could be hypnotized provided they were sufficiently "suggestible." At stake in this debate over hypnotism was a question about the limits of personal responsibility and the freedom of the will.

HYSTERIA AND DREAMS

Just as hypnotism was being revived as a tool for the study of abnormal states of consciousness, memory began to take on new powers. Pierre Janet, a philosophy professor who began his medical studies after experimenting with hypnotism, took up his research in 1889 with Charcot at the Salpêtrière. Janet emphasized the role of "subconscious fixed ideas" in hysteria. These ideas could usually be traced back to memories. Once these memories were brought to consciousness, often through the use of hypnotism, they could be analyzed, and a cure obtained.

Figure 2. Portrait of Sigmund Freud. © Hulton-Deutsch Collection/CORBIS

Sigmund Freud (1856–1939) agreed that memory was the key to hysteria. One of Breuer's patients, Bertha Pappenheim, provided Freud with the most potent combination of hysteria, hypnosis, and memory. Breuer had treated Pappenheim, whom he called "Anna O." in *Studies in Hysteria* (1895), for hysteria from 1880 to 1882. Her symptoms were the standard—paralysis, contractions, and hallucinations—coupled with a more exotic "splitting," in which she would alternate between two radically different personalities. By holding an orange in front of her face, Breuer could make Pappenheim slip from one personality into another. In her hypnotized state, Pappenheim associated each of her hysterical symptoms with an event that had occurred exactly one year earlier (her mother had kept a meticulous diary of the period, and Breuer was thus able to verify Pappenheim's statements). Every time

Scientific Biography: Sigmund Freud

Sigmund Freud, the oldest child of Jakob Freud and Amalie Nathanson, was born in 1856 in Freiburg (now Prîbor, in the Czech Republic). His father, an impoverished wool merchant, relocated the family to Vienna in 1860.

Vienna was a liberal city at the heart of a conservative, decaying empire. Despite its anti-Semitic climate, the city offered substantial opportunity for young Jewish men, particularly in the medical field. Freud was an excellent student, and despite his interest in the radical, anticlerical philosophy of Ludwig Feuerbach (1804–72) and the act psychology of Franz Brentano, he was soon drawn to medicine after enrolling at the University of Vienna in 1873.

Freud initially wanted to pursue a career in laboratory research and spent several years studying at Ernst Brücke's Physiological Institute. But he was also anxious to marry and start a family. Even in the unlikely event that a research post would be offered to a secular Jew, it would bring little financial reward. After much hesitation, Freud completed his medical degree in 1881, specializing in neurology by training under Theodor Meynert (1833–93) at the Vienna General Hospital (1882–85). In 1886, he married Martha Bernays.

Freud never gave up his desire to make a great discovery. During the 1880s and early 1890s, he published several papers on subjects as diverse as nerve physiology, morphine addiction, and neurological disorders. He was particularly interested in the use of hypnotism to treat hysteria, a disease that had recently achieved an unprecedented notoriety at the hands of the Parisian neurologist Jean-Martin Charcot. After studying under Charcot at the Salpêtrière asylum during the winter of 1885–86, Freud began seriously to investigate this mysterious disease on his own.

(Continued on next page)

Scientific Biography: Sigmund Freud (*cont.*)

Freud was not a famous asylum doctor like Charcot, but a young, unknown neurologist trying to set up a private practice. His patients were not destitute asylum inmates but well-to-do referrals wanting to consult a specialist. Most came to him by way of his good friend Josef Breuer, a distinguished Viennese physician 14 years older than Freud. Freud was in no position to wield great authority over his patients as Charcot did. So when his patients wanted to discuss, often in great detail, the strange amalgam of hysterical symptoms that plagued them, Freud obliged. His curiosity, anxiety about his professional reputation, ambition, and eagerness to please Breuer by curing his patients made him a great listener. In 1895, Freud and Breuer published their clinical observations as *Studies in Hysteria*, in which they argued that hysterics "suffered mainly from reminiscences."

As Freud's confidence increased, his relationship with Breuer suffered. In 1896, Breuer broke off the friendship because he rejected Freud's nearly-exclusive emphasis on the sexual origins of the neuroses. Blaming sex for neurotic behavior was hardly new, but it was a potentially dangerous claim for a well-regarded physician to make. Freud was also constrained by the morals of his age. When confronted with his patients' numerous stories of incest, he rejected them as infantile fantasies, largely because he found the alternative inconceivable.

Encouraged by his close friend Wilhelm Fleiss (1858–1928), an eccentric ear, nose, and throat specialist from Berlin, Freud pursued two avenues of research in the 1890s. He constructed an elaborate psychophysiological system, which he abandoned in 1895 for a purely psychological source of data: dreams. He had started to analyze his own dreams, as well as those of his patients, several years earlier. By 1899, this work culminated in the publication of his most important book, *The Interpretation of Dreams*. Although Freud was disappointed with the book's initial reception, it nevertheless established him as an original, if controversial, thinker.

Freud divided the remainder of his career between two activities: overseeing the fate of psychoanalysis by carefully managing his followers; and expanding his theories to incorporate human experience outside the clinic. His work on dreams provided the template for his psychoanalytic theories, which he and his devotees used to understand individual psychosexual development as well as the origins and growth of religion, art, and politics. Freud's husbandry of the psychoanalytic movement was sometimes ruthless but always tactical. It had to be, because Freud, a professor at the University of Vienna, never held a chair there and was thus denied the opportunity to train students in an academic setting. He lost many important allies over the next four decades, but these public conflicts served only to magnify the significance of psychoanalytic theory.

Freud was actively involved in the psychoanalytic movement up to the very end. Living in exile in London since the 1938 annexation of Austria by the Nazis, and plagued by cancer since 1923, Freud finally succumbed to the disease in 1939. Since that time, his life has been subject to a seemingly endless series of psychoanalytic biographies, a fitting tribute to one of the most contentious and influential thinkers of the twentieth century.

Pappenheim related such a memory, her corresponding symptom mysteriously disappeared, and this system of "catharsis" eventually cured her.

All the key elements of psychoanalytic therapy—repression, the symbolic conversion of memories into symptoms, and the cathartic cure—could be found in this protean tale. But the story was not true. Breuer had not cured Pappenheim. According to a letter that Freud wrote to his fiancée in 1883, Breuer had actually abandoned Pappenheim after she (falsely) claimed that she was pregnant with Breuer's child! She was committed to an asylum for several years, and only upon her departure did she find relief through an active life in social work.

Breuer was extremely reluctant to publish the case of Anna O., and probably mutilated the case history to protect his reputation. Freud, who was certainly aware of the unpublished details of the case, began to formulate a theory about Pappenheim's supposed sexual attraction to her physician. Freud called this *transference*, and made it an integral component of the psychoanalytic technique of interpreting hysterical symptoms as symbolic expressions of repressed sexual desire, an element that was not a significant component of Janet's theory of the subconscious.

In *The Interpretation of Dreams* (1899), Freud's own dreams mingled with those of his patients, demonstrating that the bizarre symptoms of hysterics and the images of dreams were both the mind's attempt to satisfy the illicit desires of the libido, or sexual drive, by substituting symptoms or images for reality. Dreams, Freud argued, were the disguised fulfilment of an unconscious wish. Their study would reveal the wellspring of desire that motivated human behavior.

Interpreting one's dreams had been a popular pastime among the literate and self-conscious European bourgeoisie for at least a century, but the peculiar genius of Freud's book was its intimacy. Freud stripped dreams of their abstract associations with the supernatural, time, or physiology, and bound them tightly to the most hidden secrets of a person's memory. His transformation of the popular (as opposed to the experimental) practice of introspection can hardly be overstated. Where dreams had once served as an exemplar of the imagination's activity, Freud reduced them to the procedural demands of memory. In the process, psychoanalysis emerged as an intensive examination of personal identity.

INSTITUTIONAL GROWTH AND EARLY DEBATES

The earliest psychoanalytic circles focused on the self. Members of Freud's Wednesday Psychological Society frequently presented papers that featured sexual "confessions," often derived from self-analysis (Freud had not yet argued that all

Scientific Institutions: Formation of the International Psychoanalytic Association (IPA)

In 1902, Freud, Wilhelm Stekel (1868–1949), Alfred Adler (1870–1937), Max Kahane (1866–1923), and Rudolf Reitler formed the Wednesday Psychological Society. They met every week at Freud's apartment at Bergasse 19 to discuss psychoanalysis. The society soon began to attract nonphysicians. Existing members granted newcomers entrance to the society by unanimous consent, based on an assessment of competence and interest rather than credentials. By 1906, the society had grown to 17 members. Otto Rank (1884–1939), an autodidact originally trained as a machinist, became the first secretary of the society and began to collect fees. Psychoanalysis also began to attract the attention of psychiatrists outside Vienna. Karl Abraham (1877–1925), Ernest Jones (1879–1958), and Carl Gustav Jung (1875–1961) formed psychoanalytic societies in Berlin, London, and Zurich, respectively. Jung organized the First Congress for Freudian Psychology at Salzburg in 1908.

Freud's original group, however, was beginning to fracture. The Viennese meetings were frequently heated, as some members, particularly Adler and Stekel, challenged Freud's authority. Freud's response was to insist that if psychoanalysis was to survive as a science, doctrinal orthodoxy was required. In 1908, he gave restless members the opportunity to leave gracefully by dissolving the society and creating a new one. Internal conflict continued to plague the new Vienna Psychoanalytic Society. Freud's demand for theoretical uniformity frustrated the ambitions and enthusiasm of new and old members. Confronted with what he called "wild" or "lay" analysis (psychoanalysis practiced by nonphysicians, such as Rank), Freud entrenched his orthodox position by naming Jung as his successor at the Second International Psychoanalytic Conference (Nürnburg) in 1910.

Jung seemed the perfect solution to Freud's problems. As the permanent president of a new international society, Jung would relocate the center of the movement to Zurich. This would strip Freud's Viennese rivals of their influence, as well as defuse the charge that psychoanalysis was strictly a "Jewish science" (Jung was the son of a Protestant pastor). Freud also hoped Jung's new position would assure the fidelity of Eugen Bleuler (1857–1939), one of the most prominent psychiatrists of his day, to the cause. Psychoanalysis could finally work its way into asylum psychiatry.

The Vienna analysts revolted, and the disgruntled members called an emergency meeting the day after Sándor Ferenczi (1873–1933), acting as Freud's deputy, announced the new plan. Freud quickly brokered a compromise. Adler, who replaced Freud as presiding officer, became, with Stekel, a coeditor of the Vienna Society's official journal. All agreed that Jung would have to face an election after two years. Freud's concessions were illusory, and Adler and Stekel left the Vienna Society within two years. Jung was reelected in 1913, but he broke with Freud and resigned a few months later. The first psychoanalytic training institute was finally founded in Berlin in 1920, but the problem of lay analysis continued to plague the IPA throughout most of the twentieth century.

psychoanalysts had first to be analyzed themselves before they could practice professionally). Several of these early analysts soon rebelled against Freud's reduction of identity to a matter of sex. Alfred Adler (1870–1937) and Carl Gustav Jung (1875–1961), among others, attempted to develop alternatives to Freud's system. Adler promoted a psychoanalysis based on his concept of "organ inferiority," which suggested that the neuroses were due to an imbalance of vitality among the organs. His individual psychology, he hoped, would ultimately restore balance to the interminable conflict between the individual and society.

Jung's analytic psychology featured a system of "psychological types," in which "introverted" or "extroverted" attitudes determined a person's personality. He argued that infants had no sexual drives and that Freud's concept of libido was nothing more than "psychical energy." For Jung, the challenge of psychotherapy was to help a person come to understand his or her own personality through the identification of a set of "archetypes" that existed in different configurations within everyone. Like Freud, Jung felt that the unconscious was a powerful source of creativity. But whereas Freud felt that this took place through "sublimation," or the redirection of the libidinal force toward artistic or literary production, Jung described the liberation, rather than the repression, of images buried in the unconscious through such endeavors.

Like its Adlerian counterpart, Jungian psychology was essentially optimistic. The future of civilization depended upon coming to terms with the unconscious, not its continued repression, as Freud so gloomily suggested in *Civilization and Its Discontents* (1930). Given this rather hopeful outlook, it is perhaps surprising that Jung was initially quite sympathetic to National Socialism during the early 1930s. But he had explicitly incorporated the concept of racial difference into his theories by arguing that the archetypes uncovered in myth were racially distinctive. This provided Nazis with a rhetorical tool through which they could dismiss the claim to universality of Freudian psychoanalysis and promote their own vision of a psychology based on the purity of race.

PSYCHOANALYSIS AND MEDICAL PRACTICE IN THE UNITED STATES AND EUROPE

When Granville Stanley Hall invited Freud and Jung to speak at Clark University in 1909, psychoanalysis was on the very fringes of medical practice in Europe. American intellectuals and medical practitioners proved more receptive to Freud's ideas. Many American neurologists, for example, quickly adopted psychoanalysis in their private practices, which usually catered to wealthy urbanites with European pretensions. The Americans' use of psychoanalysis was eclectic and pragmatic; in Europe, the debate over Freud's work was entirely polarized.

There were also structural differences. European medical schools had rejected psychoanalysis, and formation of the International Psychoanalytic Association (IPA) was, in part, an attempt to create a uniform system of training that would parallel that of the medical schools. American psychiatrists and neurologists, on the other hand, wanted to restrict psychoanalytic practice to those with a medical degree, thereby placing it outside the antiquackery legislation derived from the Flexner Report of 1910.

Thus, unlike its European counterparts, the American Psychoanalytic Association (APA) was controlled by physicians. The APA could not, however, prevent laypeople from forming their own associations and then joining independently with the IPA. This changed during the 1930s. Just as American psychiatry and neurology were formally integrating, the APA reorganized as a federation of local societies and began to enforce national standards of psychoanalytic education, which included medical training. At the 15th Congress of the IPA, held in Paris in 1938, the APA fought for a "regional agreement" by which the IPA gave its U.S. counterpart the exclusive right to register and train psychoanalysts in its region. This situation continued until the mid-1980s, when a group of nonmedical psychoanalysts launched an anti-trust lawsuit against the APA, claiming that it had an illegal (and lucrative) monopoly over psychoanalytic practice in the United States. The APA reached an out-of-court settlement with the plaintiffs and gave up its regional agreement at the 35th International Congress in Montréal, Canada, in 1987.

Events preceding World War II also influenced the development of psychoanalysis in the United States. Jewish analysts fleeing Nazism began to integrate into American psychiatry, and several came to hold prestigious positions in medical faculties that were denied them in their native Europe. Many were prolific and elegant writers, and their textbooks became part of the standard curricula of medical education that they themselves had helped to establish. Thus, during the 1950s and 1960s, psychoanalysis acquired an authoritative position in American psychiatry that was unprecedented in Europe. This situation lasted until well into the 1970s, when, for a variety of technological, educational, and economic reasons, a new form of *biological psychiatry* began to appear on the American scene.

TECHNOLOGY OF THE SELF

Mental cures form only one part of the claim of psychoanalysis to offer a comprehensive understanding of the human condition. For American psychologists, psychoanalysis became part of a technology of adjustment through the study of the *ego*, which Freud had described as a mental structure that emerged out of the conflict between the instinctual drives that made up the *id* and the moral constraints of civilization, internalized as the *superego*. In the early 1950s, psychologists in the United States jettisoned Freud's emphasis on the endless conflict between the pleasure principle and the death instinct and attempted instead to identify and correct the abnormal progress of the ego through definite stages of development.

This use of psychoanalysis dovetailed with the deep integration between psychology and educational reform. It was paralleled by another interpretation of Freudian psychoanalysis in terms of personal liberation during the 1960s and 1970s. The latter vision originated in Weimar Germany during the 1920s, with the fusion of Freudian with Marxist ideas in the work of the Frankfurt School, which counted the social theorists Max Horkheimer (1895–1973), T. W. Adorno (1903–69), and Herbert Marcuse (1898–1979) among its members. The school was ruthlessly suppressed by the Nazis, and many of its members fled to the United States and became affiliated with the New School for Social Research in New York. Marcuse's *Eros and Civilization: A Philosophical Inquiry into Freud* (1955) argued that sexual repression served the ends of a capitalistic state. More explicitly, psychoanalytic tracts such as *The Sexual Revolution: Towards a Self-Governing Character Structure* (1936, translated into English in 1945) by Wilhelm Reich (1897–1957) depicted character as "the crystallization of the sociological process of a given epoch," implying that sexual revolution would lead to the next stage of character development in the history of human civilization.

For many feminists, the nature of this sexual liberation was no more clear in the United States during the 1970s than it had been in Germany or the Soviet Union during the 1920s. In both cases, liberation frequently applied to sexual acts rather than gender roles. A debate over the relevance of psychoanalysis for women emerged among feminists in the 1970s and 1980s. Freud was a paradox for many feminists: On the one hand, he provided critical tools that could connect patriarchal social structures with hidden sexual conflicts in the individual; on the other hand, his writings merely perpetuated an analysis of sexuality based on a male model. Women were no less the "other" in this new reading of Freud than when the

French philosopher Simone de Beauvoir (1908–86) offered her critique of his work in her groundbreaking *The Second Sex* in 1949.

A somewhat different perspective on Freud arose in France, in large part through the efforts of Jacques Lacan (1901–81). Criticizing the American ego psychology of the 1950s, Lacan argued that the ego was not a thing that developed according to regular stages but the expression of a universal desire for wholeness and self-identity, originally expressed in the child's encounter with its mirror image. Psychoanalysis did not cure, but attempted to understand, human behavior in terms of a quest for unity. His one-time student, the philosopher and psychoanalyst Luce Irigaray (b. 1932), criticized Lacan for perpetuating the Freudian error that female sexuality can easily be reduced to a male model. Famously describing women as "this sex which is not one," Irigaray argued that female sexuality could not be contained within Lacan's masculine, phallocentric model. Her poetic, even erotic style of writing, claiming feminine difference on the basis that women possessed "two sets of lips," for example, infuriated many liberal feminists in the United States.

Lacan's decentering of the psychoanalytic subject was paralleled by the research of Michel Foucault (1926–84), a philosopher who had some casual training in psychiatry and psychology. In his *History of Sexuality* (1978), Foucault took Reich as a target and condemned psychoanalysis for assuming that sexuality can be understood apart from the historical forces that both create and exploit it as "biopower." Foucault's work spawned innumerable studies, and he almost singlehandedly created the subfield of historical research known as *history of the body*.

During the last decades of the twentieth century, psychoanalysis continued to retain much of its explanatory power in the human sciences throughout Europe and the Americas. The unconscious as an active force in people's lives has become part of popular parlance and late in the twentieth century, showed little sign of disappearing. But for the greater part of the twentieth century, European and North American psychologists were dedicated to other forms of investigation. Abandoning the controversies over the scientific or moral status of psychoanalysis, these psychologists turned toward behaviorism and cognitive psychology as more appropriate foundations for their discipline.

Behaviorism, 1907–60

By the time of James's death in 1910, American psychology had established itself as a professional enterprise and an academic discipline. Its primary practical focus was on education, and psychological knowledge had become an integral part of creating social equality and stability in a modern, industrialized nation-state. Yet the imageless-thought controversy precipitated a theoretical and methodological crisis. Behaviorism, which took animals as subjects and jettisoned consciousness in favor of learning as the central problem of psychology, evolved in response.

WATSON'S BEHAVIORISM

John Broadus Watson (1878–1958) studied for his Ph.D. under John Dewey (1859–1952) at the philosophy department of the University of Chicago, one of the two main centers of functionalism in the United States (the other was at Columbia). Dewey, one of the most influential philosophers of education in the United States, had published a critique of the reflex-arc concept in 1896. Dewey argued that the concept, which explained the phenomena of life as a mechanistic reaction to stimuli, completely ignored the more important question of adaptation, a problem that could not be avoided when it came to understanding the mind. His emphasis on the organism's environment resonated in Watson's theories. But Watson's mechanistic rejection of teleological, or

Science and Society: Pavlov and the Theory of Conditional Reflexes

The tale of Ivan Pavlov (1849–1936) making a dog drool by ringing a bell is one of the best known experimental parables taken from the history of psychology. But Pavlov, who won the Nobel Prize for Physiology or Medicine in 1904 for his work on digestion, never thought of himself as a psychologist. How did he come to develop a complex theory on the basis of such a simple experiment?

The answer, in part, can be found in the unique context in which the experiment was developed. Ivan Pavlov had worked at the Imperial Institute for Experimental Medicine in St. Petersburg since its creation in 1891. The Institute was designed to train Russian physicians in the new laboratory-based medicine of Louis Pasteur (1822–95) and Robert Koch (1843–1910). But Pavlov was interested in organ physiology, not bacteriology, and he taught his students to study digestion by surgically modifying a dog's stomach in order to evoke, collect, measure, and even sell, gastric secretions. These "chronic experiments" were lengthy and laborious. Keeping a single experimental dog in good health demanded more attention, space, and care than trying to create disease by culturing and injecting bacteria into guinea pigs or rabbits. But scores of army physicians, hoping to gain their doctorates by passing through Pavlov's "physiology factory" in two years or less, provided the institute with a large labor pool to conduct these experiments.

Pavlov had argued that digestion was regulated by the nerves. The discovery of secretin (a digestive hormone secreted in the duodenum) around 1902 threatened to undermine his theory. Pavlov, who was well aware of the enormous success enjoyed by psychologists in the United States (which Pavlov dubbed "the new home of science"), began to retool his factory around the observation that many of his experimental dogs would begin to salivate before they were actually fed. This *psychic secretion*, as Pavlov initially described it, was easily studied by collecting and measuring saliva, a procedure already well established at the institute. The basic experiment was easily mastered by his students (who often had little or no training in physiology), and its combinations and permutations were almost limitless.

By 1907, the study of psychic secretion had become the cornerstone of all research at the institute. Pavlov, who had once described the phenomenon in terms of the dog anticipating being fed, now rejected all such psychological language. In its place, drops of saliva signified the higher nervous activity of the cerebral cortex as a measurable quantity. Pavlov conceived of and authorized all experimental trials, which were then carried out by his near-transient students, and finally synthesized and presented by Pavlov himself. His "spitting science" (as it was derisively described by some) thus shared two key features of modern factory production: a hierarchical social structure and a reliance on measurement to coordinate large-scale production efficiently.

purposeful, explanations originated in the work of Jacques Loeb (1859–1924), a German-born physiologist then teaching at Chicago. Loeb's analysis of instincts as physicochemical "tropisms" appealed to Watson, who chose Loeb as his thesis director. In 1908, shortly after completing his dissertation on habit acquisition in animals, Watson obtained a prestigious position at Johns Hopkins, replacing Baldwin, who had been dismissed in a sex scandal.

Several psychologists were turning toward the study of animal subjects around 1900. Edward L. Thorndike's (1874–1949) study (1898) of how cats learned to escape from puzzle boxes was well known among American psychologists, as was Robert Mearns Yerkes's (1876–1956) use of primates in his program of comparative psychology at Harvard. Pavlov's method of conditional reflexes was introduced to the United States by Yerkes in 1909. In France, Alfred Binet's future successor, Henri Piéron (1881–1964), proposed a similar comparative method of studying sleep in 1907. That same year, Piéron suggested that the study of comportement (comportment or behavior) would be a fruitful direction for psychological research. Piéron's work seems to have been largely ignored by those American psychologists whom he admired, but like Watson, Piéron found biologists, not philosophers or psychologists, to supervise his doctoral research.

Thus Watson's polemical 1913 article, "Psychology as the Behaviorist Views It," did not invent behaviorism. It propagated a number of ideas already in circulation under the banner of scientific objectivity. Watson wanted to depose the last remnants of philosophical self-reflection, and to this end, behaviorists began to take animals rather than people as model psychological subjects. For early behaviorists, consciousness was an irreducible black box that could ever be only individual and private. It had no place in public science. Psychological discipline consisted of learning how stimulus and response worked with animals (which could be manipulated with few ethical consequences) and then applying these ideas to humans.

But for a psychology to be successful, people must accept it at some personal level. One of the ways in which Watson tried to bridge the gap between animals and humans was through physiological recording. Arguing that all thought was nothing but subvocal speech, Watson attempted to record the movements of the larynx during thought processes. He was ultimately unsuccessful, but retained his argument nonetheless. The fact that a behaviorist such as Watson could continue to ask subjects to introspect, and that a number of Titchener's students (Washburn in particular) were simultaneously elaborating a "motor theory of consciousness," indicates that despite the rhetorical bombast, there were still numerous connections between behaviorism and the older, introspective psychology.

Watson's infamous study of fear in a nine-month infant, "Little Albert," offers another early example of how behaviorism was extended to human experimentation. While working under the psychiatrist Adolf Meyer (1866–1950) in the Phipps Clinic at Johns Hopkins after World War I, Watson demonstrated that an infant (Albert) could be conditioned to show fear (crying) to formerly neutral stimuli (live rats, rabbits, and dogs). This was accomplished by striking an iron bar with a hammer every time an animal was presented to Albert. While this experiment undoubtedly appears unethical in retrospect, in the context of the painful, dangerous, and frequently useless treatments forced upon many psychiatric patients in this period, Watson's treatment of Little Albert was well in keeping with the psychiatric practice of the day.

Sexuality was another matter entirely. In 1920, Watson was forced to leave Johns Hopkins after his extramarital affair with Rosalie Rayner, his assistant at the Phipps Clinic, was made public. Unemployable in academe, he married Rayner and joined the New York advertising firm J. Walter Thompson, where he applied behaviorist principles to marketing. His ideas became more popular than ever. He published several best-selling books (including one on child rearing, coauthored by Rayner), and his work received considerable acclaim. The *New York Times*, for example, touted his work as marking "an epoch in the intellectual history of man."

NEOBEHAVIORISM AND RADICAL BEHAVIORISM, 1930–60

Although Watson's ideas were well received by many psychologists in the decades before and after World War II, there was no behaviorist revolution during this period. American psychology was a heterogeneous mix of pragmatic functionalism, academic introspectionism, polemical behaviorism, and professional testing. Running white rats through puzzle boxes became just one more part of the psychologist's tool kit, even as functionalism continued to supply most psychologists with an intellectual identity.

Even self-described behaviorists criticized Watson's ideas. Edward C. Tolman (1886–1959), who had been strongly influenced by James, rejected Watson's idea that learning could be reduced to a small, manageable set of mechanistic relations between stimulus and response. In his *Purposive Behavior in Animals and Men* (1932), Tolman argued that an organism always retains a set of functions or purposes that help determine the nature of any response. During the 1930s, Tolman appropriated the arguments of logical positivism (a philosophical movement with its origins in Vienna) and insisted that all observations of behavior were, unlike those of introspective psychology, empirically verifiable.

Clark Leonard Hull (1884–1952) extended Tolman's techniques of statistical inference in his *Principles of Behavior: An Introduction to Behavior Theory* (1943). Hull worked at the Yale Institute of Human Relations (IHR), which had been founded in 1929 by the Rockefeller Foundation in an attempt to unify the methodological and disciplinary array of the social sciences. From his position at the IHR, Hull was able to exercise a significant if not a lasting influence over a generation of psychologists. By the late 1940s, however, his elaborate mathematical formulations of the laws of behavior were heavily criticized, and the very idea that "learning" described a unified concept came under heavy attack.

If behaviorist theory waned in the 1950s, some of its practices endured. This was due primarily to the efforts of the Harvard psychologist B. F. Skinner (1904–90). Skinner's radical behaviorism repudiated all theory. Pavlovians and behaviorists had attempted, in varying degrees, to connect their observations to brain function. Neobehaviorists hoped to provide an empirical backbone to theories of knowledge. But Skinner took behaviorism to its logical conclusion, arguing that neither neurology nor epistemology held any relevance to psychological practice. His devotion to observation was encapsulated in what became known as a *Skinner box*. This consisted of an enclosed area containing a lever or bar that would dispense food automatically when pressed, and an animal (usually, a rat or a pigeon) that was connected to a physiograph that kept a constant record of its movements. The experiment, which practically ran itself, culminated in the animal "learning" to press the bar whenever it was placed in a similar box. Skinner contrasted his operant conditioning, which tried to understand how certain behaviors could be made more probable through reinforcement techniques, with the crude mechanism of classical conditioning, which applied stimuli in the hope of eliciting a reflexlike response.

Skinner stripped psychology down to pure technique. Its purpose, as he described in the psychological utopia of *Walden Two* (1948) and the polemical *Beyond Freedom and Dignity* (1971), was to create a rational and orderly society by constructing and applying appropriate reinforcement schedules for all facets of human activity. His uncompromising and often overextended approach opened the door for a devastating series of criticisms in the 1960s and 1970s.

DECLINE OF BEHAVIORISM

Behaviorists' intransigent refusal to tackle the question of consciousness left them with little to say about the issues of self-identity and self-image that began to dominate American culture in the 1960s and 1970s. In the left-leaning mood of that period, behaviorism ceased to appear as a benevolent attempt to achieve social order and began to look more like a prototype for manipulation worthy of the worst authoritarian regime. Not only did behaviorism deny agency to the psychological subject; it also dismissed the problem of meaning as outside the realm of scientific investigation. As reinforced "verbal behavior," language lost its ability to communicate complex internal states or experiences among its users. The idea that "the personal is political" (a popular radical slogan) made little sense if the personal could not be expressed in the first place.

One of the sharpest rebukes of behaviorism came from Noam Chomsky (b. 1928), a linguistics professor at the Massachusetts Institute of Technology (MIT). In 1959, Chomsky issued a scathing review of Skinner's *Verbal Behavior*, arguing that Skinner's categories of stimulus, response, and reinforcement were so vague as to be meaningless. In contradistinction to the behaviorist account, Chomsky proposed a "Cartesian theory of language" that drew upon developmental studies showing that children acquired language spontaneously and inductively. Comparing these with ethnographic studies,

Chomsky argued that the formal similarities of all languages indicated that there was probably a biologically based structure making human (and perhaps even primate) language acquisition possible. Chomsky's work proved to be enormously influential and established the search for and analysis of grammatical universals known as *psycholinguistics*.

Chomsky's was one of numerous critiques of behaviorist psychology that evolved in the United States in the three decades following World War II. Phenomenological and Gestalt psychology provided alternative systems for American psychologists in the 1950s and 1960s. Phenomenological psychology, originally formulated by the Würzburg school in Germany during the 1910s and 1920s, was a critique of Wundtian dualism. Inspired by the work of the German philosopher and mathematician Edmund Husserl, the Würzburgers attempted to describe the phenomena of perception without reference to either the perceiving subject or to the object perceived. Efforts at developing a new psychological language to accomplish such a task blended with Gestalt psychology in the 1920s and 1930s. Gestalt psychologists, many of whom worked out of the Berlin Institute of Psychology, devised a number of ingenious experiments demonstrating that the perception of form could not be explained in terms of sensual "elements." The ability to recognize a melody played on a number of different musical instruments, for example, indicated that perception was relational, and that it was structured around the whole rather than built up from the parts.

When such arguments took hold of the American psychological imagination after World War II, they became incorporated into an eclectic attack upon behavioral theory and practice and upon all species of scientific reductionism in general. Behaviorism did not disappear, but its scope was drastically reduced, becoming just one more psychological tradition with a narrowly defined set of tools, concepts, and practices.

Cognitive Psychology, 1950 to the present

At the end of the twentieth century, cognitive psychology was still such a new development that its historical tradition remained largely unwritten. Yet the adjective *cognitive* is now a commonplace in the language of the human sciences. In contrast to behaviorism, cognitive psychology tries to explain the phenomena of consciousness, but does so in a novel way. Its distinctive tradition emerged around 1950, when computers, electronic amplification, and information theory merged to provide tools and concepts through which a new description of consciousness could be forged.

THE FIELDS AND PRACTICES OF COGNITIVE SCIENCE

Cognitive psychology is one part of a postwar triumvirate known as *cognitive science*, the other two branches of which are *neuroscience* and *artificial intelligence*. Neuroscience continues the tradition, drawn from both neurophysiology and neurology, of investigating the brain as the organ of mind. Neuroscientists compare the performances of animal and human brains through a common visual medium such as electroencephalography (EEG) or, more recently, positron-emission tomography (PET) and functional magnetic resonance imaging (FMRI) scans. Such techniques of visualization (radically transformed by the evolution of the computer) have allowed neuroscientists to move away from the reflex-arc concept that dominated nineteenth- and early twentieth-century neurophysiology. Instead of stimulating a nerve and analyzing subsequent motor activity at the micro level, experimenters were now able to stimulate entire systems and record their electrical or biochemical responses in real time, without requiring visible or measurable motor activity. The study of dreams, for example, was radically altered by such techniques. Where a behaviorist could describe dreaming only as "verbal behavior upon awakening," cognitive scientists began to describe dreams as a form of information processing that took place during periods of rapid eye movement (REM), signified by graphical tracings first discovered in 1953.

The history of *artificial intelligence* (AI) is equally a history of technology, but in this case, the emphasis is not on recording formerly invisible brain activity. Rather, AI research relies on replicating human behavioral and cognitive performances by constructing robots that mimic human movements and by designing computing machines that model the processes of human cognition.

Cognitive psychology occupies the ground between these two fields. It draws upon the language of neuroscience and AI for new metaphors and concepts through which psychologists can describe the nature of thought and consciousness in an impersonal and objective manner. Cognitive psychology is thus opposed to the reduction of all psychology to the observation of behavior, but resists the obscure language and practices of experimental introspection.

Cognitive psychology made the brain, rather than behavior, the common denominator that brought together human and animal experiment. A cognitivist might say that the question of whether or not animals dream is insoluble for introspectionists or behaviorists. Yet REM periods, which have been detected in the sleep of almost all mammals, allow cognitive psychologists to talk about dreaming in terms of the function of memory and sensory information-processing systems in the brain while avoiding reduction of the personal and linguistic experience of having a dream.

INSTRUMENTS AND CONCEPTS

Behaviorists rejected both teleology (purpose) in evolution and intention in the individual as part of an arcane metaphysics that hindered the progress of psychology as a science. Cognitive scientists, on the other hand, embraced the idea that the mind actively intervenes in the transformation of stimulus into response. They borrowed from neuroscientists the notion of the brain as a reservoir, rather than a reflector, of energy. Similarly, the central difference between the **automata** that so fascinated Descartes and the machines created by AI is that the former were animated by external forces alone, while the latter contain programs that can transform their performance over time. Thus cognitive psychologists have identified human intention, which had for centuries been associated with the will through the phenomenon of voluntary movement, as equivalent to an internal program that directs mental performance without itself being an object of consciousness.

* **Automata.** Mechanisms that seem to be capable of self-instigated movement.

None of this would have been possible in the absence of a number of technological and social developments that began with the electronics revolution of the early twentieth century, which made possible the appearance of digital electronic computers by the late 1940s. Neurophysiology, for example, was transformed completely by the advances in electronic amplification that followed the invention of the triode, or vacuum tube, in 1905. Machines had been detecting electrical signals since the mid-nineteenth century, but the rectification and amplification of very weak signals, which could be broadcast as "waves," led to the discovery of such phenomena in the brain and nerves of living bodies by the end of the 1920s. Tube-driven technologies were at the cutting edge of both military and biological research at the dawn of World War II. To take but one example, Alfred Lee Loomis (1887–1975), a wealthy American financier and amateur scientist, was called away from his work on the EEG at his private laboratory in Tuxedo Park, New York, to help organize the Radiation Laboratory (RadLab) at MIT in the summer of 1940. Loomis's transition from brains to radar was seamless, since both investigations involved problems of generating, detecting, and amplifying signals. By the late 1940s, neurophysiologists, psychiatrists, and neurologists were beginning to describe the brain (and, by default, the mind) as a signal-processing system.

The word *feedback* dates back to 1920. It described the howling sound generated when a microphone was placed too close to a loudspeaker. In 1943, the word took on a new meaning that would prove to be enormously valuable for cognitive science. That year, Norbert Wiener (1894–1964), a mathematician at MIT working on the development of servomechanisms to help stabilize the aim of antiaircraft artillery, jointly published a paper with Arturo Rosenblueth, a neurophysiologist, and Julian Bigelow, an engineer (both dates unknown). The paper, entitled "Behavior, Purpose, and Teleology," argued that communication and control posed fundamentally identical problems, both of which could be depicted as questions about the flow of information within a system. The system could be mechanical (as in the case of a thermostat), neurological, or a combination of the two, as was the case with the interactions between people and machines that made radar detect and fighter planes fly. All systems shared the similar operation of positive feedback: They continually collected information about their performance, compared this to a plan, and then attempted to minimize the discrepancy between the two by adjusting variables. This contrasted to the negative feedback of a squealing microphone, in which a signal, left unregulated, created only noise.

Systems thus exhibited purposeful, goal-oriented behavior through a program that regulated the flow of information through a feedback mechanism. In the 1950s, cognitive psychologists used such metaphors to direct their field toward an analysis of the mind's organizational structure. This emphasis on self-regulation had already begun to bear practical and conceptual fruit in physiology. The study of regulatory hormones, for example, received an enormous boost in prestige in 1921, when insulin was discovered and subsequently mass-produced to control the symptoms of millions of diabetes sufferers. In the 1930s, several prominent American physiologists compared the physiological body to the body politic, arguing that the principles of self-regulation effectively captured the contrast between the efficient, yet freedom-preserving, nature of the scientifically managed American democracy and the totalitarian structures of Soviet communism and German fascism.

Psychologists began to use neurophysiology to criticize behaviorism during the 1940s. Karl Lashley (1890–1958) initially attempted to trace conditioned reflex pathways in the brain, but he soon rejected the idea that the chains of association that supposedly caused behavior had a material form in the brain. He argued instead that it was the brain's innate organizational structure, not environmental stimuli, that made possible complex activities such as language acquisition. Donald O. Hebb (1904–85), a psychologist at McGill University in Montreal who had worked with both Lashley and the Canadian neurosurgeon Wilder Penfield (1891–1976), took a similar stance in *The Organization of Behavior* (1949). Hebb speculated that the Gestalt psychologists' holistic approach to perception had an analog in the brain's structure. As the individual brain developed, Hebb argued, repeated behavioral patterns transformed small assemblies of neurons into complex phase sequences that directed behavior and perception by mass action, rather than passively responding to stimuli as individual units. This attempt to synthesize localist and holistic interpretations of brain structure continued in the work of the biochemist, neurobiologist, and Nobel laureate Gerald Edelman (b. 1929). His *Neural Darwinism* (1987) applied the theory of natural selection to populations of neural cells modeled in an artificial brain.

Psychologists in the 1950s also drew from their autonomous experimental traditions. Their professional involvement in World War II exposed many psychologists to information theory, which described and analyzed the performance of information-processing systems according to binary logic. This approach had already been applied to neural functioning in 1943 by the American neuropsychiatrist Warren S. McCulloch (1898–1972) and the mathematician Walter H. Pitts (b. 1924) in their paper "A Logical

Calculus of the Ideas Immanent in Neural Activity." McCulloch and Pitts argued that the "all-or-none" hypothesis of neural transmission (the notion that the nerves either fired completely or not at all) was analogous to the binary character of logical inference. In other words, neurological and epistemological systems shared an identical structure.

In the mid-1950s, the Cambridge psychologist Donald Broadbent (1926–93) applied a similar approach to the problem of mind. During his service in the war, Broadbent had taken note of how cockpit design did not take into consideration any limitations on the pilot's ability to deal with streams of data coming from several different sources. After the war, he took a degree in psychology at Cambridge and joined the Medical Research Council's Applied Psychology Unit there in 1949. Broadbent studied the phenomenon of attention and compared it to an information filter. He instructed his experimental subjects to attend to, and later repeat, the signal (spoken words) delivered to one ear while noise was delivered to the other. He found that when different sets of numbers were delivered simultaneously to both ears, subjects scored highest when they treated the information received by each ear as an independent unit and recalled each in succession. Perhaps the most noteworthy aspect of these experiments was their graphical description: Broadbent used flowcharts, already common in information theory, to represent the cognitive processes supposedly involved.

REPRESENTATION AND METAPHOR

With the increased use of metaphors drawn from information theory, computer science, and systems analysis, psychologists began to transform the concept of mental representation itself. Nineteenth-century psychologists had depicted mental representations as images, typically visual, that appeared before the mind's eye. Empiricist philosophy had insisted that all such representations were impressed upon the mind by the power of the perceived objects themselves, while an "internal observer" took notice of these images when its will was directed toward them. Several nineteenth-century psychologists, for example, compared the images of dreams to those projected on a screen by a "magic lantern." The "imageless thought" discovered by the Würzburgers around 1905 thus posed an enormous challenge to psychology, because it suggested that the public activities of seeing and hearing were not analogs to private mental processes. The vague qualities of "determining tendencies" implied that introspectionists could obtain only a rather oblique view of what the mind was doing. There was no available language to express such processes; the development of phenomenological psychology was, in part, an attempt to create a new descriptive language that could appropriately capture the subjective experience at the heart of these experiments.

American behaviorists took these results as posing a methodological rather than an epistemological problem. The idea of consciousness, Watson argued, was little better than fiction and could be eliminated from psychology by deposing introspection. Mental representation could never be captured by scientific psychology. In its place, Watson borrowed the concepts of repetition and habit to describe behavior in terms of the reflex arc. No one, after all, thought that the work of the spinal nerves in governing the knee-jerk reflex depended upon the representation of an internal image.

It was not until the 1960s and 1970s that American psychologists began to turn en masse toward the metaphor of programming as a new way of conceptualizing mental representation. What was represented in this case, however, was not an image but a set of logical rules that governed and coordinated mental activity. The Austrian philosopher Ludwig Wittgenstein (1889–1951) had already underscored the all-encompassing importance of following a rule in his observations about language. His posthumously published *Philosophical Investigations* (1953) transformed the philosophy of mind by tracing back questions about the nature of emotion, of reason, and of mind-body problems to the rule following that he felt was inherent in all language use. Wittgenstein criticized experimental psychology for treating language as an unproblematic medium through which the "imagery of the mind" could be described. In its place, he advocated the study of how language is used in practice. He did not ask, for example, "What do dream-images mean?" but rather, "What do I mean when I call this a dream?"

For some psychologists and philosophers of mind, the postwar ubiquity of computers, programs, systems, and information seemed to offer examples of rule-governed activity. But there was little consensus regarding Wittgenstein's often paradoxical formulations. Nor was there much agreement regarding the status of mental imagery within cognitive psychology itself. During the 1970s, for example, David Marr (1945–80), a psychologist at MIT, attempted to construct devices that could identify and describe objects in a picture. His scene analysis used AI to unravel the problem of mental imagery by developing programs that would allow computers to see. When the initial enthusiasm about AI began to wear off in the 1980s, however, an "ecological" approach attempted to reinstate classical ideas about mental imagery based on an organism's direct perception of objects. The American psychologist James J. Gibson (1904–79) argued in *The Ecological Approach to Visual Perception* (1979) that perception was not a passive activity and could not be properly analyzed simply by referring to the physical properties of the stimuli or the sense organ (a tradition that stretched all the way back to Berkeley). Gibson argued that subjects (particularly subjects in motion) perceive possibilities in objects called *affordances*, which directly affect the nature of cognition, something that had been a practical concern of Gibson's since he was involved in testing pilots for military service during World War II.

More recently, the AI community has touted parallel processing as a solution to these problems. Instead of building computers that model the way the mind converts a picture into functional knowledge through the "serial order" of rules, neoconnectionists try to model the mass action of millions of neurons to examine how they behave as aggregates in neural networks. Instead of using robots to show how human performances can be replicated mechanically, AI now tends to use computing technology to model the complicated phenomena that emerge out of simple units.

INTERDISCIPLINARITY

In 1977, neuroscience and AI were brought together through $20 million in funding delivered to the cognitive sciences by the Alfred P. Sloan Foundation. But neither field proved capable of dealing with human beings as knowers. They simply created models of human knowers as either brains or as machines.

Cognitive psychologists have remained hesitant about reducing their subjects to such a passive status. The question of mental imagery has remained a flashpoint between psychologists who experiment with introspecting human subjects and those cognitive scientists who do not. Although many retain the metaphors of programs and neural networks, they are frequently unwilling to describe mind as being identical to these things.

Interdisciplinarity has been described as an article of faith among many cognitive scientists. Ideally, independent and autonomous disciplines, each with its own concepts, tools, and practices, gather together around a single problem or group of problems, presumably to achieve some sort of consensus on key issues. The funding that cognitive science received from the Sloan Foundation in the late 1970s was based on just these principles. Yet despite a remarkable degree of cross-fertilization, disciplinary identities within cognitive science have proved remarkably resilient. When the first State of the Art Report was released a year after the Sloan money had started to flow, its attempt to portray research objectives in an interdisciplinary manner aroused such opposition that its publication was suppressed.

To date, there is no unified psychological paradigm. Nor does it seem likely that one will emerge in the foreseeable future. Psychologists' insistent retention of a unique disciplinary identity based on an extremely diverse set of conceptual and experimental practices is a microcosm of the condition of the human sciences in general. The human sciences do not seem to require the sort of consensus required in the natural sciences. The former are, for better or for worse, self-referential in a way that the latter are not. Every form of psychological knowledge described in this article has been so deeply rooted in its historical context that each constitutes a kind of social institution in itself. In a sense, psychology serves two masters. As a source of self-identity, it must always offer a workable conception of what it means to be human. As a science, it is obliged to borrow its authority from the natural and life sciences. Psychology has to react to change in both dimensions. It quite literally makes up its subjects as it goes along. This is not a criticism, nor could it be one. History, after all, performs a similar function, albeit in a different mode.

Bibliography

Ash, Mitchell G., and William R. Woodward, eds. *Psychology in Twentieth-Century Thought and Society.* Cambridge: Cambridge University Press, 1987.

Carroy, Jacqueline, and Régine Plas. "The Origins of French Experimental Psychology: Experiment and Experimentalism." *History of the Human Sciences* 9 (1996): 73-84.

Cooter, Roger. *The Cultural Meaning of Popular Science: Phrenology and the Organization of Consent in Nineteenth-Century Britain.* Cambridge: Cambridge University Press, 1984.

Danziger, Kurt. *Constructing the Subject: Historical Origins of Psychological Research.* Cambridge: Cambridge University Press, 1990.

Ellenberger, Henri F. *The Discovery of the Unconscious: The History and Evolution of Dynamic Psychiatry.* New York: Basic Books, 1970.

Fancher, Raymond E. *The Intelligence Men: Makers of the I.Q. Controversy.* New York: W. W. Norton, 1985.

Gardener, Howard. *The Mind's New Science: A History of the Cognitive Revolution.* New York: Basic Books, 1985.

Gauld, Alan. *A History of Hypnotism.* Cambridge: Cambridge University Press, 1992.

Hacking, Ian. *Rewriting the Soul: Multiple Personality and the Sciences of Memory.* Princeton, N.J.: Princeton University Press, 1995.

Hearnshaw, Leslie Spencer. *The Shaping of Modern Psychology.* New York: Routledge & Kegan Paul, 1987.

Herman, Ellen. *The Romance of American Psychology: Political Culture in the Age of Experts.* Berkeley: University of California Press, 1995.

Jeannerod, Marc. *The Brain-Machine: The Development of Neurophysiological Thought.* Translated by David Urion. Cambridge, Mass.: Harvard University Press, 1985.

Kusch, Martin. *Psychologism: A Case Study in the Sociology of Philosophical Knowledge.* New York: Routledge, 1995.

Leahey, Thomas Hardy. *A History of Psychology: Main Currents in Psychological Thought.* 4th ed. Upper Saddle River, N.J.: Prentice Hall, 1997.

Murray, David J. *A History of Western Psychology.* Upper Saddle River, N.J.: Prentice Hall, 1983.

Richards, Graham. *Mental Machinery: The Origins and Consequences of Psychological Ideas, Part 1: 1600–1850.* London: Athlone Press, 1992.

———. *Putting Psychology in Its Place: An Introduction from a Critical Historical Perspective.* New York: Routledge, 1996.

Smith, Roger. *The Norton History of the Human Sciences.* New York: W. W. Norton, 1997.

Woodward, William R., and Mitchell G. Ash. *The Problematic Science: Psychology in Nineteenth-Century Thought.* New York: Praeger, 1982.

Statistics and Probability Theory

Sylvia M. Svitak

From ancient musings on chance and early record keeping of seasonal floods and commerce, probability and statistics have evolved to become cornerstones of modern scientific investigation in fields as diverse as quantum theory (see below) and psychology. The theory of probability grew from the communications of Pierre de Fermat (1601–65) with Blaise Pascal (1623–62) in the seventeenth century, to its axiomatic status as a mathematical theory in the twentieth century, through the efforts of mathematical researchers, especially Andrey Kolmogorov (1903–87). Statistical laws, which depend strongly on probabilistic reasoning, were used to estimate errors in astronomical measurements in the eighteenth century and then to describe the nature of nuclear particles in the twentieth century. In social and economic contexts, the rise of statistics is why Kenneth Prewitt, director of the U.S. Bureau of Census, has called the twentieth century the "first measured century in world history."

Pre-1543 Roots

The earliest recorded problems in probability are found as far back as the third century B.C. One of the earliest recorded problems of probability is in the mathematical treatise *Sunzi Suanjing* (Master Sun's Mathematical Manual) by the Chinese warrior-philosopher Sun Zi (c. 300 B.C.). Sun Zi is known as Master Sun. His text *Ping Fa* (The Art of War) is the oldest military treatise in the world, and many people still see it as a guide to the art of life. In *Suanjing*, Master Sun asks for the probability that an expected child will be a boy or a girl. This problem was dismissed as ridiculous by a later Chinese writer. However, Western scientists and mathematicians in later centuries took it seriously as they developed a general theory of probability. Today, this problem is a standard school exercise.

In Europe, the earliest references to probability involved gambling and dice. Dante Alighieri, in the sixth canto of "Purgatory" in the *Divine Comedy*, makes a brief reference to the game of Hazard (called Zara in the canto). In 1477, the *Divine Comedy* was published with a commentary attributed to Benvenuto d'Imola (c. 1330–c. 1387) that included the game and the various throws that can be made with three dice. Luca Pacioli (1445–1517) is thought to have been the first European writer to introduce a problem on gambling into a mathematics text. He was a Franciscan monk, a highly regarded mathematics teacher, and a mentor and friend of Leonardo da Vinci (1452–1519). In the 1487 *Summa,* Pacioli's problem, which became known as the *problem of points*, asked for an equitable division of stakes between two players when a fair game is stopped before it concludes. In his example, two players were to continue playing until one won six rounds. However, the game was actually stopped when one player won five rounds and the other three rounds. Pacioli assumed this was a proportionality problem based on the number of rounds won by each player. He concluded that the stakes should be split in a ratio of 5 to 3. For the next 200 years, texts on probability repeated and expanded on the problem of points and argued its solutions. A general theoretical solution was finally given by Blaise Pascal and Pierre de Fermat in the 1650s. Thus, historians of statistics and probability say that the real history of probability theory began with the exchange of letters between these two Frenchmen. Their contributions are discussed in detail below.

In ancient civilizations and before the eighteenth century, statistics were regarded mainly as data in the daily life of agriculture and commerce. Abstract statistical theory, first developed in the eighteenth century, followed from the probabilistic ideas that began to emerge in the 1650s. (Before 1850, the word *statistics* was still used to mean information of interest about political states and countries.) The types of information in ancient records are similar to some of those that are relevant to twentieth-century societies. The information was not necessarily numerical or concerned with calculation, but was more in the nature of record keeping.

The Egyptians, Babylonians, and the Inca were among the ancient peoples who collected numerical and nonnumerical data for agricultural, commercial, legal, political, and religious purposes. The Egyptians enjoyed a fertile land fed by the Nile River, which, because of its large area, developed a complex administration. There was a system of taxation and an army to support, making it necessary to keep records for these purposes and for other computations as people traded their goods. In the last three centuries B.C., the Babylonians directed their mathematical efforts to the study of lunar and planetary motion. They collected data in tabular form to predict the daily motion of the planets. By these means, they were able to know planting times and religious holidays. The Inca did not develop a system of writing but recorded numerical information by using knots in strings called *quipus*. The quipu was not a calculator but a storage device. It played a major role in the administration of the Inca Empire. Judges used the quipu to give monthly accounts of the sentences they imposed. Each town employed keepers of the knot (*quipucamayocs*) who were appointed by the king to keep an official town census. They also kept records of production, animals, and weapons. The Inca developed a rapid delivery service across the rough terrain of their land to take quipus to their capital, Cuzco, for central recording.

1543–1700

During this period in Europe, statistical methods and probability concepts flowed separately but were beginning to converge. The English were grand keepers of concrete facts from the time of William I (the Conqueror, ruled 1066–87). He not only conducted an extensive human and economic survey, documented in the famous *Domesday Book* (1086), but also followed up with another to check the work of the first enumerators. Natural disasters proved to be another large influence on the development of statistical methods. Epidemics of bubonic plague killed almost 25 percent of Londoners from 1563 to 1665. *Bills of mortality* containing statistics about recurring outbreaks were first printed in 1532. Londoners bought copies of the bills, looking for signs of impending disease. In 1538, Thomas Cromwell ordered every parish to record weekly its baptisms, marriages, and deaths. By 1592, parish bills were produced weekly and, by 1625, printed weekly. A yearly accounting was given on what the English dubbed Doomsday, the Thursday before Christmas. In 1632, for the first time, the bills included the attributed cause of death for each listing.

John Graunt (1620–74) used the bills of mortality as the basis for his 1662 work, *Natural and Political Observations Made upon the Bills of Mortality.* On the continent, France and Italy were slow to begin keeping registers, their leading thinkers being preoccupied with gaming problems that fed their notions of probability. Eventually, Graunt's work persuaded the French to enact their own registry laws. Beyond influencing continental decisions to establish registries, Graunt's book served as the basis of vital statistical science in England, which organizes and analyzes data about the **population**. Graunt's influence reached further: His slim volume also impressed the Dutch astronomer Christiaan Huygens (1629–95). In 1669, Huygens's brother asked him about Graunt's attempt to construct a life table. In answer, Huygens formulated a mortality curve and properly defined the concepts of mean and probable life expectancy. He was first to apply

* **Population.** Any finite or infinite collection of individuals; replaces the older expressions *universe* and *universe of discourse*.

probability theory in a meaningful way to demographic statistics, and this represents the beginning of the dependence of statistical inquiry on probability ideas.

In this same period, Niccolò Tartaglia (1499–1557) and Girolamo Cardano (1501–76) continued the use of gambling and dice as a source of probability problems. In his *General Treatise on Number and Measure* (1556), Tartaglia responded to Pacioli's problem of points. Like Pacioli, Tartaglia saw this problem in terms of proportionality, but he considered Pacioli's solution to be in error. Tartaglia argued that the stakes should be divided in a ratio of 2 to 1 in favor of the player who was ahead five games to three when the game was stopped. He said the proportion for the player who won five games should be the ratio of the difference between that player's score and half the total stake, to half the total stake. Tartaglia also presented combinatorial problems in his text. In one such problem, 10 people were seated and served as many dishes of food as there were ways in which they could be seated. Each seating arrangement was considered different from the others. He obtained the solution 10! (ten factorial), and asserted that the method he used worked in general, but did not prove it.

Cardano's many works include the well-known *The Book on Games of Chance*, probably completed by 1563. Some historians claim that he anticipated the laws of large numbers, but others dispute this. It is acknowledged, however, that Cardano was definitely the first to calculate a theoretical probability correctly and, as F. N. David (dates unknown) put it, "That is possibly enough glory." In another treatise, *Practical General Arithmetic*, Cardano pointed out that Pacioli's solution to the problem of points did not take into account how many games each player had yet to win. Although he himself did not obtain a correct solution, he did use the addition rule for calculating the stakes, understanding that the rule required independent events. Cardano actually computed stakes in fair games, but they are proportional to probabilities, so his work certainly represents the beginning of probability theory.

FOUNDATIONS OF PROBABILITY THEORY: FERMAT AND PASCAL CORRESPONDENCE

Pascal and Fermat began their famous correspondence in 1654. Pierre de Fermat is probably most famous for his work in number theory and what has become known as Fermat's last theorem. He was a lawyer who became an official at the parliament in Toulouse, France. He spent his life serving in government, doing research in mathematics, and providing for his family. All accounts seem to show that he had an admirable character. Fermat was part of a circle of mathematicians and natural scientists in an informal society centered around the mathematician Marin Mersenne (1588–1648). Through his contacts in this society, Fermat came to correspond with Pascal on theoretical probability.

Blaise Pascal was the only son of a wealthy lawyer, Étienne Pascal, who was also an amateur mathematician. Étienne supervised his son's education and, when Blaise was fourteen, his father began to take him to Mersenne's meetings. Blaise had a serious interest in his mathematical and scientific research until about 1654, when religious issues, fueled by his delicate health, began to take priority. His religious interests led to his famous probabilistic argument in favor of a belief in God. According to Pascal, God either exists or does not exist. A person has no choice in life but to bet on which is true. If God does not exist, one's behavior in terms of salvation does not matter. If there is a God, betting that God does not exist invites damnation and betting that God does exist will bring salvation.

In about 1652, Pascal met Antoine Gombaud (the Chevalier de Méré, 1607–84), who used the opportunity to ask Pascal about several betting questions. The historian

Leonid Efimovich Maistrov (dates unknown) presents some evidence to indicate that the Chevalier de Méré posed purely theoretical questions. Maistrov thus asserts that the assumption that probability theory originated from questions on gambling in the seventeenth century is false. Nonetheless, questions from gaming served the theory by providing specific contexts in which to demonstrate the theoretical concepts.

In one question, de Méré asked Pascal to calculate the probability of winning a dice game if at least one six appears when a die is tossed four times. If no six appears, the opponent wins. In a second game, he wanted to know the probability of tossing a pair of sixes in 24 rolls in a game. In the first case, the odds are slightly better than one-half (0.5177 to 0.4822) in favor of throwing at least one six in four tries. Fermat and Pascal wrote these odds as 671 to 625, thereby showing that the results depend on equally likely events. In the case of the second game, the odds are 0.4914 to 0.5055; that is, tossing a pair of sixes is not favored in 24 rounds. In modern terms, the probability of one six on a toss of one die is 1/6 and the probability of no sixes is 5/6. Thus the probability of at least one six in four throws is $1 - (5/6)^4$. A similar calculation determines the odds in the second game.

The Chevalier de Méré also posed the problem of points that first appeared almost 200 years earlier in Pacioli's *Summa*. The Fermat-Pascal correspondence yielded a general solution based on general principles. From their letters, we see that Fermat arrived at the same principles as Pascal. Fermat's solution was more direct, enumerating the possible outcomes, but his method was also less efficient. In a modern form, Pascal's theorem can be stated as follows. Suppose that player A needs p games to win a match and player B needs q games, where both p and q are at least 1. If the match is stopped at this point, the stakes should be divided as follows. Player A gets a proportion of the total in the ratio $\sum_{k=0}^{k=q-1}\binom{n}{k}$ to 2^n, where $n = p + q - 1$ (the maximum number of games that are remaining). Here $\sum_{k=0}^{q-1}\binom{n}{k}$ is the sum of the coefficients of the first q terms in the binomial expansion of $(a + b)^n$. (The binomial expansion of $(a + b)^n$ is

$$\binom{n}{0}a^nb^0 + \binom{n}{1}a^{n-1}b^1 + \binom{n}{2}a^{n-2}b^2 + \cdots + \binom{n}{n-2}a^2b^{n-2} + \binom{n}{n-1}a^1b^{n-1} + \binom{n}{n}a^0b^n$$

where n is a positive integer.)

The binomial coefficients can be obtained from a mathematical construction known as an *arithmetic triangle*. Pascal developed a version of this triangle and used it in this theorem. A simplified version of the triangle is given in Table 1, the letters indicating rows and columns. The diagonal line from row D to column d indicates the binomial terms for solving Pacioli's version of the problem of points. In that case, $p = 1$, $q = 3$, and $n = 3$. The numerator of the ratio will then be the sum $1 + 3 + 3 = 7$ and the denominator $2^3 = 1 + 3 + 3 + 1 = 8$. Thus, player A should receive 7/8 of the stakes.

Pascal's Triangle

	a	b	c	d	e	f	g	h	i	j
A	1	1	1	1	1	1	1	1	1	1
B	1	2	3	4	5	6	7	8	9	
C	1	3	6	10	15	21	28	36		
D	1	4	10	20	35	56	84			
E	1	5	15	35	70	126				
F	1	6	21	56	126					
G	1	7	28	84						
H	1	8	36							
I	1	9								
J	1									

Table 1.

Fermat, Pascal, and Hugyens used the addition and multiplication rules of probability and the ideas of equally likely, dependent, and independent events. Huygens became impressed with Pascal and Fermat's work in probability during his visit to Paris in 1655 and decided to prepare a manuscript on the new problems and their solutions. He solved the gaming problems independent of the French results, and his theoretical results were equivalent to theirs. Huygens was, however, the first to present solutions in terms of expected value, a generalization of the arithmetic mean. His treatise *About Dice Games* became the fifth book of *Mathematical Exercises* published by his teacher and friend Frans van Schooten (1615–60) in 1657. Huygens's book was the most important introduction to the theory of probability for over half a century until the prolific researches of the next century overshadowed his work.

Eighteenth Century

* **Geodesy**. The mathematical science that deals with the size and shape of Earth and its gravitational and magnetic fields.

Three problems from astronomy and **geodesy** are among those that influenced the development of probability and statistics in the eighteenth century: (1) to determine mathematically the movements of the Moon, (2) to explain the observed motions of Jupiter and Saturn, and (3) to determine the shape of Earth. Scientists and mathematicians solved these problems by using probability and statistical concepts. In the process, they abstracted and generalized those concepts, elevating them to a theoretical status. The probabilistic developments of this period begin with Jakob Bernoulli (1654–1705) and the law of large numbers. In the course of the eighteenth century, Abraham de Moivre (1667–1754), Thomas Bayes (1702–61), Comte de Buffon (Georges-Louis Leclerc, 1707–88), and Adrien-Marie Legendre (1752–1833) provided ideas, experiments, and methods that are important theoretical preludes to later accomplishments in physical and social sciences.

JAKOB BERNOULLI'S *THE ART OF CONJECTURING*

Jakob Bernoulli was a member of the talented but contentious Bernoulli family, renowned for mathematical and scientific achievements. Before he settled down to teach mechanics, Jakob traveled through Europe establishing connections with the many mathematicians with whom he would correspond during his lifetime. As a student, Bernoulli was deeply influenced by the problems in Huygens's *On the Calculations in Games of Chance*. Bernoulli's *The Art of Conjecturing* is regarded as the first major work on the theory of probability. Some think it is his most original work. He gave more than 20 years to it and to the principle that Siméon-Denis Poisson (1781–1840) would later name the law of large numbers. Jakob had not completed *The Art of Conjecturing* at the time of his death in 1705, but after about eight years, his nephew Nickolaus Bernoulli (1687–1759) wrote a preface, and *The Art of Conjecturing* was released to the world. It consisted of four parts, the first three reconstructing, generalizing, and expanding on the work of previous writers. It is in the fourth part that Bernoulli made his own original contribution and discussed potential applications of probabilistic reasoning to the social sciences and economics.

In the first part, Bernoulli recounted the work of Huygens, adding his commentaries on almost all of the propositions in *Games of Chance*. One of his generalizations was the gambler's ruin problem. This classic problem was probably first hinted at by Huygens as Exercise V in *Games of Chance*. Players A and B start with 12 balls and continue to throw three dice until 11 or a 14 is thrown. Player A gives a ball to player B if 11 comes up and B gives one to A if 14 comes up. The game continues until one of the players has all the balls. Bernoulli generalized this problem to the case where player A has m dollars to risk and player B has n dollars to risk, and the game continues until one player claims all the stakes. The odds of player A winning over player B are given by the ratio $a{:}b$. Bernoulli showed that player A's chance of winning is

$$\frac{a^n(a^m - b^m)}{a^{m+n} - b^{m+n}}$$

Bernoulli wrote the second part of *The Art of Conjecturing* to provide a complete exposition of the theory of permutations and combinations that had been started by others. Part II was used as a textbook on the subject in the eighteenth century. Here Bernoulli included his work on the arithmetic triangle and a formula for the number of combinations of n things taken r at a time that is now standard. Part III is devoted to applying combinations theory to dice and other games of chance. Bernoulli solved some of the problems for the general case.

The first three parts were certainly worthy of acclaim, but it was the fourth part (*On the Use and Applications of the Doctrine in Politics, Ethics, and Economics*) that revolutionized the role probability was to play in mathematics and the sciences. Part IV's law of large numbers is the first **limit theorem** in probability. James R. Newman (1907–66) in *The World of Mathematics* states that "this theorem was the first attempt to deduce statistical measures from individual probabilities." Newman also gives a simple statement of the theorem: "If the probability of an event's occurrence on a single trial is p, and if a number of trials is made, independently and under the same conditions, the most probable *proportion* of the event's occurrences to the total number of trials is also p; further, the probability that the proportion in question will differ from p by less than a given amount, however small, increases as the number of trials increases."

* **Limit theorem.** A theorem about the nature of the sampling distribution from a population as the sample size becomes arbitrarily large.

Bernoulli's theorem was historically significant for the following reasons. All previous theoretical work on chances came from *a priori* considerations. This means that probabilities were deduced from the rules of a game and the physical nature of the game. For example, in tossing a fair die of six numbered faces, we consider the physical properties of the die and how it is tossed. If the die is symmetrical in all aspects and thrown so as not to land on an edge or point, we determine a priori that the probability of a three facing up is 1/6.

Bernoulli's law brings probability into the realm of *a posteriori* thinking, that is, inductive reasoning from experience. What would Bernoulli's theorem predict if we tossed a fair die repeatedly? Using Newman's statement of the theorem, we would say that the most probable proportion of threes facing up would be 1/6. Furthermore, as the trials increase, the chance will also increase that the proportion of threes facing up will differ from 1/6 by less than any chosen amount, however small.

Bernoulli was not the first to use empirical evidence to determine chances, but he was first to give a mathematical proof of the idea that the greater the experimental evidence one has about a proportion, the closer one is to being certain about that proportion. However, one has to be careful about a common misinterpretation. This theorem does not mean that in increasing the trials indefinitely, the limit of the empirical proportion is the theoretical probability. What Bernoulli's theorem asserts is that if one makes the number of trials n sufficiently large in each experiment and does a sufficiently large number of experiments, in the vast majority of cases, the experimental ratio will be as close to the true ratio as one pleases.

The title of Part IV promised applications to the social sciences, but those applications did not materialize beyond some general comments. Bernoulli called an event morally certain if the probability of its occurrence is 0.999 or greater. He introduced an experiment in which there is a container of 3000 white pebbles and 2000 black pebbles and where the proportion is unknown to the experimenter. The object is to determine the ratio 3:2 of white pebbles to black pebbles. He found that the number of trials necessary for moral certainty was more than 25,500, a huge number to him, greater than the population of his city, Basel. Thus he may have concluded that applications of his theorem were impractical. Nonetheless, his efforts provided ideas for the work of his contemporary, Abraham de Moivre (1667–1754).

DE MOIVRE AND *THE DOCTRINE OF CHANCES*

Abraham de Moivre was born in France to a Protestant family, and while studying at Saumur, he read Huygens's text on probability. In 1685, when the Edict of Nantes was revoked, he was imprisoned. Freed in 1688, he immediately left France and settled in London, where he remained for the rest of his life. He published *The Doctrine of Chances* in 1718 and two more editions in 1738 and 1756. He applied his ideas from the theory of series to solve problems in probability and became one of the first important modern analytic probabilists.

De Moivre answered the Chevalier de Mere's question about the odds of throwing a pair of sixes in 24 tosses as part of a more general problem that he solved. Problem V in *The Doctrine of Chances* is to find how many trials must be made to have even odds that an event will happen at least once. The solution makes use of **natural logarithms** followed by a series expansion. The result is that if the odds against the event are q to 1 and are large, the number of trials x is approximated by the first term of the series expansion and is $q(\ln 2)$. In de Méré's problem, $q = 35$, so $x = 24.5$. Thus by a very different method of analysis, the odds are shown to be against the event of throwing a pair of sixes in 24 tosses.

* **Natural logarithm.** Symbolically, x is the natural logarithm of y if $y = e^x$, where e represents the irrational number 2.71828 . . . . The constant e satisfies the property that $d(e^x)/dx = e^x$.

De Moivre was to have a profound influence on later developments through his detailed expansion of the binomial $(a + b)^n$, which first appeared in an appendix to his 1730 treatise, *Miscellaneous Analysis*. In the 1738 edition of *The Doctrine of Chances*, he translated the appendix, *A Method of Approximating the Sum of the Terms of the Binomial*, and showed for the first time in English the curve that would become known as the normal curve, a bell-shaped probability function that would later be found to represent the distribution of many variables of importance for human society. His purpose in the derivation was to estimate probability by means of experiment. Like Bernoulli, he believed that to obtain the desired probabilities one has to calculate certain binomial coefficients. He restricted his work to finding the probability of $n/2$ successes in n trials, where a trial had two possible outcomes. This meant that he was looking for the ratio that the middle term of $(1 + 1)^n$ has to the sum of all the terms, 2^n. De Moivre again used logarithms and series expansions to produce his results. He next produced a generalization to determine the probability of successes other than $n/2$. Writing those successes X in terms of distances t from the middle term, that is, $X = n/2 + t$ successes, de Moivre showed the first instance of the normal probability curve. In modern notation, he showed that the probability is

$$P(X = n/2 + t) = \frac{2}{\sqrt{2\pi n}} e^{-(2t^2/n)}$$

De Moivre did think of $P(X = n/2 + t)$ as ordinates on a curve that has two **points of inflection,** one on each side of the maximum. He calculated those inflection points to be at a distance $1/2\sqrt{n}$ from the maximum. He used the integration formula

* **Points of inflection.** Points on a function's graph where the direction of curvature changes from upward to downward, or vice versa.

$$\frac{2}{\sqrt{2\pi n}} \int_0^k e^{-(2t^2/n)} dt$$

to approximate the sums of the terms of the binomial expansion. For $k = 1/2\sqrt{n}$, he determined that the series converged quickly enough for him to ascertain that the sum was equal to 0.341344. Today, we recognize that his calculation showed a part of the empirical rule for normal distributions. He showed that in a binomial experiment for a large number n of trials with equally likely events, the probability of the number of occurrences of the event in the interval between $-1/2\sqrt{n} + 1/2n$ and $1/2n + 1/2\sqrt{n}$ will be 0.682688. That is, approximately 68 percent of the occurrences fall within one **standard deviation** of the mean. Today, this is known as the *margin of error*. De Moivre did outline a generalization of his method to approximating terms of $(a + b)^n$. He was also able to calculate the sums of large numbers of such terms and improve on Bernoulli's number of experiments needed for moral certainty. Although de Moivre did not show it, his methods would reduce the 25,500 cases that Bernoulli needed for moral certainty to about 6500 cases.

* **Standard deviation.** The spread of data values around the mean of a data set.

BAYES AND STATISTICAL INFERENCE

As significant as de Moivre's work was, it did not solve actual problems posed by science. Empirical scientists wanted to know what the facts that they observed indicated about the process that produced the facts. This is a fundamental question of statistical inference, and the theoretical considerations had to wait for Thomas Bayes and others.

Bayes was the son of one of the first nonconformist ministers to be ordained in England and was educated privately. He also became a minister. Bayes set out his theory of probability in "Essay towards Solving a Problem in the Doctrine of Chances," which was published in the *Philosophical Transactions of the Royal Society of London* in 1764. The paper was sent to the Royal Society by Richard Price (1723–91), who believed that it had great merit. Bayes undertook his study because he wanted to find a method of determining the probability that an event will occur in a single trial under given conditions with the following assumption. Nothing is known about the event except that, under the same conditions, the event occurred a certain number of times in a given number of trials, and it failed to happen a certain number of times. Furthermore, Bayes wanted to show that the required probability can be determined within any two given limits.

In modern terms, X is the number of times the event has occurred in n trials, x the probability of its occurrence in a single trial, and u and υ the two given probability limits. Thus, Bayes wanted to calculate $P(u < x < \upsilon \mid X)$, that is, the probability that x is between u and υ, given X. To obtain his result, Bayes needed to prove two propositions, which he numbered Proposition 3 and Proposition 5. Today, Proposition 3 would read: If E is the first event and F is the second event, the probability of both occurring is the product of the probability of E with the probability of F given E or, symbolically, $P(E \cap F) = P(E)P(F \mid E)$. Proposition 5 in today's notation is: The probability of E given that F occurred is the quotient of the probability that both occurred and the probability of F or, symbolically,

$$P(E \mid F) = \frac{P(F \cap E)}{P(F)}$$

Hence Bayes's basic problem is that of determining $P(E|F)$, where E is the event that the probability of success in a single trial x is greater than u and less than υ and F is the event that X successes occurred in n trials.

It may be easy to comprehend the problem Bayes attempted to solve, but his solution is very difficult to read. However, in the process of his proof, Bayes invented a clever geometric model based on de Moivre's use of area in his analyses. The model involved the rolling of two balls on a flat rectangular surface to determine the two events. Bayes's physical model has the characteristic of symmetry that Bernoulli's pebble experiment does not. The symmetry is one of the reasons that

* **Inverse probability.** A form of probabilistic reasoning that reasons from observed events to the probabilities of the hypotheses that may explain these events.

Bayes had the success with statistical inference and **inverse probability** that had eluded Bernoulli. Bayes's contributions are significant. He was the first to derive the binomial distribution curve and obtain all its properties. Bayes's formula provided a partial answer to the question of conditional statistical reasoning, and his conclusions were accepted by Pierre-Simon de Laplace (1749–1827) in a 1781 memoir. In 1854, Bayes's results were challenged by George Boole (1815–64), and since then have been subject to controversy.

BUFFON'S NEEDLE EXPERIMENT

➢ **Differential calculus.** See *Calculus*.

➢ **Integral calculus.** See *Calculus*.

Comte de Buffon (Georges-Louis Leclerc, 1707–88), was born in Montbard, France, to a wealthy family, and was educated at the Jesuit college in Dijon. He became the leading theoretical biologist of his time. Buffon enjoyed studying mathematics as a student, and it led him to become a scientist. Buffon introduced **differential** and **integral calculus** into probability theory and became a well-known example of the eighteenth century's natural scientists who applied probabilistic ideas to their research. In his scientific observations, he would use the following type of probabilistic reasoning. For example, suppose that the movement of any one of the planets is equally likely in either direction, east to west or west to east. Then the probability that all six planets would rotate around the Sun in the same direction is small, $(1/2)^6 = 1/64 = 0.015625$. He observed that the six planets all rotated in the same direction. Then he said that this event is not random but subject to some law that should be determined.

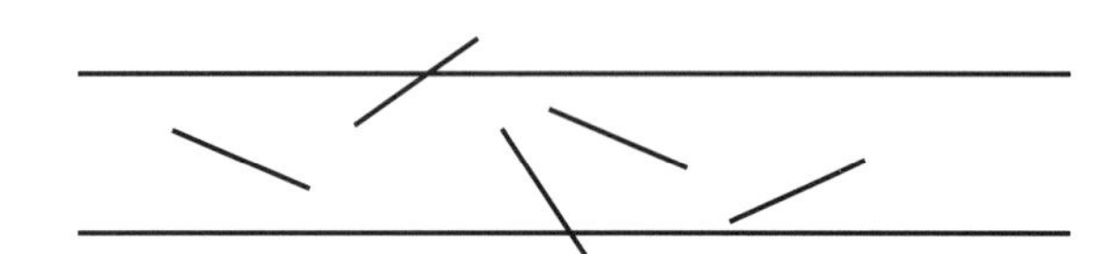

Figure 1. Buffon's needle problem experiment.

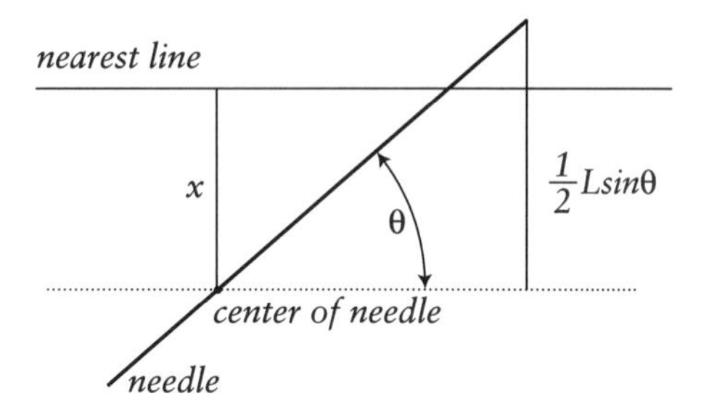

Figure 2. Buffon's needle problem solution.

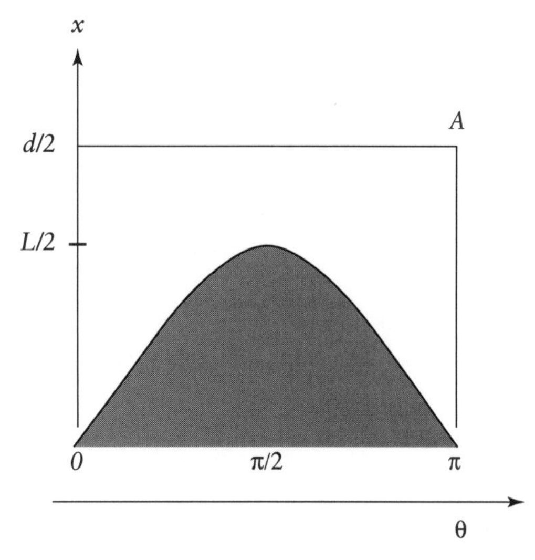

Figure 3. Buffon's determination of the desired probability in the needle problem.

Buffon's famous coin and needle problems are considered to be among the first problems in geometric probability. He used a statistical experiment to estimate π, a number that, as Petr Beckmann (dates unknown) states in *A History of π,* shows up in all branches of higher mathematics and quite frequently in probability theory. The problem can be stated as follows. Suppose that a plane is covered with a series of parallel lines d units apart. Toss onto the plane a needle L units long. The length L of the needle is less than d. Determine the probability that the needle will intersect one of the lines on the plane (Figure 1, 2, and 3).

Buffon showed that the probability that a tossed needle would cross a line is $2L/\pi d$. He let x be the vertical distance of the center of the needle to the nearest line and θ be the angle of its position, as shown in Figure 2, which shows that the solution of the problem is equivalent to finding the probability that $x < 1/2L \sin \theta$. Figure 3 shows how to determine that probability. The area under the sine curve is determined by a simple integration. The desired probability is the ratio of the shaded area under the sine curve to the entire rectangle.

Buffon also tried to verify this result by experiment, throwing a needle many times onto lined paper. However, the problem and solution were largely ignored until the mathematician Pierre-Simon de Laplace (1749–1827) gave it his attention and a new perspective.

LEGENDRE AND THE METHOD OF LEAST SQUARES

There are few details of Adrien-Marie Legendre's early life. His family was wealthy, and he had an excellent education in mathematics and physics at the Collège Mazarin in Paris. For five years (1775–80), he taught with Laplace at the Military Academy in Paris. Legendre was appointed to two French commissions to work on measurements of Earth. One commission worked with the Royal Observatory at Greenwich, and the other was assigned to measure the meridian arc from Barcelona to Dunkirk. That arc was the basis on which the length of the meter was defined. This is important to the history of statistics because it produced Legendre's method of least squares. It first appeared in 1805 in the appendix of *New Methods for the Determination of the Orbits of Comets* and is a model of clarity. The three problems from astronomy and geodesy mentioned earlier motivated the method of least squares. Astronomers and geographers were concerned with the errors that existed in their measurements and wanted methods to assess their magnitude. They also wanted mathematical models that would predict as accurately as possible the relationships among the quantities they observed. The models were usually linear, and that is where Legendre began his description of the method of least squares.

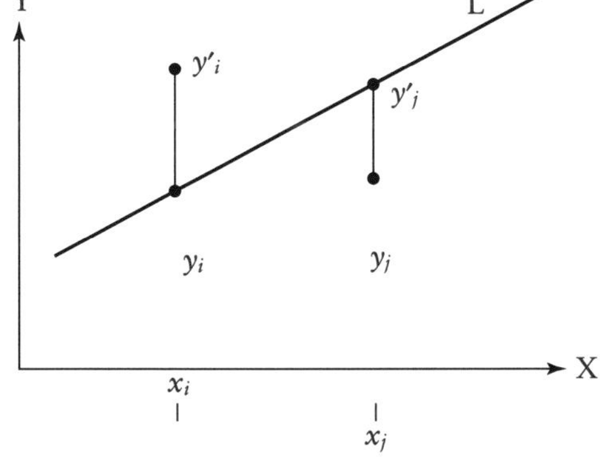

Figure 4. Illustration of Legendre's least squares concept in the case of two variables.

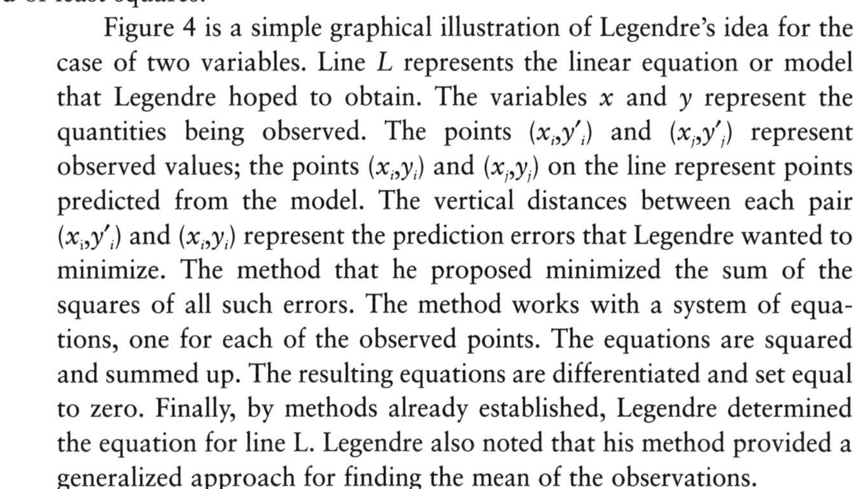

Figure 4 is a simple graphical illustration of Legendre's idea for the case of two variables. Line *L* represents the linear equation or model that Legendre hoped to obtain. The variables *x* and *y* represent the quantities being observed. The points (x_i,y'_i) and (x_j,y'_j) represent observed values; the points (x_i,y_i) and (x_j,y_j) on the line represent points predicted from the model. The vertical distances between each pair (x_i,y'_i) and (x_i,y_i) represent the prediction errors that Legendre wanted to minimize. The method that he proposed minimized the sum of the squares of all such errors. The method works with a system of equations, one for each of the observed points. The equations are squared and summed up. The resulting equations are differentiated and set equal to zero. Finally, by methods already established, Legendre determined the equation for line L. Legendre also noted that his method provided a generalized approach for finding the mean of the observations.

Carl Friedrich Gauss (1777–1855) also laid claim to the method of least squares, saying that he had been using it since 1795. This claim deeply offended Legendre, and the two mathematicians fought over the primacy of this discovery. What is clear from Legendre's description of the method is that he was well aware of its generality. Although Gauss's work surpassed that of Legendre, he did not communicate it before 1805, and he did not seem to indicate that he saw the potential generality of the method as Legendre had. Perhaps the edge should be given to Legendre.

The method of least squares was a standard technique in astronomy and geodesy on the European continent by 1815, and in England by 1825. By the latter half of the nineteenth century, it would be adopted in biology and in the study of human characteristics. The method of least squares would serve to underpin modern-day concepts of **regression** and would eventually lead to modern statistical techniques in multivariate and factor analysis.

* **Regression.** The mathematical model that is used to represent the dependent relationship of a variable in terms of the other variables in a statistical study.

Nineteenth Century

The stage was now set for the theoretical synthesis by Laplace and Gauss. Bernoulli's theorem and the method of least squares opened the way for probability to be applied to statistics. The motivational emphasis for developing probability theories and statistical methods switched from dice and gambling problems to the problems of error in large-scale measurements in astronomy and Earth sciences.

LAPLACE-GAUSS SYNTHESIS

In the first 20 years of the nineteenth century, the theoretical foundation for a theory of errors was laid down by Laplace, Gauss, and Legendre. Astronomers eagerly exploited the principles and the least squares methods these mathematicians gave them. Paper after paper in astronomy began with a pronouncement that the work was based on the ideas of these celebrated mathematicians. It was about 1810 when the ideas of Laplace and Gauss converged to the synthesis that was necessary for modern science. By that time, Laplace had produced an impressive set of probability techniques for the theoretical analysis of binomial distributions. He also contributed useful methods of combining measurements in complex situations. His work was a powerful current in a stream of efforts that reached practical scientific application with Legendre's 1805 publication of the least squares method. However, there was a missing link between his two lines of work. That link would be provided by Gauss in 1809.

Gauss was born in Brunswick, a duchy that is now a part of Germany. Throughout his school years and beyond, he discovered many mathematical theorems on his own. One of these was the method of least squares. A few years before 1801, Gauss had come to know the astronomer Franz Xavier von Zach (1754–1832). In June 1801, Zach published the orbital positions of Ceres, a new "small planet" that was discovered in January by the Italian astronomer Giuseppi Piazzi (1746–1826). Piazzi was able to observe only 9 degrees of its orbit before it disappeared behind the Sun. Zach asked Gauss and several others to predict Ceres' position. Gauss's prediction differed greatly from the others, and when Ceres was rediscovered by Zach on December 7, 1801, it was almost exactly where Gauss had said it would be. Gauss did not disclose his methods at the time but said that he had used a least squares approximation method.

Gauss discussed statistical estimators in his *Determination of the Accuracy of Observations*, published in 1816. In 1823 (and again in 1828 with a supplement), he devoted *Theory of the Combination of Observations with Small Random Errors* to mathematical statistics and the least squares method. This treatise gave his most complete exposition of the theory of errors, which was his primary contribution to probability. His methods are still in use today, and the normal law is commonly referred to as the *Gaussian law of error*. However, his work on probability was scattered throughout his many writings, and he never brought his ideas together in one volume as Laplace did.

Laplace's father expected him to have a career in the church. However, with support from two perceptive teachers at the University of Caen, Laplace realized his talent for and love of mathematics. He left the university after two years with a letter of introduction to Jean le Rond d'Alembert (1717–83). He was soon appointed professor of mathematics at the Military Academy in Paris. At the same time, he quickly began producing the many noteworthy mathematical papers that have earned him his high position in the history of mathematics and science. In 1795, Laplace taught probability at the Normal School that was founded for educating schoolteachers. Although the school survived for only four months because the 1200 pupils found the level of teaching well beyond them, Laplace later wrote up the lectures of his course as *A Philosophical Essay on Probabilities*. It was published in 1814 and greatly influenced the intellectual climate in Europe at that time.

It seems that after 1786, Laplace focused on astronomical studies, and probability theory took a back seat. After he completed the fourth volume of *Celestial Mechanics* in 1805, he returned to probability and in 1810 presented a major result known in modern statistics as the *central limit theorem*. His result generalized de Moivre's limit theorem. The theorem in the case of a finite population can be stated in simple modern terms as follows. Draw a simple random sample of a given size n from a population with mean μ and standard deviation σ. For large n, the sampling distribution of the sample mean $\bar{x}$ approximates a normal curve.

In 1809, just before Laplace's paper on the central limit theorem, Gauss published his second book, *Theory of Motions of the Heavenly Bodies*. It was a major treatise that used Laplace's ideas of probability. In the main part of the work, he showed how to estimate and then refine the estimation of a planet's orbit from any number of observations. In his text, Gauss provided a number of important concepts for modern statistics. Among them, he defined a measure of precision, which in today's symbols would be $1/\sqrt{2}\sigma$, where σ represents the standard deviation of the set of measurements. He described briefly the principle of least squares. Most important, he provided the idea that would lead Laplace to complete his theory of probability and present the synthesis in his 1812 treatise.

Gauss's results of 1809 were certainly significant, but he reasoned incorrectly. He argued in a circle, assuming the conclusion he set out to prove. He started his development in much the same way as Laplace had done before him, but then moved in a new direction with his approach to the error curve. This move is what motivated Laplace to envision the synthesis that truly laid down the theoretical foundations for probability and statistics. Laplace had started by reasoning a priori to a specific curve and then used the curve to find a method of combining observations. Gauss turned that process around. Gauss said that at the start, he could state only general characteristics about the error he needed. He stated that the curve should be a maximum at an error value of zero, should be symmetric about the value zero, and should be zero outside the range of possible errors. He assumed that the arithmetic mean of the observations is the error with the most probable value. He then showed that this assumption forced the curve to be normal and gave the curve symbolically in the following notation: $\varphi(\Delta) = (h/\sqrt{\pi})e^{-h^2\Delta^2}$. The constant h is a measure of the precision of the observation, Δ represents the error values, and $\varphi(\Delta)$ is a measure of probable error for a given Δ. Gauss went on to show how this error curve led to the arithmetic mean of the observations as the most probable value for the true value of an observed quantity. Of course, he later realized his faulty reasoning and revised his work. Nevertheless, history shows that his 1809 paper became the celebrated reference on the method of least squares.

Laplace saw past Gauss's faulty reasoning to the idea that would complete his own development of the theory of probability. What Laplace saw in Gauss's work was the connection between his central limit theorem and **linear estimation**. He quickly produced a short supplement in time for it to be included in his 1810 paper, where he referred to Gauss's work and showed how the central limit theorem provided a better basis for Gauss's error curve. Now Gauss's proof of the least squares method made sense. The connection that Laplace made produced a torrent of intellectual activity on the subject in the period after 1810. Laplace went on to show that in certain situations, the least squares estimates were best because they had the smallest probable error. He, Gauss, and scores of other mathematicians and scientists went to work devising methods of determining the accuracy of the estimates. By the time Laplace died in 1827, the theoretical synthesis was well developed. Throughout the middle of the century, numerous researchers continued to present

* **Linear estimation.** An estimation that is a linear function of the observations.

their versions and refinements. Science saw the synthesis of two major conceptions, the use of least squares to combine observations and the use of mathematical probability to make inferences about true values and to determine degrees of certainty for those inferences. The physical sciences now had strong theoretical and practical tools for understanding the natural world.

The first edition of Laplace's *Analytical Theory of Probability* was published in 1812 in two volumes and contained the synthesis that Gauss's 1809 work inspired. It brought together in one place most of the numerous memoirs that Laplace wrote on probability and included new material. Two years later, the second edition came out, its length increased by about 30 percent. *Analytical Theory* may be considered the greatest single work on the subject of probability, but even highly regarded mathematicians of Laplace's time admitted that it was extremely difficult to read. However, many of them, including Isaac Todhunter (1820–84), greatly improved on his proof, extended it, and provided valuable commentary.

The first edition of *Analytical Theory* contained generating functions and approximations to various expressions occurring in probability theory. Laplace defined probability in the second edition and discussed Bayes's theorem and moral and mathematical expectation. He gave methods for finding probabilities of compound events when the probabilities of their simple components are known. He discussed the method of least squares, Buffon's needle problem, and inverse probability. Finally, Laplace provided applications to mortality, life expectancy, and the length of marriages, and examined moral expectation and probability in legal matters. Supplements in later editions of *Analytical Theory* addressed the application of probability to errors of observation. Laplace also applied probability theory to calculate the masses of Jupiter, Saturn, and Uranus, to deal with triangulation methods in surveying, and to determine the meridian of France. Laplace did much of this work between 1817 and 1819, and it appears in the 1820 edition.

EXTENSION OF PROBABILITY AND STATISTICS TO SOCIAL PHENOMENA

In the last part of the nineteenth century, Francis Galton (1822–1911), a cousin of Charles Darwin (1809–82), developed statistical concepts for his work on the inheritance of human characteristics. He was, above all, attracted to the idea that human characteristics were normally distributed. Galton followed the lead of Adolphe Quetelet (1796–1874) in applying the normal curve to social phenomena and human characteristics.

Quetelet learned probability theory from Laplace around 1823. The experience prepared him for the work he was to do at an observatory that he established in Brussels in 1833, where he developed methods for the comparison and evaluation of data. In the middle years of the century, Quetelet turned his probabilistic and statistical reasoning to social issues, steering the course of social science in Europe and influencing intellectual and political thought. Quetelet became the first to use the normal curve for something other than error analysis. He collected and analyzed statistics on crime, mortality, and so on, and devised improvements in census taking. His statistical work on the constancy of crimes provoked debate on free will and determinism.

He took his work further, believing that probability and statistics, and the normal curve in particular, is the basis for studying human physical, mental, and moral characteristics. He proposed some of these ideas in *A Treatise on Man, and the Development of His Faculties* (1835), where he presented his concept of the "average man." He believed that the idea of a central value for measuring physical or mental characteristics in a population was not just an abstract idea but a real property of those in the population. He also believed that measurements of a human trait are grouped around the central value according to the normal curve. This concept, he thought, would help discover differences among populations and so provide a scientific rationale for the ideas of racial differences that were afloat in nineteenth-century Europe. His reasoning can be illustrated by his analysis of the chest size of more than 5000 Scottish soldiers from data that he found in an 1817 issue of the *Edinburgh Medical Journal*. He reasoned that if the chest size of one soldier was measured repeatedly, a set of slightly differing measurements clustering around one size would be obtained. The normal law of error could be applied here in a similar way as for measuring errors in astronomical quantities. He then reasoned that the measurements of all 5000 chests would group around one central size according to the normal curve. This idea was so strong for Quetelet that he concluded that it would be impossible to judge from a clustering of the measurements whether one chest or 5000 chests had been measured. Therefore, as long as care was taken to obtain accurate measurements of individuals in a particular race or population, it would be possible to measure physical and intellectual characteristics for the whole population. This would, of course, lead to comparisons among populations. Such reasoning has led to the infamous "bell curve" debates. [See, e.g., Richard J. Herrnstein (dates unknown) and Charles Murray's (b. 1941) *The Bell Curve, Intelligence and Class Structure in American Life* (1994) and Stephen Jay Gould's revised and expanded *The Mismeasure of Man* (1996), in which he criticizes the arguments in *The Bell Curve*.]

Science and Technology: The Normal Curve

The probability curve $y = (2/\sqrt{2\pi})e^{-x^2/2}$ became known as the *normal curve* in the 1870s. Figure 5 represents the normal distribution curve in standard form, that is, where the mean has been translated to zero and the scores are shown in standard deviation units. The population mean is designated by μ, the standard deviation by σ, and a particular score by x_i. The score x_i is then expressed in terms of standard deviations from the mean by $z_{x_i} = (x_i - \mu)/\sigma$, known as a *z-score*. An area under the bell curve between two z-scores measures the probability or proportion of scores in the interval defined by those z-scores. For instance, the shaded gray area in Figure 5 represents the probability that a score will be between one and two standard deviations from the mean value.

The first appearance of the normal law is attributed to Abraham de Moivre. In 1756, de Moivre produced an English translation of "A Method of Approximating the Sum of the Terms of the Binomial $(a + b)^n$ Expanded into a Series" for the third edition of *The Doctrine of Chances*. This rare document shows that some of the historical roots of the normal distribution belong to games of chance. As probability theory continued to develop into the eighteenth century, many mathematicians and mathematical astronomers contributed noteworthy papers on the law of error to deal with the variable results of astronomical measurements. This work enabled them to assist in the analysis of census data and records of social and economic variables. Different ways of thinking from economics, mathematics, and political statistics converged through those who became multiple experts in these areas. The result was the science of statistics and the towering presence of the normal law at its center. Quetelet's influence was passed on by his great admirer Francis Galton. Expanding the application of statistics and the normal distribution to the social sciences, Galton argued that if they were to be sciences, they must subject their phenomena to measurement. In turn, Galton had a great influence on twentieth-century statistics and its use in education. His publication on the method of correlation seems to have attracted Karl Pearson (1857–1936) to his life work and to the mathematical theory of correlation.

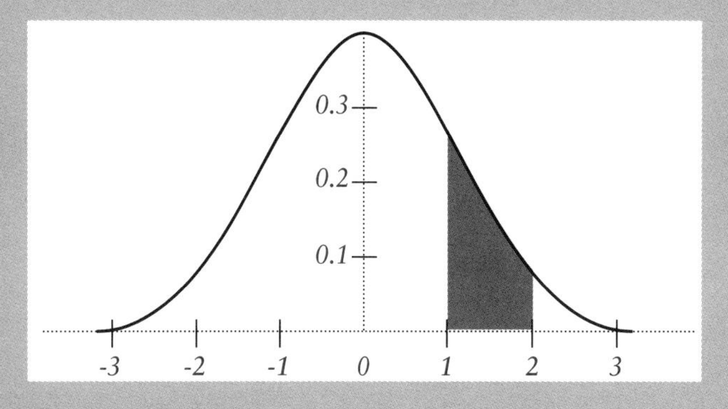

Figure 5. Graphical representation of a probability given the normal distribution.

STATISTICS AND THE SOCIAL SCIENCES

By adopting the new statistical methods and the normal curve in particular, social and psychological disciplines have become known as sciences. Francis Galton had a wide range of scientific interests, and his independent wealth enabled him to pursue them. He was not a strong mathematician, but was good at generating new ideas: He left it to more able mathematicians to work out the details. Many statistical concepts are owed to Galton, among them the **percentile system,** including quartile deviations and the median, the idea of regression analysis, and **correlation.** Through Pearson, Galton's influence extended to the field of psychology and theories of intelligence.

* **Percentile system.** A system that divides the distribution of an ordered set of scores into 100 equal parts.

* **Correlation.** The statistical study of the associations between two or more variables.

Galton's ideas needed to be expressed with mathematical precision to bear fruit as science. The person for that job was the mathematical economist Francis Edgeworth (1845–1926). He translated Galton's conceptions into mathematical form, where they eventually produced more applications than even Galton foresaw. Edgeworth provided the mathematical theory but not the empirical evidence necessary to entice the science community to embrace Galton's visions. This was left for Karl Pearson. Pearson had great ambitions as a scholar and, with the assistance of George Udny Yule (1871–1951), he turned the mathematical theory into methods that sold Galton's ideas to the world. Pearson originated many technical statistical terms. The historian of statistics Helen Mary Walker (1891–1983) lists the origins of approximately 120 technical statistical terms in a brief dictionary in her 1929 text and attributes fully a third of them to Pearson.

Pearson's greatest contributions to applied statistics, in particular to biological problems of heredity and evolution, are found in the 18 papers he wrote from 1893 to 1912, entitled *Mathematical Contribution to the Theory of Evolution.* They include his contributions to **inferential statistics**. For example, Pearson produced a family of curves that he used for describing biological data, and he wanted the answer to the question of how well one of those curves represented the data. He developed a method still in use today, called the ***chi-square test***, to answer his question.

* **Inferential statistics.** The branch of statistics that seeks to answer the question of what a data set implies about more general conditions: for example, what a sample implies about the population it represents.

* **Chi-square test.** A test used in statistics to determine the degree to which an observed set of probabilities fits the theoretical or predicted values.

THE PROBABILISTIC REVOLUTION

As the authors of *The Empire of Chance* (Gigerenzer et al.; see Bibliography) wrote: "For much of its history, probability theory *was* its applications." The world did not see probability theory in its pure mathematical form until the twentieth century. Applications determined the scientific and mathematical formulations of probability, and probability changed the scientific understanding of phenomena across a multitude of disciplines. During the seventeenth and eighteenth centuries, probability grew to embrace problems in law and commerce. The insurance industry benefited through data analysis and inductive reasoning. The revolution spread to sociology, physics, biology, and psychology in the nineteenth century, when probability came to be a major tool of scientific inquiry. In the twentieth century, the revolution invaded the fields of agriculture, economics, medicine, politics, and even sports. Baseball is frequently referred to as "the game of statistics." The writer and baseball historian Ernest J. Lanigan once confided: "All my interest in baseball is in its statistics."

The probabilistic revolution profoundly affected human thinking about the natural world by altering our conception of the physical nature of matter and energy. Isaac Newton (1642–1727) provided the first unified theory of modern physics, which was generalized to provide a deterministic mechanical view of nature. Once certain conditions are known at a given time, all other properties of a physical system can be determined uniquely and precisely in the future and in the past. Probability and statistics are necessary only to improve the precision of the results and to adjust for errors in measurement. Probability was not considered an actual characteristic of matter. The serious questions of the relationship between phenomena observed and the methods used to observe them were postponed until the twentieth century. Until 1860, the classical view in physics centered on the idea that matter was composed of tiny particles known as atoms and molecules. Much of the work at that time was aimed at determining the quantitative and qualitative properties of these particles. However, in 1869, James Clerk Maxwell (1831–79) noted that the velocities of the molecules in a gas are distributed according to the Gaussian law of errors. This observation ushered statistical physics into the laboratory of scientific methods.

The probabilistic revolution arose from gaming theory and from astronomy and spread to the social and economic sciences. However, the interactions among the sciences were not unidirectional. Maxwell and Ludwig Boltzmann (1844–1906), independent of each other, used analogy to the probabilistic reasoning of Quetelet and the historian Henry Thomas Buckle (1821–62) to justify their statistical model for a theory of gases. The authors of *The Empire of Chance* call this "a striking instance of the importance of social science for the natural science."

Twentieth Century

The twentieth century opened with major new ideas in two particular disciplines. At the end of the nineteenth century, Karl Pearson became excited over Galton's ideas of correlation. In 1901, he published a paper in the *Philosphical Magazine* entitled "On Lines and Planes of Closest Fit to Systems of Points in Space." This appears to be the earliest paper on principal component analysis, a form of **multivariate analysis** heavily dependent on the idea of correlation. In 1904, the English psychologist Charles Spearman (1863–1945) used the ideas of correlation to construct a theory of intelligence that continues to have influence in studies of mental ability. Concurrently, he initiated the pervasive multivariable technique known as *factor analysis*, which has found application in many fields.

* **Multivariate analysis.** The study of the relationships in data sets in which many variables are measured on each individual in a sample or population.

Quantum theory, with its statistical and probabilistic nature, revolutionized both the science and the worldview of the twentieth century. The study of the interaction of light and matter led to the idea that matter can be thought of as both a particle and a wave. Max Planck (1858–1947), Albert Einstein (1879–1955), and Werner Heisenberg (1901–76) not only brought profound changes to the prevailing perspectives of physics, but also drew the world's attention to the idea that at many levels, nature and human knowledge operate on a principle of uncertainty. The way in which statistics addressed quantum theory and the **uncertainty principle** is explored in detail below.

* **Uncertainty principle.** The principle that accurate measurement of an observable quantity necessarily results in uncertainties in the observer's knowledge of the values of other observables; also known as the indeterminacy principle.

The twentieth century was also a time for major advances in general statistical methods. Mathematical statistics was born as a separate discipline in the early 1930s. The 1930s also saw the emergence of probability theory as a formal mathematical system. These developments are also discussed in detail in the final section, "The Reign of Statistics."

ADVANCES IN STATISTICAL METHODS: FISHER, NEYMAN, AND PEARSON

Three researchers are credited with the development of the modern practice and theory of statistics: Ronald Aylmer Fisher (1890–1962), Jerzy Neyman (1894–1981), and Karl Pearson's son, Egon Pearson (1895–1980). They put statistics and its applications on firm mathematical ground. Two of the three, Neyman and Pearson, became colleagues and collaborators. Fisher had a long-standing dispute with Karl Pearson that eventually extended to Pearson's son and to Neyman. When Karl retired, his laboratory at University College London was split between Egon and Fisher. Tension between the two sections was the prevailing mood. When Neyman first came to England, he established a friendly relationship with Fisher. However, in 1935, there was a bitter dispute over Neyman's paper on agricultural experimentation, which called into question the effectiveness of tests that Fisher had devised 10 years earlier in *Statistical Methods for Research Workers.* Fisher took great offense at Neyman's analysis of his work, and thereafter, the tense atmosphere between the two departments became hostile.

Fisher earned his bachelor's degree in astronomy from Cambridge in 1912. He developed an interest in error theory in astronomical observations that led him to investigate statistical problems. In 1915, he published a paper on the distribution of the sample correlation coefficient, a work that initiated studies of the exact distributions of various sample statistics. He was greatly aided by his impressive geometric intuition, which helped him to arrive at results that were only later proved mathematically. In 1919, Fisher began work as a biologist at the Rothamsted Agricultural Experiment Station, where he made many contributions to both statistics and genetics. He introduced randomization, the analysis of variance, and experimental design procedures into the experiments he conducted. Two of his books, *Statistical Methods for Research Workers* (1925) and *Design of Experiments* (1935), are classics in twentieth-century statistics.

In 1921, Fisher provided criteria for deciding when a sample statistic is a good estimator for a population parameter: It must be consistent, efficient, and sufficient. At the same time, he introduced the concept of likelihood. The likelihood of a parameter is proportional to the probability of the data. Likelihood is a function that usually has a single maximum value, which he called the *maximum likelihood.* Fisher also developed statistical methods suitable for small samples, as William Gossett (1876–1937) had done. Dale Varberg (dates unknown), in *The Development of Modern Statistics*, asserts that Gossett's results were really guesses, and he credits Fisher with rigorously verifying those results around 1926. In 1922, Fisher defined statistics in terms of its purpose, which is the reduction of data. In doing so, he identified three fundamental problems that statistics must address: (1) to identify the type of population from which the data came, (2) to estimate the parameters of the population, and (3) to determine the distribution of the population.

As a young boy, Egon Pearson was aware of his father's work and his efforts to start the journal *Biometrika.* Many of the lectures he attended on astronomy involved him in statistics. In 1921, Egon joined his father's Department of Applied Statistics at University College London and began research on statistics. In 1924, he became an assistant editor of *Biometrika*, and he met Neyman in 1925. In June 1926, he and Neyman undertook their first joint research project. The goal of this project was to develop a mathematical theory of tests that could be used to minimize the number of incorrect conclusions regarding the hypotheses being tested. During this period, Pearson corresponded with Gossett, whose ideas were significant to his work with Neyman. He also revised his father's two-volume work, *Tables for Statisticians and Biometricians*, completing Volume 1 in 1954 and Volume 2 in 1972. The tables are seen as models of their kind.

Neyman studied at Kharkov University in the Ukraine, where he met Sergey Bernstein (1880–1968), who influenced him to read Karl Pearson's *Grammar of Science.* In 1925, he was awarded a fellowship to work with Karl Pearson in London. His interest in statistics was ignited anew by Karl's son, and they soon began their fruitful collaboration, which lasted from 1928 to 1938. Neyman and Pearson developed the logical foundation and mathematical rigor for the theory of testing that it had lacked previously. Their work was disputed by some mathematicians, including Fisher, but eventually their ideas were accepted.

In 1927, Neyman returned to Poland, and by 1928, he had managed to set up the Biometric Laboratory for Experimental Biology in Warsaw. It was extremely difficult

* **Survey sampling.** A systematic sampling procedure in which all elements in a population are placed on the same footing.

to obtain support from the Polish government, which was deep in financial crisis. Eventually, he left Poland for England and a post in Egon's laboratory, until 1938, when he immigrated to the United States. He settled at the University of California at Berkeley for the remainder of his life. Neyman's work on hypothesis testing, confidence intervals, generalized chi-square testing, and **survey sampling** revolutionized statistics. His efforts led him to contribute significantly to many fields, including biology, cosmology, medicine, survey sampling, and weather modification.

The two important papers that Neyman and Pearson wrote on hypothesis testing in 1933 were "On the Problem of the Most Efficient Tests of Statistical Hypotheses" and "The Testing of Statistical Hypotheses in Relation to Probabilities a Priori." In particular, Neyman and Pearson addressed the types of error that are encountered in hypothesis testing. Error can occur in two ways. An error occurs if the hypothesis being tested is rejected but is actually true. Neyman and Pearson called this a *type I error*. The other error, a *type II error*, occurs when the hypothesis is accepted but in reality is false. The relationship between these two types of error is complicated, but Neyman and Pearson were able to produce a guiding principle for handling these errors. They said that among the tests that have the same type I errors, one should choose a test for which the size of the type II error is as small as possible. This guiding principle and the related concept of a power function became important statistical concepts.

QUANTUM MECHANICS AND UNCERTAINTY

Quantum theory was developed to deal with the properties of very small quantities: molecules, atoms, and elementary particles. With its uncertainty principle and the statistical nature of its predictions, quantum theory shook the prevailing belief in scientific determinism. Classical physics up to that time had considered energy to be a continuous phenomenon and assumed that matter occupied a very specific region in space and moved in a continuous manner. Quantum theory holds that energy is emitted and absorbed in tiny, discrete packets of energy called *quanta*, which behave in some respects like particles of matter. However, sometimes these particles seem to have wavelike properties. They appear not to be localized in a region but rather to be somewhat spread out [see *Atomic and Nuclear Physics*].

Quantum theory can be traced back to 1859 and a theorem that Gustav Robert Kirchhoff (1824–87) proved about blackbody radiation. A *blackbody* is an object that absorbs all the energy that falls upon it: Because it reflects no light, it appears black to an observer. A blackbody is also a perfect energy emitter, and Kirchhoff proved that the energy emitted is a function solely of the temperature and the emitted energy's frequency. Before 1990, several scientists proposed solutions for the blackbody spectrum, some incorrect, some not totally adequate. In 1900, Planck guessed the correct formula for the function by making the unprecedented assumption that the total energy is made up of quanta. In 1905, Einstein proposed that radiation itself is quantized according to Planck's formula, and he used probabilistic notions to explain the **photoelectric effect**.

* **Photoelectric effect.** The liberation of electrons by electromagnetic radiation falling on a substance.

* **Photon.** An elementary particle composed of a quantum of electromagnetic radiation.

* **Bose-Einstein statistics.** The statistical mechanics of a system of indistinguishable particles for which there is no restriction on the number of particles that may exist in the same state simultaneously.

In 1924, Niels Bohr (1885–1962), working with several colleagues, made proposals regarding the interaction of light and matter that rejected the notion of a **photon**. This stimulated significant experimental work but produced a number of paradoxes, one of which puzzled Einstein: How does the electron know when to emit radiation? In the same year, another fundamental paper on these issues appeared. Satyendranath Bose (1894–1974) sent his manuscript, *Planck's Law and the Hypothesis of Light Quanta*, to Einstein, who immediately saw its importance and arranged for its publication. Bose's work and Einstein's extension of it led to the development of the quantum statistical concepts of **Bose-Einstein statistics** and

Fermi-Dirac statistics. Bose-Einstein statistics govern systems of quantum particles called *bosons*, which have the property that an unlimited number of bosons can exist in exactly the same state, and Fermi-Dirac statistics apply to systems of elementary particles called *fermions*, which have the property that no two fermions can exist in the same state.

The final mathematical formulation of quantum theory, quantum mechanics, was developed during the 1920s. Louis de Broglie (1892–1987) proposed in 1924 that light waves sometimes show particlelike properties, as in the photoelectric effect, but also show wavelike properties. His propositions were confirmed by experiment, and two different formulations of quantum mechanics followed. Erwin Schrödinger's (1887–1961) wave mechanics (1926) involves a mathematical wave function, which is related to the probability of finding a particle at a given point in space. Werner Heisenberg's (1901–76) matrix mechanics (1925) made no mention of wave functions but was shown to be mathematically equivalent to Schrödinger's theory. Actually, 1926 turned out to be an important year. Schrödinger also gave his equation for the hydrogen atom, and Paul Dirac (1902–84) provided the complete derivation of Planck's law. Max Born (1882–1970) abandoned the causality of traditional physics. Speaking of collisions, Born wrote that physics no longer asks the state after a collision but how probable is a given effect of the collision.

In 1927, Heisenberg pronounced his *uncertainty principle*, which places an absolute theoretical limit on the accuracy of certain measurements. It states that one cannot measure both the position and the velocity of a particle exactly. For example, light must shine on a particle if one is to see it. But Planck's theory does not allow arbitrarily small amounts of light, and at least one quantum must be used. This will disturb the particle and change its velocity in a way that cannot be predicted. To measure the particle's position accurately, light of short wavelength must be used. Again, Planck's theory shows that quanta of these short waves have higher energies than those of visible light, so they will disturb the velocity even more. The more accurately one tries to measure the position of a particle, the less accurately one can know the velocity, and vice versa. Heisenberg's principle sums this up in an equation which states that the uncertainty in the position of a particle multiplied by the uncertainty in its speed is always greater than a quantity called Planck's constant divided by the mass of the particle.

Not everyone accepted the uncertainty principle. Einstein was a most outspoken opponent. He challenged Bohr with one of his famous "thought" experiments in 1930. Einstein suggested a box filled with radiation with a clock fitted in one side. The clock is designed to open a shutter and allow one photon to escape. If the box is weighed again some time later, the photon energy and its time of escape can both be measured with arbitrary accuracy. Within a day, Bohr had a response. The mass is measured by hanging a compensation weight under the box. This is turn imparts a **momentum** to the box, and there is an error in measuring the position. Time, according to special and general **relativity theories**, is not absolute, and the error in the position of the box translates into an error in measuring the time. Although Einstein was never happy with the uncertainty principle, he was forced to accept it after Bohr's explanation.

* **Momentum**. A property of a moving body equal to its mass times its velocity.

* **Relativity theory**. A theory that describes the effects on mass, time, and distance when physical objects move at very high velocities or when a gravitational field influences them.

THE REIGN OF STATISTICS

In the twentieth century, statistics began to reign over determinism in the physical sciences. In biology, chance rules the evolution of life. Statistics presides over educational research; no study on pedagogy or learning is deemed worthy unless supported by data processed and analyzed by statistical methods. Statistics figures in economics, business, and politics, as witnessed by the daily appearance of charts and graphs in newspaper

Axiomatic Mathematical Foundations of Probability

In the twentieth century, a number of attempts were made to formalize probability and produce a mathematical foundation for it. Sergey Bernstein (1880–1968) attempted to axiomatize the theory in 1917 but did not succeed. Bernstein is known for solving two of David Hilbert's (1862–1943) famous 23 problems. Hilbert presented them in 1900 as important mathematical problems to be solved in the coming centuries. For the nineteenth and twentieth problems in the list, Bernstein provided analytical solutions to a class of equations known as *elliptic differential equations* (differential equations include the first or higher derivatives of the function whose solution is being sought).

In 1927, Aleksandr Yakovlevich Khinchin (1894–1959) published *Basic Laws of Probability Theory.* Richard von Mises (1883–1953) developed foundations for probability from a limiting frequency approach. Émile Borel (1871–1956) regarded probabilities as analogous to the measurement of physical magnitudes and said that they can be measured only within a certain approximation.

Andrey Kolmogorov was influenced by a number of outstanding mathematicians early in his education and showed remarkable talent as a researcher. He published eight papers in 1925, all written while he was still an undergraduate. His first paper on probability appeared in that year, and in 1933 he published his monograph *Foundations of Probability Theory*. In this work, Kolmogorov constructed probability theory in a rigorous way from fundamental axioms. He succeeded in living up to his words on page 1 in that text: "The theory of probability as [a] mathematical discipline can and should be developed from axioms in exactly the same way as Geometry and Algebra." In doing so, Kolmogorov solved Hilbert's sixth problem, "To treat in the same manner, by means of axioms, those physical sciences in which mathematics plays an important part; in the first rank are the theory of probabilities and mechanics."

In addition to setting out the axioms upon which probability theory rests, Kolmogorov made a substantial contribution to the concepts of mathematical expectation: in particular, conditional mathematical expectations. He set down the axioms in paragraph 1 of his monograph and proved them consistent. However, he also stated that the system of axioms is not complete because different fields of probability have to be studied for the various problems in the theory of probability. He also discussed the relationship of the axioms to experimental data and gave an empirical deduction of the axioms.

accounts. Statistics has dominion over public health policies and medical practices, winning the case against smoking and deciding on the efficacy of various cancer treatments. Statistics plays a prominent role in the testing of persons for placement in schools and colleges and for educational opportunities such as scholarships. Statistics is found with increasing frequency in the legal process, where statisticians and others who use statistical methods serve as consultants and expert witnesses.

Beginning in the latter half of the twentieth century, the computer has been supporting the reign of statistics through its ability to handle large quantities of data quickly and efficiently. Data collection, measurement of variables, descriptive statistics, inferential prediction, and structural modeling of the relationships among variables require enormous amounts of organizational and computational labor, especially for large data sets involving multiple variables. For instance, in the 1930s, Leon Thurstone (1887–1955) parceled out the calculations required by the centroid method of factor analysis to a cadre of human calculators. He then reassembled the results for analysis and interpretation. The process took weeks and months. Today, statistical programs churn out a factor analysis in a matter of minutes. As our computational power continues to grow, it seems certain that probability and statistics will continue their reign.

Bibliography

Beckmann, Petr. *A History of π*. New York: St. Martin's Press, 1971.

David, F. N. *Games, Gods and Gambling: The Origins and History of Probability and Statistical Ideas from the Earliest Times to the Newtonian Era*. New York: Hafner, 1962.

Fisher, Ronald Aylmer. *Statistical Methods for Research Workers*. 12th ed., rev. Edinburgh, Scotland: Oliver and Boyd, 1925.

———. *The Design of Experiments*. New York: Hafner, 1953.

Gigerenzer, G., Z. Swijtink, T. Porter, L. Daston, J. Beatty, and L. Krüger. *The Empire of Chance*. Cambridge: Cambridge University Press, 1989.

Gould, Stephen Jay. *The Mismeasure of Man*. New York: W. W. Norton, 1996.

Herrnstein, Richard J., and Charles Murray. *The Bell Curve, Intelligence and Class Structure in American Life*. New York: Free Press, 1994.

Laplace, Pierre-Simon. *Philosophical Essay on Probabilities*. Translated from the 6th French edition by F. W. Truscott and F. L. Emory, with an introductory note by E. T. Bell. New York: Dover Publications, 1951.

Legendre, Adrien-Marie. *New Methods for the Determination of the Orbits of Comets*. Paris: Courcier, 1805. A portion of the appendix, "On Least Squares," was translated by H. A. Ruger and H. M. Walker in D. E. Smith's *A Source Book in Mathematics*. New York: McGraw-Hill, 1929.

Maistrov, L. E. *Probability Theory: A Historical Sketch*. Translated by Samuel Kotz from *Teoriia Veroiatnostei Istoricheskii Ocherk*. New York: Academic Press, 1974.

Stigler, Stephen M. *The History of Statistics: The Measurement of Uncertainty before 1900*. Cambridge, Mass.: Harvard University Press, 1986.

———. *Statistics on the Table: The History of Statistical Concepts and Methods*. Cambridge, Mass.: Harvard University Press, 1999.

Todhunter, Isaac. *A History of the Mathematical Theory of Probability from the Time of Pascal to That of Laplace*. London: Macmillan, 1865. Reprint, New York: Chelsea, 1949, 1965.

Walker, Helen Mary. *Studies in the History of Statistical Method: With Special Reference to Certain Educational Problems*. Baltimore: Williams & Wilkins, 1929.

Systematics

Joe Cain

Open any biology textbook and names jump out. Some are familiar: rice, dog, virus. Others are formal and imposing: hydrozoa, Hominidae, *Homo sapiens*. Names provide an information infrastructure within the life sciences and compress large amounts of information into compact bundles. Underlying all names are systems for organizing nature. These systems guide biologists as they make choices about what names to use. Systematics is the study of these systems.

Key Distinctions in Ideas about Systematics

Several key distinctions are important to an understanding of the history of the ideas involved in the study of systematics. One such distinction is that between the sister concepts of taxonomy and systematics. Crudely, systematics involves systems—the overarching frameworks used to describe relationships and patterns. It focuses on the rules, plans, or theories governing the construction of categories. By contrast, taxonomy involves the day-to-day process of identifying what a thing is and placing it within any one system. Systems are at the heart of all taxonomy, and taxonomists will disagree (sometimes sharply) about which system is best.

Another key distinction separates artificial and natural systems. Some taxonomists want their systematics simply to provide a quick means for storage and retrieval. They care little about why a system was constructed, only that it stores things efficiently. Such systems are understood to be artificial because their rationale is more or less arbitrary and they use distinctions decided by convention alone. Their inventors simply chose a system that seemed a useful and reasonable instrument, then moved ahead.

Extreme kinds of artificial systems are easy to imagine. Museum objects are cataloged by their date of acquisition. Patent numbers are determined by their sequence of award. Each system provides easy retrieval of information. Importantly, artificial systems are judged on pragmatic criteria, such as efficiency of storage, speed of retrieval, or value for identifying conventional associations. Although artificial systems may be preferred for specific uses, they cannot be preferred as inherently better.

In contrast, those building natural systems seek something more: arrangements that reflect real underlying patterns. They want systems to reveal the world's architecture "as it really is," and they speak more of "uncovering" categories than of creating arbitrary divisions. Those proposing natural systems usually claim special status. The underlying realism, they argue, makes their system inherently better than mere convenient artificial systems. A core theme in the history of systematics is the search for a natural system.

A good example of a proposed natural system was quinarianism. In the first half of the nineteenth century, British entomologist William Sharp Macleay (1792–1865) claimed that the number five sat at the very center of nature's order. He proposed five primary groups for animals, each with five classes of five orders, then five families, and so on. Geometry gave him a means of expressing similarities and differences down to small details. Quinarianism had a small group of advocates at the time. Some used the same rationale but thought different numbers were the fundamental units. This emphasis on key numbers was not broadly adopted, but it illustrates an attempt to organize nature using "real" and "underlying" features.

Controversies in systematics tend to focus on competing claims to the discovery of natural systems, or on the extent to which taxonomists are forced to compromise the principles of a natural system for the practical realities of clear, useful organization. In practice, artificial and natural systems blur together. Those deeply committed to a natural system regularly accept artificial arrangements as provisional solutions. Moreover, what counts as "natural" varies quite radically between groups of taxonomists. Several examples of these controversies and compromises are presented below.

A third key distinction is between group and rank. Naming creates groups by distinguishing between this thing and that. Groups have boundaries, definitions, and memberships. Groups may also be ranked, ordered using a scheme for priority or inclusiveness. Ranks can be hierarchical: Combining lions and whales into "mammals" not only provides a more inclusive group but has also been used to imply that lions and whales are instances of a more fundamental quality of "mammalness." In the history of systematics, taxonomists have built systems with many different relationships between group and rank. They have also thought a great deal about which kinds of groups and ranks are natural and which are artificial.

Useful Classifications

The most frequently used systems for classifying nature are artificial ones, relying on pragmatic distinctions between useful and useless. This approach is common in the history of medicine and pharmacology, where classifications were used to sort plants by pragmatic criteria such as their curative or pharmacological value. Recorded in the form of catalogs known as *herbals* or *pharmacopoeias*, these classification systems tended to include only local materials of practical value, plus perhaps a few exotics. Usually, such systems encompassed several hundred herbs, shrubs, and trees. These systems used local names, although sometimes they attempted to list synonyms used elsewhere. As practical guides, they normally grouped plants according to the maladies they were meant to treat. Descriptions usually emphasized where to find the plant, useful parts, handling or preparation instructions, and general descriptions of the plants themselves.

Herbals and pharmacopoeias are known from every culture from which written records have survived. Extensive pharmacopoeias are known from several Egyptian dynasties. These mention curatives imported from the Near East and the eastern Mediterranean. Pharmacopoeias from the Babylonian royal library in Nineveh dating from approximately 650 B.C. describe hundreds of plants known to the Babylonians's Sumerian predecessors. The Babylonians worked to assimilate this information with other information about useful plants collected through their own medical traditions. By 500 B.C., practitioners in India had access to herbals listing nearly 700 plants sorted by the disease or condition they might be used to treat. Chinese herbals known from as early as 300 B.C. clearly classify plants according to their pharmacological value. Herbals are also one of the most common types of text surviving from medieval Europe, derived from Greek sources such as Pedanius Dioscorides' (fl. A.D. 50–70) *De materia medica* (c. A.D. 60). When printing was introduced in the fifteenth century, herbals were among the first and most common texts converted to this new format and sold in vernacular languages. Herbals continue to be published today, although the format tends not to be so direct.

Catalogs of plants organized for other purposes (e.g., for basic foods, dyes, and textiles) are known across ancient and medieval civilizations. As with herbals, their underlying classifications emphasized practical value rather than the properties of the objects themselves. The frequent use of pragmatic classifications should be a reminder about the broad context for natural history investigations. People organize nature for many purposes, and scientific classifications exist alongside these alternatives. Distinguishing the scientific systems from these others sometimes destroys the broader point.

Bestiaries were the analog to herbals and pharmacopoeias for information about animals; lapidaries were the same for minerals or rocks. The choice of "animals" was broadly interpreted in bestiaries: Mythological beasts (such as the dragon, unicorn, griffin, and phoenix) and composite monsters (such as the mermaid and ant-lion) sat comfortably alongside common animals (such as the bear, bull, boar, cat, deer, dog, fox, goat, hedgehog, and horse) and beasts that were exotic to Europeans (elephant, lion, camel, and ape).

Medieval bestiaries had as their basic core some version of ancient Greco-Roman catalogs. Best known was the *Physiologus* from an unknown author before A.D. 200. The number of creatures described in bestiaries grew with increasing time and distance away from those Mediterranean experiences. Information was sometimes based on observations and sometimes copied uncritically and haphazardly from ancient texts or repackaged from fantasy and fable. Names and the relative order of presentation changed, usually for no obvious reason. Supplemental information was added in an unsystematic fashion. Bestiaries were like scrapbooks. Lapidaries followed the same pattern.

During the Renaissance, naturalists dismissed medieval bestiaries and lapidaries for their poor quality. These naturalists made reputations for themselves by constructing alternative catalogs of nature derived only from "reliable" reports or first-hand observation. This focus on accuracy obscured two aspects of how bestiaries and lapidaries functioned in medieval literature. First, medieval authors gave priority to ancient texts over direct observations because their notion of *fact* was intimately tied into the authority of those texts. For complex reasons, this textual authority trumped eyewitness reports. Renaissance naturalists were trying to shift the relative importance of texts and observations, so they strongly criticized the older style.

More important, the underlying organizations of bestiaries had little to do with the actual animals that these books described. To medieval scholars, creatures in the bestiaries functioned as vehicles for moral and allegorical messages transmitted through them. For instance, information about foxes focused attention on personality traits that people associated with this animal (such as cunning and cleverness, as in "sly like a fox"). Falcons and eagles symbolized nobility; lions represented courage; unicorns represented holiness and chastity. This symbolism functioned as a shorthand—reminders of moral duties and a moral code—in religious and moral education. It also formed part of the complex language of heraldry. Bestiaries functioned as storage systems for moral tales, not natural history.

Encyclopedic Natural History

Systems grounded in practical needs (physical or moral) can be compared with those grounded in a desire for comprehensive coverage. Eighteenth-century French philosophers sought systematic collections of knowledge to serve as a monument to reason. Calling themselves *encyclopadists*, they published the *Encyclopédie, ou dictionnaire raissoné des sciences, des arts et des métiers* (Encyclopedia, or Systematic Dictionary of Sciences, Arts, and Trades, 1751–72). Scottish philosophers followed suit with the *Encyclopaedia Britannica* (1768–71). Many others followed over the next century.

Attempts at comprehensive collection did not begin with the encyclopedists; they are known in every literate culture. The Romans set the model for later Europeans. In natural history, the oldest known encyclopedia is the *Naturalis historia* (Natural History, A.D. 79) of Pliny the Elder (c. A.D. 23–79). Pliny assembled knowledge about the known world into themes such as anthropology, botany (which included agriculture, forestry, and horticulture), zoology, and mineralogy. He divided zoology into basic categories: humans, mammals and reptiles, fish and other marine animals, birds, and insects. Most of his zoological information came from Aristotle (384–322 B.C.), most of the botany from Theophrastus (371–287 B.C.); both are discussed below.

Pliny's *Natural History* embodies a Roman tradition of compilation and presents a mountain of facts—more than 33,000 items—arranged loosely within its general categories. For Pliny, *comprehensive* meant the exhaustive collection of statements from all available sources. He was uncritical in organizing information within categories. Information was not ranked, and the arrangement was haphazard. Instead of synthesis, Pliny's goal was accumulation, producing a kind of one-stop shop for all recorded knowledge. To create *Natural History*, he surveyed more than 2000 manuscripts to extract facts and observations.

Science and Society: Art and Renaissance Natural History

Renaissance encyclopedias showed their real innovation in illustrations. This shifted illustration from idealized and imaginary images toward realism. The goal was accuracy, although this was manifested in several different ways.

One early example was *Herbarum vivae eicones* (Living Portraits of Plants, 1530–36) by Otto Brunfels (1489–1534). His illustrator, Hans Weiditz (fl. 1516–22), produced images drawn directly from nature. Weiditz's technique produced exquisite detail and accuracy. Ironically, he took literalism too literally. His engravings represented plants exactly as he saw them, blemishes and all—broken stems, withered leaves, and bits eaten away by insects. Critics said this was too real and failed to provide a sense of the ideal or typical.

Leonhard Fuchs (1501–66) also produced illustrated encyclopedias, such as his *De historia stirpium* (History of Plants, 1542). Compared with Weiditz's literalism, Fuchs's engravings extrapolated to an ideal state. His images showed plants as a composite, placing everything necessary for identification on a single engraving. Typically, this included showing several angles of view for key parts and drawing key parts as they appeared at several seasons of the year or in several habitats.

This approach set a new standard in scientific illustration. It was made possible not only by a change in thinking about the role of images, but also by technical developments in printing, such as the development of engraving with copper plates rather than wood blocks.

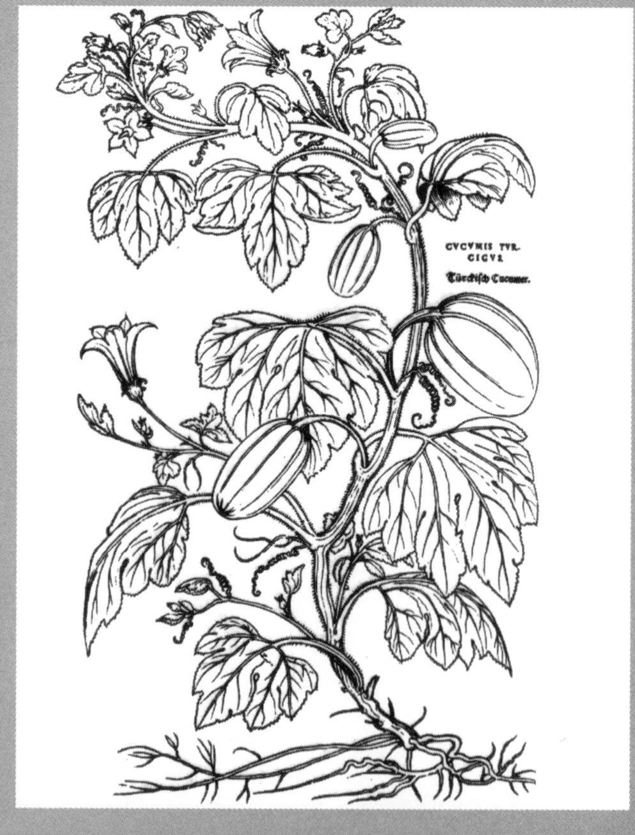

Figure 1. Fuch's illustration of the vegetable marrow, a relative of the pumpkin. The binomial name for this plant, Cucumis turcicus, *came only after Linnaeus. Note the effort to depict the plant as it might appear in nature. From Leonard Fuchs,* De historia stirpium *(Basel, 1542).*

Pliny's encyclopedia was widely copied through the medieval period, when scholastic tradition gave authority to the ancients in both content and method. The same can be said for herbals, such as Dioscorides' *De materia medica*, or the bestiary *Physiologus*. As mentioned above, reformers in the Renaissance relied more heavily on direct experience than on information found in texts from ancient authorities. Regardless, most Renaissance systems followed the ancient model of the encyclopedia: comprehensive data collection instead of synthesis or analysis.

For instance, although the German naturalist Otto Brunfels (1489–1534) extended the herbal's coverage and redefined its accuracy using direct observations, his organizing system preserved ancient priorities for cataloging, as shown in *Herbarum vivae eicones* (Living Portraits of Plants, 1530–36). In his training, Brunfels prepared for scholastic medicine. He studied in the herbal tradition. His book began as an edition of *De materia medica*, with additions from local fauna. Respecting the authority of classical sources, Brunfels added new plants cautiously and referred to them symbolically as *orphans*. The catalog approach simply followed the long-established tradition. In his *De historia stirpium* (History of Plants, 1542), the German Leonhard Fuchs (1501–66) furthered this trend toward comprehensive encyclopedias with increasingly realistic information. His descriptions were based on firsthand knowledge of specimens. His illustrations provided important taxonomic information and were meant to substitute for textual descriptions.

The Swiss Konrad Gesner (1516–65) did the same for animals in his encyclopedia *Historiae animalium* (History of Animals, 1551–58). Gesner included material on many European animals that were not known to classical writers. He also included information brought back from explorations of Africa and the Americas. Nevertheless, Gesner was a traditional encyclopedist in his organizational system. He recorded every bit of information, from whatever sources he could assemble. Following Pliny's style, his catalog provided vernacular names, habitats, diseases, behaviors, diet, practical uses, folklore, fables, emblems, relations to heraldry, and nearly anything else he had heard about a particular animal.

System Building by the Ancients: Aristotle and Theophrastus

Herbals, bestiaries, and encyclopedias used organizational systems defined sometimes by practical criteria (useful for one task or another), sometimes by alphabetical arrangement of names, sometimes at random. Such systems rely on relatively low-level organizational principles or simply copy a pattern from predecessors.

The ancient Greeks provide a crucial point of focus for the creation of natural systems in Western science. Greek philosophers in the fifth and fourth centuries B.C. concerned themselves not only with compiling facts about nature, but also with investigating underlying causes and mechanisms. In this analytical tradition, Aristotle and his disciple Theophrastus developed a framework for systematics that influenced nearly all later generations. Through philosophical reflection and firsthand experience with natural history material, they sought not only natural systems, but also explanations for why the system existed that way. Their conclusions began with the concept of *parts*.

Born in Macedonia, Aristotle moved to Athens in 367 B.C. to study in Plato's (427–348/7 B.C.) Academy, remaining there until just after Plato's death. Although initially sympathetic to Plato's theory of forms, Aristotle rejected this approach in later work. In his alternative, Aristotle turned the relationship around. Ideal forms exist as abstractions, he suggested, only because they are manifested in particular things. The material had primacy over the ideal. This approach developed from Aristotle's theory

Science and Philosophy: Plato's Ideal Forms

Aristotle's views were shaped by his disagreements with the theory of forms developed by his mentor, the Athenian philosopher Plato. Plato drew a distinction between the material world around us and the world of the ideal. The material world was an ephemeral place, constantly changing. The world of ideals contained only eternal and changeless forms. Each form was the pattern of a particular concept: stone, sponge, chair, or dog. Plato suggested that ideal forms also existed for nonphysical things, such as beauty, justice, love, and good. Crucially for Plato, the ideal took precedence over the material. Everything in the material world was an imperfect manifestation of perfect forms.

Because the material world was imperfect, Plato reasoned, knowledge created through studies of physical things was bound to be unreliable. Our senses cannot be trusted. True, changeless knowledge came only from understanding forms. This required contemplation and reason. Plato's approach to definition set the rules for his taxonomy. When constructing definitions, he predictably focused on forms and their distinctions. He might divide the form "animal" into those that are tame and those that are wild, those that live in herds and those that are solitary, those that are aquatic and those that are terrestrial, and so on, until enough divisions were produced to create a class containing only the thing he wanted to define. If these divisions were true—that is, if tame, wild, terrestrial, and so on, were genuine forms—definitions should require knowledge of the complete set of forms in which a thing participates. Importantly, this routine allowed only two-element divisions (i.e., either-or divisions). It was also cumulative but not progressive: Only the sum of divisions was key, not the sequence by which they were made. Plato's definitions produced necessary and sufficient conditions for grouping. For an object to count as being one of a kind of thing, all necessary forms had to be present.

Sometimes Plato's approach is called *essentialism*, because the composite of forms creates a thing's essential nature. More accurately, it is a kind of formalism or idealism.

of cause. He divided causes into four types: efficient, material, formal, and final. Efficient causes describe the agency by means of which something comes to be. Material causes describe the substance that constitutes the thing. Formal causes describe the configuration and conception of a thing as well as the necessary conditions of its existence. For living things, he called this "soul." Final causes describe the reason, goal, or purpose for which a thing exists. To claim complete understanding of a thing, Aristotle argued, required an understanding of its four causes together.

When thinking about a system for organizing animals, Aristotle concentrated on two problems. First, how do we define natural categories for judging similarities and differences? Second, how can single organisms be divided into natural parts? Aristotle approached the latter question first. Building a clear idea of parts could help solve his questions about categories.

But parts is a difficult concept. Organisms can be divided in many ways. For instance, they could be organized into segments randomly. However, segments divided in this way, Aristotle suggested, gave little information about the organisms as living entities. Instead, he proposed, organisms should be divided along the lines of integrated functional systems, such as circulation, propulsion, reproduction, and digestion. Functional systems consisted of parts, and parts were identifiable by their role in the functional system. For example, feathers, wings, and feet were parts for producing propulsion. General terms such as these gave Aristotle a vocabulary for describing functional systems within his theory of causes: Feathers aid flight, feet aid walking. Of all the functional systems, Aristotle ranked reproduction (generation) the highest.

Thinking of organisms as sets of integrated systems (with each system composed of parts), Aristotle was able to balance studies of similarity and difference. Broad similarities in systems and parts bound kinds of organisms together as natural groups called *genera* (*genera* is the plural of *genus*). Singular differences in one system or one part divided animals into different species of the same genus. Genus and species were Aristotle's only two formal categories of classification. Features used to draw distinctions or similarities were called *differentiae*. Generic differentiae had to be naturally divisible: footed into two-footed, four-footed, and lacking feet; or feathers into barbed and not-barbed. In practice, distinguishing one genus from another might require many differentiae to be listed. Once a genus was defined, more differentiae could be provided to distinguish one species of the genus from another.

Aristotle's system for classification began by recognizing broad natural groups, such as birds or fishes, then locating differentiae within their functional systems and parts. Animals were either blooded or bloodless. Birds possessed either barbed or nonbarbed feathers. Plants either produced seeds or they did not. Next, Aristotle would further refine the differentiae: blooded animals were either warm- or cold-blooded, seeds exhibited some of many natural shapes, and so on. Refinement of differentiae continued until only individuals of the same specific kind remained.

Aristotle's system—based on functional systems and their parts, their relative importance, and studies of their different incarnations—brought two innovations to classification systems. First, sequence was important. Abandoning Plato's notion of definition by combining many independent forms, Aristotle insisted on specific order in his definitions so that he could produce ranks. This solved the problem of groupings that defied intuition. For instance, both squids and dogs have complex eyes, but they do not form a natural group because one is blooded, the other is bloodless. In Aristotle's system, bloodedness took priority over vision in the hierarchy of functions, so it took priority in the sequence. Second, Aristotle's sequencing led to stability. For Plato, the discovery of a new form added an entirely new dimension to the matrix of definitions. All classification would need to be redone in light of this new option. Aristotle's approach allowed additional information to be added without disturbing the system as a whole. New differentiae could be added to the end of a sequence. Distinguishing a new species from a known one simply required one more differentia.

Aristotle's approach was followed by Theophrastus. Born on the island of Lesbos, he studied in Plato's Academy. Sixteen years younger than Aristotle, Theophrastus became his disciple, followed him while in exile from Athens, and became head of Aristotle's Lyceum when the master left. Aristotle's influence on Theophrastus was profound. Of his writings that survived into the Renaissance, two are key to systematics. *Causes of Plants* (c. 300 B.C.) considered development, reproduction, and growth. *Enquiry into Plants* (c. 300 B.C.) examined classification and morphology.

As with Aristotle, Theophrastus emphasized functional systems, parts, and differentiae. His identification of plant parts made use of familiar concepts: root, stem, branch, twig, leaf, flower, fruit, and so on. Theophrastus also used absence of a differentia as a property. Thus fungi became a class because they lack the organs of higher plants; seedlessness was a character of ferns. As with Aristotle, Theophrastus's practical procedure was to consider large numbers of parts common to many plants, noting their presence or absence and differences in size, number, quality, and function, as well as their mutual relationships in space.

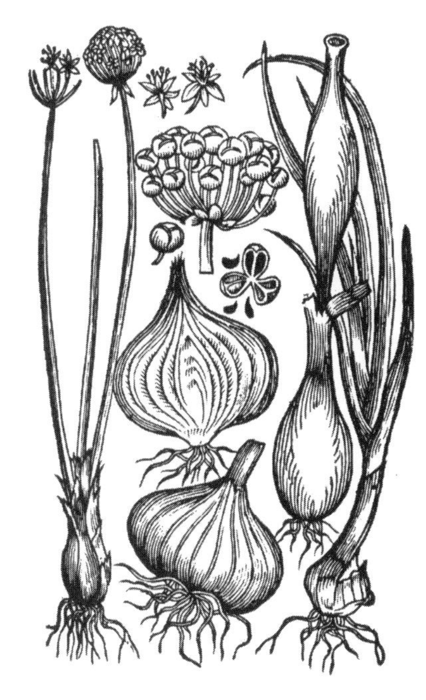

Figure 2. Members of the genus Allium, *which includes onions. This plate combines several species of this genus. From Theophrastus,* Historia plantarum *(1644).*

The analytical approach championed by Aristotle and Theophrastus went into decline during the Hellenistic period (late fourth to late first century B.C.) and into Roman times, coinciding with a general decline in analytical philosophy. Instead, the basic conclusions drawn by Aristotle and Theophrastus were embedded in the encyclopedic tradition as matters of fact. Revivals of ancient texts in early modern and Renaissance Europe brought a revival of the parts-and-differentiae systematics. New Latin editions of Aristotle's *Parts of Animals* and his other biological works, as well as Theophrastus's botanical texts, resurfaced in Rome by the mid-fifteenth century A.D.

Andrea Cesalpino (1519–1603) placed special emphasis on these two ancient philosophers in his *De plantis libri XVI* (Book of Plants, 1583). Cesalpino taught botany, medicine, and philosophy at the University of Pisa, combining knowledge of the classics with an interest in field observations. Cesalpino rejected the pragmatic classifications of herbals derived from Dioscorides, and the alphabetical arrangements of others. These artificial systems, he argued, made virtues of trivial features (such as scent, taste, and utility to humans) or accidental qualities (such as those varying with conditions of soil, climate, and season). Cesalpino argued for a return to Greek traditions and natural systems. He thought systematics should be built on fundamental natures (Aristotle's differentiae). For differentiae in plants, Cesalpino emphasized organs of growth and nutrition (roots, stems, and leaves) for the broadest categories, and organs of reproduction (then called *fruitification*) for subclasses. Joachim Jung (1587–1657) pursued this program as well, although he lacked Cesalpino's passion for promoting Aristotle's system.

Other system builders in the sixteenth and early seventeenth centuries used Aristotle's notion of parts to develop the principle of homology. Parts in different organisms that are identical in their inner nature, their argument went, are homologous no matter what their shape or function: a bird's wing compared with a human's arm, for example. Comparative anatomists in this period identified themselves as Aristotelians and rushed to locate homologies in fundamental systems and their parts. This research program encouraged standardized nomenclature and fueled comparative morphology across Europe for more than a century. Examples included French anatomist Guillaume Rondelet's (1507–66) comparisons among fishes and between dolphins and humans (1554), French naturalist Pierre Belon's (1517–64) comparisons of birds and humans (1555), and Italian anatomist Guilio Casserio's (1552–1616) comparison of the auditory bones in **vertebrates**.

* **Vertebrate.** Any animal with a backbone.

Natural Theology

Religion and science have enjoyed cordial relations for much of science's history, contrary to what most people today think. This is best illustrated in the history of systematics by the cases of the English botanist John Ray (1627–1705) and the Swedish naturalist Carl von Linné (1707–78), better known as Linnaeus.

John Ray attempted to describe all known plant species in his *Historia plantarum* (History of Plants, 1682) using a combination of natural history and anatomical description. This covered more than 6100 descriptions. He also compiled descriptive catalogs of birds, fishes, mammals, and reptiles with a view to creating a world catalog. This global project was the culmination of projects of lesser scope. Living in Cambridge for many years, Ray first produced a catalog of local plants (1660). After exploring Wales and Cornwall in 1662 with the naturalist Francis Willughby (1635–72), he described the flora of England and Wales. About this time, Ray and Willughby agreed to produce a natural history of all living things, Ray focusing on plants, Willughby on animals. For this, Ray traveled across Europe (1663–66), funded by wealthy patrons connected through Willughby. He also made considerable use of patrons and friends such as Hans Sloane (1650–1753) for access to rarities in botanical gardens near London.

This was an exciting time to be a naturalist. Ray and Willughby shared an enthusiasm common among northern European naturalists in the second half of the seventeenth century. As naturalists in southern Europe had done a century before,

Science and Society: Natural Theology

Natural theology begins with an analogy to engineers and architects. We happily judge the skills of these craftspeople by studying their designs. We go further, too, inferring aspects of their character and personality: thrifty, extravagant, fashionable, graceful, smart, irresponsible, and so on. Natural theologians studied nature as a way to understand qualities of the divine in precisely the same way. Order and adaptation in nature were the product of a caring and skillful designer. Nature's many "intricacies" and "contrivances" led natural theologians to reflect on the creator's personality: wisdom, intelligence, benevolence, and foresight.

Linking the existence of purposes in nature with the presence of design and designer was an intellectual synthesis of Aristotle and Christianity that had been popular since Thomas Aquinas (1224/5–74). Natural theology complemented revealed theology, the divine word as set down in sacred texts.

In natural theology, systematics became the search for the original organization set down in the design. Comparative anatomy became the study of clever fits between form and function, or the unity found in anatomical plans. Ecology became the study of integrated natural economies and mechanisms for their self-regulation.

Two approaches to natural theology developed. One focused on individual objects and sought to understand how they functioned and why they possessed a particular design: a bee's stinger, a plant's flower, an elephant's trunk. This emphasized the craftsmanship and contrivances in creation and encouraged thinking in terms of particular divine actions. British theologian William Paley (1743–1805) encouraged this style of natural theology. A second approach stepped back from individual cases to focus on the design of underlying laws. This sought nature's clockwork mechanism and thought of natural laws (such as gravity) as the divine machinery for administration and governance.

Natural theology was not universally accepted. French satirist Voltaire (1694–1778) savaged it in *Candide* (1759). Scottish philosopher David Hume (1711–76) attacked it in *Dialogues Concerning Natural Religion* (1779).

British naturalist Charles Darwin (1809–82) was influenced by both traditions of natural theology. When writing his *On the Origin of Species* (1859), he went out of his way to appeal to natural theologians. Common descent explained the appearance of common anatomical structures. Natural selection explained the close fit between form and function. Stepping back, Darwin presented his ideas as laws of nature set into the original clockwork mechanism. Closing his famous book, Darwin explained: "Thus, from the war of nature, from famine and death, the most exalted object which we are capable of conceiving, namely the production of the higher animals, directly follows. There is a grandeur in this view of life. . . ."

northerners were receiving large amounts of new materials imported from Africa, Asia, China, Japan, and the Americas that came with economic and colonial expansion. Ray's *Historium* was one effort to absorb this new material into systematics.

Unlike many of his contemporaries, Ray was explicit about methods. His *The Wisdom of God Manifested in the Works of the Creation* (1691) combined Aristotle with natural theology. Ray learned Aristotle through Cesalpino and focused on fundamental differentiae defined by the purposes of things (their final causes). This emphasis made sense to Ray for two reasons. First, he undertook systematic collecting of single species from all types of habitats and localities. Ray argued that this helped him to distinguish between true differentiae and accidental qualities. Collecting in different habitats and working with living plants also led Ray to emphasize ecology, life history, and physiology in understanding plants. It encouraged him to reflect on the fit between form and function as well as on what varied and what remained constant. Second, Ray's methodology was grounded in natural theology. He was a devout Puritan, and he sought a natural system that arranged nature according to its original design. This focused his interest on divine purposes and the functions of parts, just as Aristotle had done on other grounds. Ray's natural system was based on fruitification because this anatomy was most stable across different habitats and because it related directly to purposes in nature.

Several technical improvements to classification showed in Ray's earliest work. He produced descriptions in a standard format that noted habitat, morphological characteristics, timing of flowering, medical properties, and so on. Ray compared each species found in nature with similar species in encyclopedias and herbals. (Ray had the luxury of travel, as well as access to botanical gardens and private collections.) He also set some basic rules for classification and naming. New classifications should seek minimal change to previous catalogs. He discouraged changes to commonly accepted names. He sought definitions that used clear, distinct, and precise anatomical details. He insisted that differentiae be obvious and easily observed.

Linnaeus

Born in rural Sweden and trained in medicine at Lund and Uppsala, Carl von Linné latinized his name to Carl Linnaeus as part of a lifelong campaign to manipulate the judgment of history and to cultivate a self-image as the "prince of Botany." As with Ray, Linnaeus sought a natural system for taxonomy and introduced important technical innovations. He grounded his natural system in two concepts: natural theology and sex.

Nothing seemed more natural for Linnaeus than connecting nature with the divine. In later editions of his *Systema naturae* (System of Nature, 1735), he explained: "I tracked His footsteps over nature's fields and found in each one, even in those I could scarcely make out, an endless wisdom and power, an unsearchable perfection." Studying nature's original organization was a form of worship; it made systematics the most noble science. With a confident and driven personality, Linnaeus claimed for himself the role of chosen interpreter of nature, a prophet selected by God to reveal the original method for creation. Because God had selected him, Linnaeus argued, his views enjoyed special protection from criticism. Linnaeus was also a compulsive organizer and brilliant observer. His slogan was "to know the thing itself." To this end, he surrounded himself with herbaria, collector's reports, and specimens imported from around the world.

As did so many others, Linnaeus's systematics began with Aristotle and the parts-and-differentiae approach. Like Aristotle, he gave priority to the reproductive system, although this was for a different reason. Linnaeus was a convert to a new theory about plant reproduction. The theory of plant sexuality—that sexual generations existed in plants, that reproductive anatomy was divisible into multiple sexes, that pollen was functionally equal to sperm and seeds to fertilized eggs, and so on—was developed in the generation immediately before Linnaeus. German natural philosopher Rudolph Camerarius (1665–1721) persuasively argued this point in *De sexu plantarum* (On the Sex of Plants, 1694). Linnaeus learned this theory at university, referring to it as the "nuptial of the flower." It forced comparative anatomists to rethink the structures of flowers in terms of duties to be fulfilled by "husbands" and "wives" in nuptials that were either "public" or "clandestine," an analogy to the kinds of marriages permitted in their own societies.

As early as 1731, Linnaeus used a sexual system as the key to systematics. Although the reasoning was new, Linnaeus's emphasis on the anatomy of reproduction (fruitification) was not. Many of his predecessors experienced practical difficulties when sorting according to these features because of an insufficiency of variety: there simply were not enough parts for useful divisions. Linnaeus solved this problem simply by expanding the list of characters relevant to fruitification. He used their number, form, proportion, and situation to produce considerable potential for variety. Linnaeus put so much emphasis on fruitification that he excluded other features from classification except at the species level. Critics complained that this went too far, but Linnaeus had a trump card. He argued that God had determined that plants should be distinguished by the details of their reproductive system;

thus these could serve as the only genuinely natural differentiae. The remainder of the plant was of secondary importance.

Linnaeus built a hierarchical ranking for plants using five layers: class, order, genus, species, and variety. Classes were defined by a plant's anatomy for the function of "husband" (i.e., number, arrangement, and features of the stamen); Linnaeus constructed 24 plant classes in this fashion. For any one kind of husbandly anatomy, orders were defined by a plant's anatomy for the function of "wife" (i.e., the pistil). Because many flowering plants carry both functions on the same flower, Linnaeus identified 35 orders using wifely functions. Further differentiae of the sexual anatomy produced genera. This hierarchy set the genus as the fundamental unit in systematics because it represented the precise expression of the reproductive anatomy for that kind. Other functional systems provided parts and differentiae to distinguish genera, species, and varieties. Linnaeus also extended these general principles to the classification of animals (and in a modified way to minerals) with the goal of creating one global arrangement of all known genera. In theory, this overall system was simple and natural. In converting this into a practical system, however, Linnaeus was forced to make many compromises and simplifications. It left him with an artificial system that only approximated the divine plan.

Linnaeus first used the sexual system for dividing groups and ranks in the 1730s. His pamphlet *Systema naturae* (System of Nature, 1735) and *Genera plantarum* (Genera of Plants, 1737) describe this system. Later editions of these works added more groups, translating their descriptions into the Linnaean format. Linnaeus first used trivial names for species in *Species plantarum* (Species of Plants, 1753) and in the tenth edition of *Systema naturae* (System of Nature, 1758). When

Linnaeus and Binomial Nomenclature

Linnaeus's most enduring legacy is his use of binomial nomenclature. This broke species names into two parts: genus terms and trivial terms. Linnaeus was not the first to develop this technique. He adopted the idea in the 1750s and spent the rest of his life heavily promoting it.

Linnaeus was reacting against a long-established but troublesome practice for naming species. Traditionally, formal names provided complete descriptions of the kind. The first known species of a kind received a single, generic name (imagine describing the first specimen of a kind as "blob"). Similar species of this kind found later would be identified using this generic name plus descriptive adjectives to identify what was peculiar to the new species ("blob with silver colors"). The name of the original species would need to be changed (from "blob" to "blob original" or "blob with blue colors") to distinguish it as now only one species among several. If another species was described ("blob with silver colors found only under rocks"), this would require more differentiation and further changes in the formal names of the previous species, and so on.

In this system, names and their qualifying adjectives became ever longer and more cumbersome. They also changed constantly. This instability destroyed the standardization expected from formal taxonomy. With the pace of discoveries rapidly increasing in the seventeenth and eighteenth centuries, names were being changed at such a rate that no one knew which names applied to which specimens. Linnaeus used the traditional approach early in his career but grew frustrated with its cumbersome and unstable nature. He was not alone.

The binomial approach draws a crucial distinction between names and descriptions. Arbitrary names serve as placeholders for the detail of descriptions. Thus species of the same genus are given a common genus term (e.g., "blob") plus an arbitrarily assigned supplemental term ("blob one," "blob two," and "blob three"). Both names remain constant regardless of changes in descriptions. Distinctions might need to be refined when new species are discovered, but this has no effect on the nomenclature. Thus the onion (*Allium cepa*), leek (*Allium ampeloprasum*), and shallot (*Allium cepa*) could be distinguished clearly and permanently while still understood as specific forms of a single genus. Should a new species of *Allium* be discovered, it could be set into the system with little disruption.

The binomial system had immense practical value. It also appealed to Linnaeus on theoretical grounds because it defined the genus as the fundamental unit of creation. Species were trivial variations on a theme.

botanists refer to 1753 or the tenth edition of *Systema* as the origin of "real" taxonomy, they are pointing to the adoption of trivial names and the consistent use of binomial nomenclature as the key development.

There are several important points to make about Linnaeus. First, some claim that he had an evolutionary concept before Charles Darwin (1809–82). Although many people had evolutionary concepts before Darwin, this is not a simple issue for Linnaeus. He gave priority to the genus in his hierarchy. He believed that it represented the fundamental unit of life's variety. Early in his career, Linnaeus spoke of species being constant and created; later, his view became more complex. Although he kept genera constant and fundamental, Linnaeus allowed for crossing between species to form hybrids, just as might happen with varieties. This later view was not a true evolutionary concept, for two reasons. Fluctuations between species constituted a kind of noise in the system—nothing fundamental changed in the generic final or formal causes. Moreover, Linnaeus always rejected claims that change occurred across higher categories. This excluded evolution of new genera, orders, or classes. Linnaeus's systematics was static in its fundamentals.

Linnaeus was an aggressive promoter of his own work. He published many pamphlets, guides, and catalogs, and he applied his ideas to an ever-increasing range of material. He was also a popular teacher. However, claims for his originality should be read with skepticism. The underlying basis of his systematics and the methods of his taxonomy were grounded in traditions established before him. For instance, Swiss botanist and anatomist Gaspard Bauhin (1560–1624) used binomials in his *Pinax theatri botanica* (Illustrated Exposition of Plants, 1623). His brother Jean Bauhin (1541–1613) spent 40 years arranging a botanical encyclopedia that made use of this approach. Linnaeus's emphasis on the genus was Aristotelian and probably drew directly from French botanist Joseph Pitton de Tournefort (1656–1708), who traveled more extensively than Ray and who served as a professor in the Jardin des Plantes in Paris from 1683 to his death. Tournefort argued for establishing the genus as the universally recognized unit of taxonomy. He also established methods for distinguishing genera later used by Linnaeus, although he rejected the theory of plant sexuality. Overall, Linnaeus combined passionate advocacy, clever synthesis, compulsive organization, and a determination to understand nature's system.

The Linnaean system appealed to many audiences, from learned scholars to pharmacists, from explorers to the growing number of local amateur collectors (many of them women) interested in "botanizing" in the countryside. It provided an orderly aesthetic for the universe. The use of arbitrary names in binomials also allowed for the sending of diverse signals. Patrons could be recognized, and discoverers could be named. Special features of the organism could be identified, and ideas about the theoretical importance of a species could be embedded. The interpretation of such signals tells important stories in the history of taxonomy. Groups and ranks were relatively easy to insert. With every new edition of his fundamental works, Linnaeus expanded the reach of his system, adding new discoveries and transforming formal names from other systems into his own. By the 1758 edition of *Systema*, Linnaeus had named or renamed 7700 plants and 4400 animals.

Linnaeus ensured that collecting expeditions were sent across the globe to expand the coverage of his system. Swedish botanist Daniel Solander (1733–82) and English botanist and aristocrat Joseph Banks (1743–1820) carried the Linnaean system on James Cook's (1728–79) first circumnavigation (1768–71). Collectors were sent with the Dutch East India Company to the Cape colonies, Java, Sri Lanka, and Japan during the 1770s. More than 20 Linnaean disciples were dispatched across Arabia, North America, South America, northwestern Russia, the Caucasus, Lapland,

Surinam, China, Senegal, and Sierra Leone. Later, wealthy and well-connected Linnaeans, Joseph Banks among them, sponsored many more. Linnaeans working in botanical and medical gardens also converted their garden catalogs to this new operating system. On Linnaeus's death, British disciples purchased manuscripts and collections from his estate, founding the Linnean Society of London in 1783.

Parisian Alternatives to Linnaeus

Linnaeus met his coldest reception in Paris, where generations of biologists had also struggled to create a natural system. Among Linnaeus's contemporaries were Comte de Buffon (Georges-Louis Leclerc, 1707–88), Louis-Jean-Marie Daubenton (1716–1800), Jean-Baptiste de Lamarck (1744–1824), Antoine-Laurent de Jussieu (1748–1836), and Augustin-Pyramus de Candolle (1778–1841). Their views varied significantly, but together they focused on the compromises Linnaeus made to create a practical system. For example, they lambasted his emphasis on a few features of the reproductive system. It might have been convenient and easy to use, but this approach failed to express the true relationships of species one to another. It was simpleminded and utterly failed for organisms other than flowering plants. These critics represented Europe's elite life scientists; they had good reason to see little value in the Linnaean system.

Buffon was the strongest critic. He attacked Linnaeus for foolishly thinking that small sets of characters could yield distinctions at so broad a level as a class or family. Organisms were complex biological wholes that needed to be considered in their entirety, Buffon complained. Using a few trivial parts of flowers was like using paint colors to classify architecture. At the very least, Linnaeus needed to consider far more than static morphology. Buffon insisted that taxonomic schemes, even artificial ones, must also consider facts of development, generation, overall organization, habits, and so on. Linnaeus also brought a false sense of simplicity to nature and a false sense of humanity's power to understand creation. Buffon deeply respected divine power and the nuances of design. He argued that natural order was simply too complex and subtle to be understood by humans or to be captured in taxonomic systems. Linnaeus had been ridiculously overconfident, Buffon asserted.

To be fair, Buffon was less interested in the details of taxonomy than in animal habitats, reproduction, senses, and their relationships with the environment. His greatest work began as a simple catalog of the royal cabinet of natural history in Paris. At its completion (after Buffon's death), *Histoire naturelle générale et particulière* (Natural History, General and Particular, 1749–89) had grown into a 50-year, 30-volume comprehensive history of nature. *Histoire naturelle* was not an encyclopedia but a comprehensive system for understanding geology and biology. Buffon's goal was to uncover the natural laws guiding all of life's phenomena. He relied on causes that were constant, ever-present, and acting without interruption. His concept of a creator was a geometrician who manufactured universal laws, then set the world in motion and left the system to work out its own details. This made the divine into an architect rather than a micromanager. As natural theology, it differed from a focus on single instances of creation, individually manufactured at creation. Instead, it focused on system building and laws. Buffon challenged the notion of perfection in nature, suggesting that laws were not always entirely obeyed. Thus despite harmony and beauty in these laws, monstrosities sometimes occurred. These were nature's mistakes. Buffon included extinctions among these mistakes. He also supposed that species could degenerate in some environments, especially in warmer climates. His *Époques de la nature* (Epochs of Nature, 1778) presented a history of nature in which occasional catastrophes decimated entire regions. Life was restored by migrations from elsewhere.

The French botanist Antoine Laurent de Jussieu introduced Linnaeus's binomial system to Paris. At the same time, he harshly criticized the sexual system as a basis for the systematics of plants. Jussieu cited all the reasons presented by Buffon, and more. Linnaeus used too simple an approach for fundamental categories. Reliable taxonomy required the synthesis of exhaustive studies of all aspects comprising an organism. Linnaeus misunderstood the difference between convenient or external characters for identifying species, on the one hand, and internal, fundamental criteria that supported a natural system, on the other. Most important, Jussieu complained that the Linnaean system ignored a sense of deep continuity cutting across the whole of nature. Any method claiming a natural system, Jussieu demanded, had to represent this continuity. Jussieu attempted such as system in *Genera plantarum secundum ordines naturales disposita* (Genera of Plants Arranged According to Their Natural Orders, 1789).

When Jussieu considered the whole of creation, he imagined continuous chains of form along broad themes. Each chain was made complete and continuous in the creation; links currently missing in any one chain simply represented forms not yet discovered or identified. This understanding of nature was known as the *great chain of being* and was a view Jussieu shared with Buffon, Lamarck, and many others. It relied on three basic concepts. First, it assumed that the universe was created with everything possible made into an actual organism somewhere on Earth. Second, it assumed continuity, with each species sharing overlapping qualities with its neighbours. Third, it assumed linear gradation that sometimes demonstrated progress.

Aristotle and Plato Reborn in Paris

Opposition to Linnaeus continued in later generations of Parisian naturalists. Important alternatives were championed by Georges Cuvier (1769–1832) and Étienne Geoffroy Saint-Hilaire (1772–1844). Both claimed to be providing natural systems for classification, although their systems were fundamentally different.

Cuvier emphasized Aristotelian functional anatomy in which purpose and final cause took priority. Structures were means by which organisms carried out functions, he argued. These structures were organized into functionally integrated systems: locomotion, sensation, digestion, circulation, respiration, generation, and so on. As Cuvier explained in *Le règne animal distribué d'après son organisation* (The Animal Kingdom, Distributed According to Its Organization, 1817), systematics should organize life according to these functional systems because function determines structure. Whereas Aristotle emphasized reproduction when building fundamental natural classes, Cuvier showed his Enlightenment prejudice for the brain by emphasizing nervous systems. He proposed four basic plans or embranchments: vertebrates, **articulates** (worms, crustaceans, spiders, insects), mollusks, and **radiates** (urchins, starfish, nematodes, corals, **infusorians**). There were no transitions between these basic groups, no structural affinities or homologies. Embranchments could be divided into classes, Cuvier proposed, based on respiration and circulation systems. Other functional systems were subordinate, useful for organizing at lower levels. Thus a hierarchy of functions (and functional systems) determined arrangements within Cuvier's systematics.

Underlying each animal's functional integrity, Cuvier argued, was a divine design built upon an anticipation of the conditions of its existence—those conditions necessary for the animal to survive and reproduce in a given environment.

* **Articulates.** A branch of the Brachiopoda (a phylum of marine animals) having hinged valves that usually bear teeth.

* **Radiates.** Eumetazoa (animals with tissues or showing signs of tissue formation) that have a primary radial symmetry.

* **Infusorians.** Living things found by looking through a microscope at water, even if resulting from the soaking with water (infusing) of an object such as hay or other vegetation. The term was later restricted to the active, one-celled forms called protozoa, but both uses are now obsolete.

Cuvier suspected that organisms were created with precisely those organs that were needed for meeting its established conditions. He explained extinction by hypothesizing changing environments and periodic catastrophes in which organisms could not survive. Cuvier's system allowed a synthesis of comparative anatomy, taxonomy, and paleontology. In reconstructing fossils, Cuvier used his assumptions of functional integrity to rebuild organisms from isolated parts. A carnivore's tooth, for instance, implied a digestive system for eating flesh, a sensory system for predation, and a locomotion system for pursuit and capture.

Geoffroy was influenced by Plato's theory of forms. In books such as *Philosophie anatomique* (Anatomical Philosophy, 1818), Geoffroy argued for a system of underlying plans in nature. Organisms or species were instances of some plan; it determined their fundamental structure and organization. These plans bound groups together as natural units. Geoffroy charged systematics with sifting through the complex modifications of organisms to locate homologies. Vertebrate limbs provided a simple example. From the trunk, the basic scheme follows: one bone (humerus in the arm), which connects to two (radius and ulna), which connect to a set (carpals then metacarpals), which extend to digits (phalanges). This common abstract plan has elements minimized and maximized along various lines (compare the horse, bat, and whale), but the theme transcends the differences. Geoffroy's taxonomy labeled these connections by the term homology: Parts of different organisms were essentially the same structure. (Homology also related to the life of one organism, as when a caterpillar and butterfly were compared.)

Defenders of Cuvier's functionalism and Geoffroy's structuralism (also called *transcendentalism*) fought a bitter dispute during the 1820s. Each claimed fundamental insights into anatomy and rejected their rival's alternative basis for a natural system. Geoffroy's supporters argued that unity of plan helped anatomists explain seemingly functionless anatomy such as wings and wishbones on flightless birds or the appendix in humans. These existed as part of a structural constancy or symmetry in the plan. Cuvier's supporters responded that nothing in nature existed without function. They criticized efforts by structuralists to locate homologies across the basic embranchments, such as a proposed unity between insects and mollusks, or between cephalopod mollusks (the squid and the nautilus) and vertebrates. The latter suggestion especially infuriated Cuvier. This dispute drew a great deal of attention across Europe but produced no clear victor. Instead, an uneasy balance was established in which extreme views (allowing only one or the other approach) were replaced by a broader view in which the benefits of both approaches were incorporated into compromise positions.

The British zoologist Richard Owen (1804–92) followed this controversy from London. Owen began his career at the Hunterian Museum at the Royal College of Surgeons; three decades later, he became superintendent of natural history collections of the British Museum. Initially, Owen followed Cuvier, but moved toward transcendentalism as the debate unfolded. His *On the Archetype and Homologies of the Vertebrate Skeleton* (1848) and *On the Nature of Limbs* (1849) investigated the unity of plan across all vertebrate groups and attempted to construct the basic structural plans—archetypes—for the broad vertebrate groups. Owen's mature view attempted a reconciliation of Cuvier and Geoffroy in which the archetype served as the starting point for a species. Basic plans, he suggested, were modified variously to meet different functional needs. Other British naturalists also participated in this debate. Robert Knox (1793–1862) defended Geoffroy and transcendentalism. Peter Mark Roget (1779–1869), whose thesaurus first appeared in 1852, also defended Geoffroy. Unlike the Parisian tradition, which placed divine action behind the scenes and focused on "secondary" causes, British and American transcendentalists argued that the manifestation of unity was direct evidence of design. Roget, for instance, did this in his *Animal and Vegetable Physiology Considered with Reference to Natural Theology* (1834).

A similar approach was adopted by the Swiss-born American naturalist Louis Agassiz (1807–73), founder of the Museum of Comparative Zoology at Harvard University. In his *Essay on Classification* (1857), Agassiz defended the search for basic plans in nature. These provided evidence of design and the direct role of the divine in the creation of individual species. Hierarchy in systematics, he argued, provided a key to the layers of divine thinking during creation. Basic plans for organisms were manifested in the architecture reflected in different embranchments. Classes expressed the ways and means used to execute those plans. Orders demonstrated the degrees of complication in those ways and means. Different genera showed alternative choices in the execution of specialized parts. Individual species varied in the relationships of individuals to each other and to the world, the proportions of their parts, ornamentation, and so on. Agassiz layered these concepts together in both his approach to classification and his understanding of the processes of creation.

Darwin and Genealogical Units

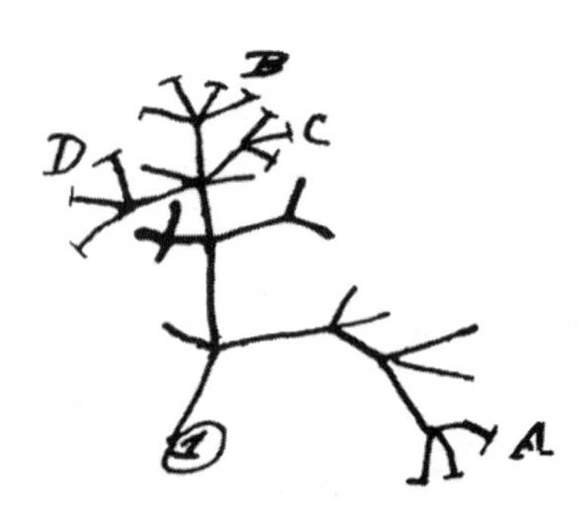

Figure 3. One of Darwin's first sketches of the idea of evolutionary branching, from one of his notebooks in the 1830s. It shows how descent connects species into ever more inclusive groups as lines are followed down in evolutionary time. Darwin used this thinking to suggest that taxonomy should be built from evolutionary relationships. By permission of the Syndics of Cambridge University Library.

When British naturalist Charles Darwin published his theory of evolution via natural selection in his *On the Origin of Species* (1859), he included a discussion of its implications for systematics. Darwin's core thesis about evolution was simple: Species today are modified descendants of species in the past, linked via parent-offspring relationships just as organisms are connected over generations. Both form lineages. Darwin confidently proposed genealogy as the only true basis for natural classification. Community of descent was the hidden bond, he explained. The one diagram he produced in the *Origin* reinforced this point, offering a theoretical sequence of divergence: one species diverged into two, those two each diverged into more, those descendants diverged into more, and so on (see Figure 4). Taxonomic groups must provide a sense of proximity in descent. Splits in lineages provided the foundation for subgroups. Darwin argued that the purpose of classification must be to represent family relationships in evolution, just as a pedigree chart does for prized animals or families.

Darwin was well aware of the competition between systematic theories in the preceding generation. In describing the advantages of his descent theory, Darwin absorbed these alternatives. Darwin's tactic was to represent each as presenting important empirical generalizations, then show how evolution could explain those generalizations. Unity of type came from unity of descent: All vertebrates have the same basic forelimb structure because they inherited it from a common ancestor. Cuvier's emphasis on function and the conditions of existence was subsumed by emphasizing the power of natural selection to modify structure for functional reasons. Darwin also emphasized the value of translating concepts in comparative anatomy, such as homology, away from Aristotelian parts or archetypes. For him, homologies were parts whose structure and function may be completely different but are linked by common descent. Provided that their lineages could be traced to a common ancestral structure, homology was established. Darwin's research on orchids was rich in such examples, some surprising even to orchid specialists.

Darwin's emphasis on continuity led him to argue that formal boundaries between species were arbitrary. In the *On the Origin of Species*, he supposed, a

Figure 4. Darwin's illustration of descent (time moves upward). This is the only diagram to appear in Darwin's On the Origin of Species *(London, 1859).*

long series of gradual and slight changes accumulated along a continuum of divergence. If every form that had ever lived on that continuum were to reappear, he suggested, taxonomists would find it impossible to draw taxonomic distinctions because the steps would be as fine as those that exist between living varieties. Gaps in museum collections or the fossil record provide arbitrary break points for taxonomists. Darwin had direct experience of grappling with the problems of documenting descent and drawing such distinctions. For nearly a decade before writing the *Origin*, he studied the comparative anatomy of barnacles. His first book after the *Origin* was a study of orchids and the evolutionary history of their adaptations for fertilization. Each shows him struggling to understand and trace communities of descent.

Thomas Henry Huxley (1825–95), more than anyone, attached his future to Darwin's rising star. At the first opportunity, Huxley jumped ahead to defend the principle of descent and to demonstrate its importance in systematics. In the *Origin*, Darwin avoided many controversial topics, especially the origin of humans. In a series of lectures and essays from 1860, culminating in *Evidence as to Man's Place in Nature* (1863), Huxley focused directly on the "ape theory" and built a detailed series of arguments from comparative anatomy for setting humans and apes within the same community of descent (see Figure 5). The physical differences between humans and gorillas, he suggested, were less than those between gorillas and gibbons. By any reasonable standard in comparative anatomy, Huxley argued, humans were apes, and an evolutionary genealogy must group them as such. In *Descent of Man* (1871) and *Expression of Emotions in Man and Animals* (1872), Darwin pressed further. He argued that natural selection explained the origin of the morality and intellect that seemed to make humans special.

Depending on the historian, the introduction of evolution into systematics either changed everything or nothing. How could this be so? The answer depends on the

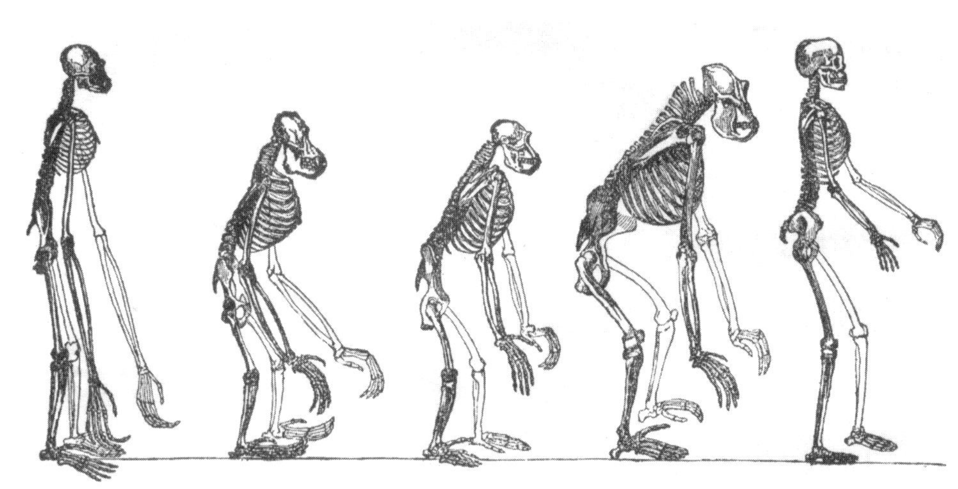

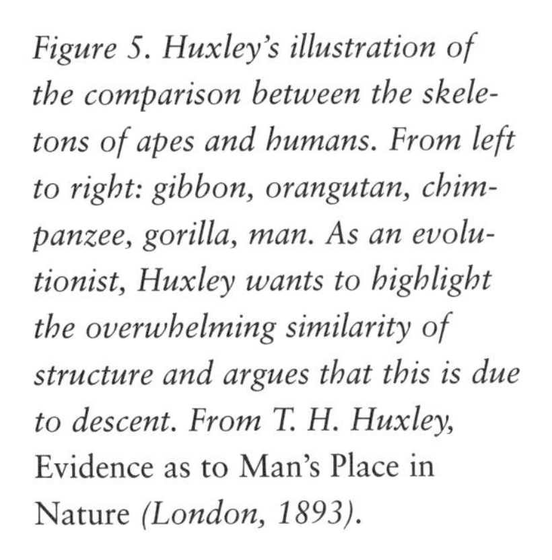

Figure 5. Huxley's illustration of the comparison between the skeletons of apes and humans. From left to right: gibbon, orangutan, chimpanzee, gorilla, man. As an evolutionist, Huxley wants to highlight the overwhelming similarity of structure and argues that this is due to descent. From T. H. Huxley, Evidence as to Man's Place in Nature *(London, 1893).*

particular way in which change is defined. Evolution changed little at the level of day-to-day practical decisions about classification. Many systematists who accepted the descent theory made little use of it in practice. They used the notion of relatedness only in general terms. (A claim that all species in a genus are sister species is easier to prove than a claim that C evolved specifically from B, and B evolved specifically from A.) For instance, after the *Origin*, Joseph Hooker (1817–1911) and George Bentham (1800–84) produced a comprehensive taxonomy for more than 97,000 species of seeded plants in *Genera plantarum* (Genera of Plants, 1862–83). Hooker was in Darwin's inner circle. He defended evolutionary principles in many settings. Although comprehensive, *Genera plantarum* made no use of the Darwinian emphasis on communities of descent.

Darwin's descent theory, however, forced many changes at the level of the theoretical underpinnings of systematics. The redefinition of homologies as features of different species derived from a single character found in a common ancestor is a good example. The notion of diverging lines of species gives an explanation for both grouping and hierarchy in taxonomy, as Darwin expressed in his diagram of species diverging into two, then each of these into two more, and so on. He encouraged thinking about higher categories in taxonomy in terms of evolutionary relationships. Common ancestry explained rudimentary and vestigial organs. Natural selection explained how homologous structures evolved to serve vastly different functions.

The most significant theoretical change related to the problem of defining the boundaries between taxonomic groups. The naming of a new species, genus, or other group in the Darwinian approach carried with it an implicit claim that a split occurred in a lineage, with one line becoming two. Groups in this approach are defined by historical relationships alone. There is nothing essential in their nature, no fixed qualities. To further blur definitions, Darwin heavily emphasized variation within groups. This variation provided the raw material for natural selection. The tendency to vary was ever-present. Darwin's evolutionary concept of groups provided moving targets for taxonomists whereby a species could be defined at one time by qualities absent at another time.

Another key point in Darwin's approach is the issue of instability in names. Consider the problem of assigning rank within Darwin's diagram of diverging species. Suppose that the most recent split helps to group species, the next most recent groups the genus, the next groups the family, and so on. What happens to these decisions about ranks when one of the species splits into two new lines? Must the "species" group now be redefined as a genus, and must the genus now be thought of as a family? Or suppose one branch undergoes many subsequent splits and the other undergoes no new splits. How is a taxonomist to rank these different branches? More practically, with every split, how are names to be assigned? Does each new branch earn a new species name (meaning that the parent species effectively ceases to exist)? Without careful consideration, evolutionary systems threatened the stability of nomenclature. Darwin gave no solution to these kinds of problems, relying on the judgment and experience of individual taxonomists.

A second complication related to the problem of proof for lineages and evolutionary relationships. How was a taxonomist to establish genealogy for a group of species? Darwin gave no solution to this problem. One was later offered by a group of German comparative anatomists, among whom Ernst Haeckel (1834–1919) was especially vocal. Haeckel was an enthusiastic supporter of the principle of descent. In books such as *General Morphology of Organisms* (1866) and *History of Creation* (1868), he asserted evolution's priority in systematics. A single connected series was the only way, he argued, to express the true relationships between organisms. To solve the problem of proof for evolutionary relationships, Haeckel emphasized comparative embryology. The historical record of a species, he claimed, was preserved in the developmental history of individuals. Thus, by following the stages of development (ontogeny) in a species, evolutionists could observe a retracing or recapitulation of evolutionary history (phylogeny). Comparison of developmental sequences for different species allowed taxonomists to identify communities of descent and to locate moments of evolutionary divergence. Thus Haeckel offered comparative embryology as the solution for problems of taxonomy.

Evolution and Taxonomy after Darwin

At its best, Haeckel's embryological approach offered enough evidence to draw connections between major groups, but it failed to provide working taxonomists with the fine degree of resolution needed for their day-to-day problems. Haeckel's approach lost momentum in systematics by the 1880s. At the same time, embryologists themselves were shifting focus away from comparative anatomy and toward the biochemical mechanisms of development. This was part of a broad shift in emphasis in the life sciences away from explanations grounded in comparative and inductive methods and toward those grounded in experimentation and physicochemical causes.

Systematics responded in two ways. Many shifted away from natural systems and toward merely convenient arrangements. This avoided the need to engage in

Science and Society: The 1925 Scopes Trial

In July 1925, Dayton, Tennessee, saw the trial of John Thomas Scopes for teaching evolution. Earlier that year, Tennessee had made unlawful the teaching of "any theory that denies the story of the divine creation of man as taught in the Bible, and to teach instead that man has descended from a lower order of animals." Scopes agreed to be tried so that the American Civil Liberties Union could challenge the law's constitutionality. He also knew that the town's businessmen were planning a publicity event to put Dayton "on the map." Attention ran beyond their wildest expectations when populist hero William Jennings Bryan volunteered for the prosecution and civil rights campaigner Clarence Darrow entered for the defense.

In the trial, the prosecution argued that states have the power to prescribe a curriculum and to decide what should or should not be taught in the classroom. The defense emphasized free speech and the right to worship according to one's own conscience. More important, the defense argued that the antievolution law falsely supposed a contradiction between evolution and the Bible. Many scientists, they argued, understood evolutionary processes as the mechanism that God used to carry out Creation. The trial lasted nearly two weeks. Much of the defense's strategy was ruined when the court prevented scientists from testifying on current evolutionary theory. Scopes was convicted and fined, although his case was dismissed on appeal. That dismissal prevented further legal challenges because the prosecution failed to retry Scopes's case. The antievolution law remained valid until repealed in the 1960s.

Press coverage shaped popular images of the Scopes trial. The American satirist H. L. Mencken emphasized Dayton's carnival atmosphere, calling it "Monkeytown." Mencken called Bryan a fraud and opportunist who preyed on genuine faith for his own profit. Southern commentators complained about outsiders, especially northerners, imposing inappropriate values on their communities.

Recent impressions about the trial derive from the film *Inherit the Wind* (1960), which used Dayton as a vehicle for criticizing anti-communist campaigns of the 1950s. In these campaigns, free speech was a key issue. Bryan's character in the film is a McCarthy-like brute who persecutes the innocent.

The Scopes Trial was only superficially about evolution. Bryan used antievolutionism to criticize modernism. Evolutionists, he supposed, preached "survival of the fittest," an ethic that justified destroying the weak and defenseless. Bryan connected this with thinking that produced the horrors of World War I and the dehumanization of assembly lines. He also thought science promoted skepticism and disrespect for authority. Bryan packed a great deal into the tags "atheist," "agnostic," and "evolutionist," and his defense of the Bible had more to do with these wider concerns than with biblical literalism. Many later antievolution campaigns shared Bryan's deeper concerns about society.

debates with the new brand of experimental biologists over how they might prove these connections. Other taxonomists responded by shifting toward the new methodologies. They focused on extensive fieldwork to catalog variation across many habitats. Geographic and ecological distributions became the key theme for these studies. Variables in the environment might be found to correlate with the distribution of varieties. Experimental gardens might be created so that environmental variables could be varied systematically and the responses of organisms monitored.

This experimental taxonomy pressed systematists once again to consider the basic relationship between morphology and the categories of taxonomic systems. Of all the ways in which organisms vary, which are trivial and which are fundamental for classification? How can the vocabulary of taxonomy reflect these patterns of variation? Should systematics attempt to relate these patterns to biological processes, such as adaptation to local circumstances? One result of this reflection was the polytypic species concept, used by many naturalists in the first several decades of the twentieth century. This accepted the presence of geographical variations—sometimes with substantial differences between varieties—within a single species. In some cases, taxonomists could not agree whether a cluster of populations should be counted as a set of varieties within a species or as a set of "true" species within a closely related genus. Such studies pushed the limits of terminology within classification theories.

New Systematics

In the 1930s, taxonomists spoke widely about a revolution in their approach, trumpeting change in volumes such as *The New Systematics* (1940), edited by Julian Huxley (1887–1975). Although different participants in these changes emphasized different points, their activity clustered around three basic themes.

One theme was a renewed interest in mechanisms for species formation: What processes bring new species about? Interest developed from a growing emphasis on polytypic species and studies of clusters of closely related species within a genus. These seemed to offer snapshots of groups undergoing evolutionary divergence, caught close to the crucial step of species formation. Could these give insights into evolution in action? Interest in speciation put great strain on the nomenclature of lower taxonomic categories—such as variety, race, subspecies, and species groups—as taxonomists searched for ways to express dynamic processes within formal categories. In the process, most argued that the species represented a fundamental evolutionary unit. It had an objective reality in nature. Whether other levels in systematic hierarchies shared this quality was a matter of intense debate.

The focus on speciation fed wider interests among taxonomists in understanding organisms as living, functioning things. The idea was to move away from studying organisms simply as dead anatomical specimens and toward studying them as living, biological entities complete with physiology, behavior, ecology, geography, development, and so on. As biologists explored these different subjects, they could provide comparative data for taxonomy. Alternative species concepts arose during the period of the new systematics. Each was meant to offset the long-entrenched morphological approach by emphasizing some other aspect of living organisms. The most durable of these alternatives, now known generally as the *biological* or *genetic species concept,* emphasized gene flow from one generation to the next: Membership in a species was to be defined by interbreeding.

The broad shift in interest toward biological processes also brought a range of operational and theoretical challenges into taxonomy. It demanded interdisciplinary data collection and discussions about the relative importance of different kinds of data. Taxonomists were forced to broaden their concept of taxonomic characters to include behaviors, ecology, and physiology as well as cytology (the study of cells) and karyology (the study of chromosome number and shape).

A good example of the challenge that this brought to taxonomy was the case of *Drosophila miranda*. In 1935, two geneticists decided that one local population of a well-known species of fruit fly needed to be described as a new species. This population was anatomically identical to other populations of the common species. Yet members of this group failed to interbreed with members of other groups because of certain physiological differences. Based on this physiological evidence, the geneticists argued that this block to gene flow meant they had identified a new species. Some taxonomists complained about the practical implications of this case: How can identical forms be counted as different species? Others understood it to be a case of recent speciation, accepted the weakness in the nomenclature, and used *Drosophila miranda* to explore evolutionary mechanisms.

A third theme of the new systematics also emphasized methodological changes. The placement of specimens in taxonomic groups often requires difficult judgments about the importance of subtle differences. Two experts can sort the same set of specimens into radically different patterns based on their impressions about the relative importance that certain features should have in their weighing up of similarities and differences. Taxonomists in the 1910s and 1920s suffered considerable abuse for what appeared from the outside to be subjective decisions based on aesthetic opinions. Colleagues wanted the certainty of facts, not the vague impressions of artists.

With the rise of the new systematics, taxonomists looked for techniques that improved their sense of objectivity. For example, some developed quantitative approaches to data, representing a species description not as a single type specimen but as an average or range of variation. Others pressed the statistical approach further by introducing confidence tests. Thus the significance of differences between populations could be judged numerically. Testing became more valuable, replacing a general reliance on craft and expert's intuition. For instance, gene flow, and thus genetic species concepts, could be tested for living organisms by breeding experiments and studies of mating habits. The effectiveness of isolating mechanisms and **hybridization** could be observed both in nature and in laboratory conditions.

* **Hybridization.** The process of producing offspring through intercrossing members of different species.

The goal of increasing objectivity in taxonomy also motivated interest in comparative biochemistry, **serology**, and **immunology**. In later decades, these approaches added techniques to compare protein structures, then amino acid sequences, then DNA and RNA sequences. In 1958, the British chemist Francis Crick (b. 1916) proposed that a protein taxonomy should be used in systematics, although he did not explore how this might work practically. With the appearance of each technique, proponents claimed a new objectivity. Although always advertised as revolutions in classification, these techniques relied on traditional principles of comparative anatomy: homology and divergence. Their long-term value remains a subject of considerable discussion.

* **Serology.** The branch of science concerning the properties and reactions of blood serums.

* **Immunology.** Study of the immune system.

With its range of changes, the new systematics created a genuine sense of progress among taxonomists by the 1940s. However, as World War II ended and interest in the life sciences clearly shifted toward physiology, biochemistry, and molecular biology, panic again set in among systematists worried about marginalization. They pressed hard for recognition during the following decade. In Britain, the Systematics Association reorganized in 1946. In the United States, the Society of Systematic Zoology formed in 1947. In the decade of political maneuvering that surrounded the launch of the (U.S.) National Science Foundation in 1950, systematists competed with other biologists for center stage—each claimed to be an essential cornerstone for the study of life.

Evolutionary Systematics after World War II

In the second half of the twentieth century, three approaches to systematics competed for primacy. This competition divided the systematics community into distinct groups that sometimes criticized one another with considerable aggression. Each approach has strong advocates today, and their relative merits continue to be a source of debate.

Evolutionary systematics has its origins in the polytypic species concept and the focus on processes of speciation adopted in the new systematics. Its most vocal advocate has been the German-born American zoologist Ernst Mayr (b. 1904), with books such as *Systematics and the Origin of Species from the Viewpoint of a Zoologist* (1942) and *Methods and Principles of Systematic Zoology* [1953, coauthored with E. Gorton Linsley (b. 1910) and Robert Usinger (1912–68)]. Another major proponent was the American paleontologist George Gaylord Simpson (1902–84), whose 1945 "Principles of Classification" was revised into *Principles of Animal Taxonomy* (1961). This approach is often presented as the "establishment" in the second half of the twentieth century because its principal advocates were also major contributors to the period's predominant school of evolutionary theory.

The basic approach of evolutionary systematics was to use classification to express evolutionary relationships, with a special focus on groups close to the species level. Evolutionary theory set the principles guiding the taxonomist's formulation of these groups. For example, because gene flow became the currency of evolutionary studies, species became the fundamental unit of attention. Geographic and ecological varieties were recognized as subspecies and presented either as candidates for speciation or incipient species, depending on circumstances. Species clusters were set neatly within a genus. Variation within populations was understood to be an essential element of population structure. Characters thought to be adaptations to local conditions were highlighted within taxonomic distinctions as the features making a group distinct. The presence of geographic barriers was deemed sufficient to draw boundaries in classifications. In the overall scheme of classification, taxonomic groups were built around a common ancestor and its descendants, following Darwin's emphasis on continuity and descent with modification.

In his approach to systematics, Mayr emphasized the genetic and geographic aspects of organisms. His nomenclature focused on "biological" species, subspecies, and genera. In contrast, and as might be expected from his background, Simpson worked to relate these concepts to the taxonomic needs of paleontologists. In an evolving lineage, how is one to draw boundaries between species or other taxonomic groups? What kind of living group corresponds to the paleontologist's species?

Not all taxonomists accepted evolutionary systematics. American zoologists Waldo Schmitt (1907–77) and Richard Blackwelder (b. 1909) denied the necessity of placing evolutionary relationships at the heart of taxonomy. They did not deny evolution as a phenomenon; they simply disputed the need for a natural system. Instability was their principal concern. Phylogenies were difficult to test. Similarly, changing theories about evolutionary relationships demanded changes to taxonomic arrangements.

More important, Schmitt and Blackwelder complained that classifications based on ancestor-descendant relationships sometimes obscured more than they clarified. For instance, strict use of genealogy would require that reptiles be destroyed as a taxonomic group, because this group seemed to contain several independent evolutionary lineages that split very close to the original transition from amphibians to reptiles. But dissolving the class of reptiles and replacing it with groups according to these different lineages would have destroyed an extremely convenient and useful approach, leaving little more than a mess in its wake. Schmitt and Blackwelder thought rather little of the genealogical system for classification. Their opposition to the evolutionary systematics of Mayr and Simpson added an important subtext to the formation of the Society for Systematic Zoology in 1947; they wanted to promote their own approach over others.

Numerical Taxonomy

Debates in the 1940s and 1950s over the value of evolutionary systematics demonstrated the continuing tension in systematics over the relative importance of natural and artificial systems in classification. In theory, evolutionary arrangements might be preferred, but in practice, they seemed difficult to produce, test, and confirm. They also did not always seem particularly useful.

Numerical taxonomy grew out of complaints such as Blackwelder and Schmitt's about evolutionary systematics. It sought to focus taxonomy on methods for assessing overall degrees of similarity and difference independent of claims about evolutionary relationships. Its rules for grouping made use of straightforward counts of similarities. The underlying idea was to build classifications that stood the test of time, regardless of the success or failure of particular evolutionary scenarios or particular attempts to draw evolutionary lineages. Although many people were involved in pressing this alternative, a key contribution was *Numerical Taxonomy* (1973) by Austrian-born American microbiologist Peter Sneath (b. 1923) and British biologist Robert Sokal (b. 1926).

On the surface, numerical taxonomy was a straightforward search for increased objectivity through measures of similarity and difference represented as cold, hard facts. Numerical taxonomists claimed to follow traditional techniques of comparative anatomy with a few important extensions. First, this approach encouraged the simultaneous comparison of many characters in measures of similarity and difference. Sometimes this included hundreds of characters at a time. To assist such comparisons, numerical taxonomists made use of new computational techniques for the analysis of variation. They took advantage of new computing technologies to process these large databases. Critics complained that such data sets failed to capture how organisms exist as integrated wholes. This, they said, was simply playing with numbers.

Second, numerical taxonomists discouraged the practice of weighting the value of characters in these calculations. Aristotle and Linnaeus gave priority to reproductive systems. Cuvier emphasized the nervous system. Evolutionary systematics gave priority to adaptations. Numerical taxonomists sought to emphasize nothing about organisms more than anything else. This relied on the principle that taxonomists had no a priori grounds for such preferences. Rather than emphasize a few diagnostic characters, numerical taxonomists sought comparisons over large numbers of characters in the search for clusters. Assessments of overall similarity and differences, using as many characters as possible and not restricting themselves to only morphological ones, was the goal. (Later versions of numerical taxonomy reintroduced weighting, although this remained controversial.) Importantly, numerical taxonomists made no claims about evolutionary or phylogenetic relationships: Morphological similarity did not imply evolutionary relatedness.

Cladism

Cladism (also cladistics) has its origin in German zoologist Willi Hennig's (1913–76) *Phylogenetic Systematics* (1966), an English revision of his *Grundzüge einer Theorie der phylogenetischen Systematik* (1950). Hennig's ideas developed during the 1940s. Because little exchange occurred between Allied and Axis countries immediately following World War II, he was not widely read in Allied countries until the 1960s.

Like those defending evolutionary systematics, Hennig wanted to revive interest in underlying theories of systematics. Nothing could be worse for taxonomy, he argued, than to ignore underlying principles, be they physiological, ecological, or phylogenetic. Hennig argued that phylogeny should take precedence as a theme in classification. The sequence of evolution, he argued, had fundamental implications for all relationships between organisms. Criticizing Haeckel's evolutionary morphology for circular reasoning and weak methodology, Hennig proposed an alternative. Taxonomic groups should be formed according to evolutionary *clades*, portions of a phylogenetic tree resulting from a common ancestor. Species-to-species relations were the basic unit of phylogeny. Branching events, in which one species gave rise to two new species, represented the creation of new evolutionary lineages and new taxonomic groups. (Hennig's basic intuitions closely paralleled those described by Darwin in the *Origin*.)

Hennig charged taxonomists with the task of reconstructing the history of evolutionary branching. This could be done by identifying common traits shared by all members of a clade as well as unique traits possessed only by certain of its branches.

Scientific Institutions: Taxonomic Codes

No government agency controls taxonomic names or holds the official database of valid names. Nomenclature is normally negotiated among professionals through the way names are used in published classifications. Sometimes multiple systems for the same group coexist in the literature. Advocates of one approach usually ignore or criticize schemes proposed by others. The community relies on the principle of attrition: Specialists will use good systems and ignore poor ones; in time, the better systems will prevail.

When irresolvable conflicts occur, taxonomists appeal to systems developed by their own community. One example is the *International Code of Zoological Nomenclature*, administered since 1961 by the International Union of Biological Sciences. This approach has its origin in nineteenth-century efforts to standardize decision making. These codes emphasize principles of priority and standardization. They give rules for structuring nomenclature and publication. They define standard practices for documenting and preserving type specimens. They also provide rules for legitimate exceptions: how a researcher might overrule core principles.

One of the first codes was Hugh Edwin Strickland's (1811–53) *Series of Propositions for Rendering the Nomenclature of Zoology Uniform and Permanent*, which the British Association for the Advancement of Science supported in 1842. Codes became an important feature of international scientific conferences through the nineteenth and twentieth centuries. They remain important today.

International codes evolve following conceptual or methodological developments. In 1999, the fourth edition of the zoological *Code* acknowledged the need to reconcile fundamental differences in approach between the Linnaean hierarchy used by evolutionary systematics and concepts of branching developed in cladism. Cladists had been complaining for years that Linnaean ranks (class, order, family, etc.) were arbitrary and too few to adequately capture the notion of cladistic branching. They also complained that the Linnaean hierarchies allowed groups to contain more than one evolutionary branch. Rather than attempting a resolution, the zoological *Code* instead took refuge in artificial systems, appealing to the need for stability in nomenclature and procedures as well as for rapid information retrieval. Natural systems may be preferred for aesthetic reasons, the zoological *Code* implied, but taxonomists still need to grapple with daily problems of identifying materials. Cladists strongly criticized this conclusion, and future revisions of the zoological *Code* are sure to be contentious events.

Common traits were said to be *ancestral*, shared with the common ancestor and passed on as species branched. Unique traits were later additions, appearing only in one lineage and "derived" from the species in which they evolved. Hennig argued that the sequence of branching events could be inferred once the series of new traits was set in accumulating sequence. This was a study of layering one new trait upon another. As the taxonomist added derived traits to branches of a clade, the overall sequence of phylogeny resulted. In contrast to evolutionary systematists, cladism did not identify one particular species as a common ancestor. Claiming that "two forms share a common ancestor" was not the same claim, they argued, as "this form evolved from that form." Instead, cladists used their taxonomies to produce lists of features expected in that common ancestor.

Cladism gained popularity among British and American systematists critical of numerical taxonomy. Artificial systems were a weak compromise, they complained. Besides, processes such as parallel evolution easily distorted statistical measures of similarity and left the taxonomist with groups of dubious value. Fundamental features of groups were easily swamped by the massive, unweighted data sets. Cladism offered a means of returning systematics to evolutionary lineages and natural systems while providing more rigorous and direct methods for lineage construction. At the same time, defenders of cladism criticized the natural system of evolutionary systematics as overreaching from the evidence at hand. Taxonomists could assess the degree of relatedness, cladists argued, but they could not draw connections between groups as specific ancestors or descendants. Thus they agreed on the natural system but disagreed on what could be proved or known for certain.

As it came to be used, Hennig's approach had several problems. It assumed that all evolutionary branching resulted from the splitting of one lineage into two. Critics complained that this made other kinds of evolution invisible, such as when a single lineage never branches but acquires new features, or when a branching splits into more than two lineages. Another complaint focused on Hennig's use of parsimony as an underlying principle when reconstructing sequences of derived traits. Hennig preferred sequences invoking the fewest evolutionary changes. Critics complained that this preference might lump together cases where similar traits evolved independently in different lineages.

Predictably, the strongest criticisms focused on implications for the stability of names for higher categories. In Hennig's system, every branching event produced a new clade, and each new clade required a unique taxonomic name. Ongoing research regularly altered the specifics of phylogenies and the sequences of branching events. This changed the nested sequences of clades, which in turn required changes to taxonomic names.

Worrying about the underlying assumptions of cladistics, especially parsimony, some cladists proposed a disconnect between the construction of clades and discussions of evolutionary processes. In fact, detaching evolutionary theories from taxonomy minimized the theoretical assumptions at work in classifications, they argued. This made them less vulnerable to criticism when changes occurred in evolutionary explanations. These cladists did not abandon evolution; they simply sought ways to insulate taxonomy from changes in evolutionary theory by minimizing the reliance of classifications on particular evolutionary scenarios. This approach, dubbed *transformed cladism*, was defended in the 1970s by American systematists Gareth Nelson and Norman Platnick (both dates unknown) and in the 1980s by Colin Patterson (1933–98).

Today, systematics continues to be a contentious field of research, with advocates of the three most recent schools arguing over fundamentals. Cladism has proven to be the most robust approach. This is due partly to the ease with which it absorbs new kinds of data, such as DNA sequencing, and also to its methodological caution on evolution: It allows claims of common ancestry (identifying sister groups) but not specific claims of ancestry. Cladistic principles also increasingly organize museum displays and textbook accounts of the evolutionary history of life. It would be foolish, however, to expect the evolution of systematics to stop. The contest between artificial and natural systems is bound to continue far into the future.

Bibliography

Allen, David. *The Naturalist in Britain: A Social History.* Princeton, N.J.: Princeton University Press, 1976.

Appel, Toby. *The Cuvier-Geoffroy Debate: French Biology in the Decades before Darwin.* New York: Oxford University Press, 1987.

Hennig, Willi. *Phylogenetic Systematics.* Urbana: University of Illinois Press, 1979.

Hull, David. *Science as a Process.* Chicago: University of Chicago Press, 1988.

Huxley, Julian. *The New Systematics.* Oxford: Clarendon Press, 1940.

Koerner, Lisbet. *Linnaeus: Nature and Nation.* Cambridge, Mass.: Harvard University Press, 1999.

Mayr, Ernst. *Systematics and the Origin of Species from the Viewpoint of a Zoologist.* New York: Columbia University Press, 1942.

Mayr, Ernst, E. Gorton Linsley, and Robert L. Usinger. *Methods and Principles of Systematic Zoology.* New York: McGraw-Hill, 1953.

Morton, A. G. *History of Botanical Science.* London: Academic Press, 1981.

Rehbock, Philip. *The Philosophical Naturalists: Themes in Early Nineteenth-Century British Biology.* Madison: University of Wisconsin Press, 1983.

Simpson, George Gaylord. *Principles of Animal Taxonomy.* New York: Columbia University Press, 1961.

Sneath, Peter H., and Robert R. Sokal. *Numerical Taxonomy.* New York: W. H. Freeman, 1973.

Interdisciplinary Timeline

Date	Astronomy and Cosmology	Chemistry	Earth Sciences	Life Sciences
c. 580 B.C.	Thales postulates that the world originated from water.			
c. 532 B.C.				
c. 500 B.C.				Xenophanes studies fossils and speculates on the development of living things.
c. 450 B.C.				
c. 400 B.C.				Hippocratic book Nature of Man postulates the existence of four humors (blood, phlegm, black bile, and yellow bile).
c. 387 B.C.				
355 B.C.				
c. 350 B.C.				Aristotle carries out detailed observations on marine organisms and expounds an epigenetic theory of development.
c. 306 B.C.				
c. 300 B.C.				Diocles composes the first known anatomy book and coins the term anatomy. Herophilus and Erasistratus dissect human cadavers in Alexandria. Erasistratus hypothesizes that the arteries contain pneuma, and the veins contain blood.
c. 230 B.C.	Archimedes composes *The Sand-Reckoner*.			
c. 200 B.C.	Eratosthenes gives a rough estimate of the radius of Earth.			
130 B.C.	Hipparchus discovers the precession of the equinoxes.			
56 B.C.				
46 B.C.	Julius Caesar orders the development of the Julian calendar.			
A.D. 150	Ptolemy completes the *Almagest* (Greatest).			Galen investigates urine and blood experimentally.
c. 180				Galen proposes his system of the human body, based on four primary qualities and four elements.

Note: We've added English titles when such editions are reasonably available, except in some cases when the work is commonly known by its foreign-language title.

Mathematics	Physics	Technology	Culture
Pythagoras states the theorem that bears his name.			
	Empedocles postulates the existence of four elements (air, water, earth, and fire) that combine to form all physical bodies.		
	Democritus asserts that all things are made up of tiny, indivisible particles or atoms.		
			The Academy at Athens is founded by Plato.
			The Lyceum at Athens is founded by Aristotle.
			The School (the "Garden") in Athens is founded by Epicurus.
Euclid presents geometry as a deductive system in the *Elements*.	Euclid states the law of reflection in *Catoptrics*.		
Apollonius composes *On Conic Sections* and names the ellipse, parabola, and hyperbola.			
	Lucretius composes *De rerum natura* (On the Nature of Things).		

Date	Astronomy and Cosmology	Chemistry	Earth Sciences	Life Sciences
1054	Native American and Chinese astronomers observe the Crab supernova explosion.			
1303				
c. 1330				
1347				
1451				
1455				
1477		Thomas Norton recognizes the importance of color, odor, and taste to chemical analysis.		
1485				
1490				Leonardo da Vinci describes capillary action.
1496	A Latin translation of Ptolemy's *Almagest*, started by Georg Peurbach and completed by Johannes Regiomontanus, is published.			
1514	Nicholas Copernicus presents the central ideas of his heliocentric theory in *Commentariolus*.			
1517				Girolamo Fracastoro postulates that fossils are the petrified remains of organisms that lived at the bottom of a sea that once covered the land.
1540	Georg Joachim Rheticus prints Narratio prima.			Michael Servetus postulates pulmonary circulation.
1543	*De revolutionibus orbium coelestium* (On the Revolution of the Heavenly Spheres) presents Copernicus's heliocentric system in detail.			Andreas Vesalius publishes *De humani corporis fabrica* (On the Architecture of the Human Body).
1545				
1546				Georgius Agricola introduces the word *fossil* for anything dug from the ground.

Mathematics	Physics	Technology	Culture
			The use of spectacles is first recorded.
	Jean Buridan proposes the concept of impetus.		
			William of Ockham advances the principle now known as Ockam's razor.
		Nicholas Cusa invents concave-lens spectacles to treat nearsightedness.	
			The Gutenberg Bible, the first European printed book, is published.
	Leonardo da Vinci advocates the use of lenses for the study of small objects.		
Girolamo Cardano publishes a formula (discovered by Niccolò Tartaglia) that will solve any cubic equation.			

Date	Astronomy and Cosmology	Chemistry	Earth Sciences	Life Sciences
1551	Erasmus Reinhold publishes astronomical tables based on Copernicus's theory.			
1551–1571				Konrad Gesner publishes *Opera botanica* (Botanical Works) and *Historia plantarum* (History of Plants), works that influence Carl Linnaeus and Georges Cuvier.
1556			Agricola's *De re metallica* becomes the standard text on all aspects of mining and mineralogy.	
1559				Matteo Realdo Columbo demonstrates the pulmonary circulation of blood in detail.
1560				
1569			Nicolaus Mercator produces the first projection map.	
1572	Tycho Brahe discovers a supernova in Cassiopeia. Brahe publishes *De nova stella* (The New Star).			
1576	Thomas Digges modifies the Copernican system by removing its outer edge and replacing the edge with a star-filled unbounded space. Brahe builds an observatory at Hven.			
1577	Brahe uses parallax to prove that comets are distant entities, not atmospheric phenomena.			
1582	Pope Gregory XIII introduces the Gregorian calendar.			
1583				
1584	Giordano Bruno argues that stars form planetary systems and the universe is infinite.			
1585			Mercator publishes *Atlas*, the first great atlas of the world.	
1590				

Mathematics	Physics	Technology	Culture
			The Academia Secretorum naturae, the first scientific society, suppressed by the Inquisition, is founded by Giambattista della Porta.
	Galileo Galilei notes the isochronism of the pendulum.		
Simon Stevin presents a systematic account of how to use decimal fractions.			
		Zacharias Jansen combines two convex lenses within a tube, constructing the forerunner of the compound microscope.	

Date	Astronomy and Cosmology	Chemistry	Earth Sciences	Life Sciences
1591				
1592				
1595	Johannes Kepler publishes *Mysterium cosmographicum* (Mystery of the Cosmos).			
1596				John Gerard's *Herbal* contains the most complete survey of botanical knowledge at this time.
1597		Andreas Libavius publishes *Alchemy*, a treatise on chemistry as used in medicine.		
1600	Kepler becomes Brahe's assistant at his Prague observatory and composes one of the first histories of science, *A Defense of Tycho against Ursus*.			
1602	Brahe's *Astronomiae instaurata progymnasmata* (Introduction to the New Astronomy), containing a star catalog and a description of the 1572 supernova, is published posthumously.			
1603				Johannes Fabricius describes the valves in the veins.
1604	Kepler observes a supernova in Serpentarius (Ophiuchus).			Fabricius publishes *De formata foetu* (The Formation of the Fetus), one of the first important studies of embryology.
1605				
1606	Kepler's *De stella nova* describes the nova observed in 1604.			
1608	Thomas Harriot draws lunar maps with the assistance of the telescope.			
1609	Kepler publishes *Astronomia nova* (New Astronomy), containing a physical explanation for planetary motion and the first two laws that bear his name.			

Mathematics	Physics	Technology	Culture
François Viète uses letters of the alphabet to represent unknown quantities.			
		Galileo builds a crude thermometer using air in a tube.	
	Galileo admits in correspondence with Kepler that he accepts the Copernican system.		
Viète finds a solution to the problem of constructing a circle touching three given circles.	William Gilbert publishes *De magnete*, the first book in physics based entirely on experimentation.		Giordano Bruno is burned at the stake.
			The Bodleian Library opens at Oxford.
	Galileo discovers the laws of pendulum and free fall by experiment. Kepler publishes *Astronomiae pars optica* (The Optical Part of Astronomy), outlining a new theory of vision.		
			Francis Bacon publishes *Advancement of Learning*.
		Hans Lippershey builds his first telescope (3×).	
		Galileo makes substantial improvements to the telescope (9×, 30×).	The first Accademia dei Lincei (listing Galileo as a member) is created in Rome.

Date	Astronomy and Cosmology	Chemistry	Earth Sciences	Life Sciences
1610	Galileo publishes *Sidereus nuncius* (Starry Messenger), reporting telescopic observations of four of Jupiter's moons and the rugged features of the lunar surface. Kepler uses the dark night sky to argue for a finite universe.			
1611				
1613	Galileo uses sunspots to demonstrate the rotation of the Sun (first reported by Fabricius in 1611).			
1614				Santorio Santorio advances the first study of metabolism.
1615				
1616				
1618				
1619	Kepler's third law of motion is stated in *Harmonice mundi* (Harmony of the World). Kepler's *Epitome astronomiae Copernicanae* (Epitome of Copernican Astronomy) is a defense of the Copernican system published in three parts (1619–21).			
1620		Johannes Baptista van Helmont coins the term *Gas* to describe substances that are like air.		
1621				
1622				Gaspare Aselli discovers the lymphatic system.
1623				

Mathematics	Physics	Technology	Culture
	Kepler publishes *Dioptrice*, containing the theoretical principles of the Keplerian telescope (built in 1630 by Scheiner).		
John Napier invents logarithms.			
Kepler uses infinitesimals to calculate the volume of various solids of revolution, such as wine casks, anticipating the integral calculus.			
			The Church issues an injunction against Galileo not to "defend" or "support" the Copernican doctrine. Copernicus's *De revolutionibus* is placed on the Church's *Index of Prohibited Books*.
	Francesco Maria Grimaldi discovers the diffraction of light waves.		
			René Descartes has a celebrated dream in which he is instructed to work out the unity of the sciences on a purely rational basis.
			Bacon analyzes the scientific method in his *Novum organum*, containing a vigorous defense of the experimental method.
Diophantus's *Arithmetica* is published in Greek with a Latin translation, launching the revival of number theory.	Willebrord Snell states his law of refraction, discovered independently by Descartes in 1629.	William Oughtred invents the slide rule.	
			Galileo publishes *Saggiatore*, in which the distinction between primary and secondary qualities is elaborated.

Date	Astronomy and Cosmology	Chemistry	Earth Sciences	Life Sciences
1625				Joseph of Aromatari hypothesizes that the chick is present in the egg before incubation.
1626				The Jardin des Plantes is inaugurated in Paris (later became the Musée d'Histoire Naturelle).
1627	Kepler publishes the *Rudolphine Tables* (completed in 1623).			
1628				William Harvey publishes *De motu cordis* (Movement of the Heart), in which he describes the function of the heart as being like a mechanical pump, pushing arterial blood into the veins, contrary to Galen.
1632				
1634				
1635				
1637				
1638				Harvey works on *De Generatione* (On Generation), advancing a theory of oviparous reproduction.
1639				
1640				
1641				
1642				

Mathematics	Physics	Technology	Culture
			Bacon publishes *New Atlantis*, which contains a description of the "House of Solomon," in which scientists perform collaborative experiments.
	Galileo's *Dialogue Concerning the Two Chief World Systems* introduces the principle of relativity to physics. He is immediately summoned to trial (1633).		
Giles Personne de Roberval gives a mathematical description of the cycloid.			
Bonaventura Cavalieri develops his method of calculating volumes by using infinitely small sections, an important precursor to the integral calculus.			
An appendix to Descartes's *Discours de la méthode* (Discourse on Method) contains the first published development of analytic geometry.			
	Descartes introduces the concept of an aether filling all of space, to explain the tides.		
Girard Desargues begins the study of projective geometry.			
Blaise Pascal proves Pascal's theorem, also known as the mystic hexagram.	Evangelista Torricelli applies Galileo's laws of motion to fluids, creating the science of hydrodynamics.		
		Ferdinand II, the Grand Duke of Tuscany, invents a thermometer that has one end sealed.	
	Pascal advances the principles of hydraulics.	Pascal builds an adding machine.	

Date	Astronomy and Cosmology	Chemistry	Earth Sciences	Life Sciences
1643				
1644				
1647	Johannes Hevelius produces the first map of the side of the moon observable from Earth.			
1648		Van Helmont publishes *Ortus medicinae* (On the Development of Medicine), containing a description of carbon dioxide.		
1649				
1650		The first chemistry laboratory is established at Leyden University.		
1651				Harvey describes organ differentiation in the developing embryo.
1655	Giovanni Cassini discovers Jupiter's Great Red Spot. Christian Huygens identifies Saturn's rings as rings.			
1657				
1658				Jan Swammerdam observes red blood cells under a microscope.
1660		(to 1678) Robert Boyle conducts experiments on gases and studies the effects of combustion and respiration on the atmosphere.		
1661		Boyle publishes *The Sceptical Chymist*, defending an atomic theory of matter.		Marcello Malpighi describes blood capillaries in the lungs of frogs.
1662				

Mathematics	Physics	Technology	Culture
	Torricelli measures atmospheric pressure with his mercury barometer.		
	Descartes publishes *Principia philosophiae* (Principles of Philosophy), containing the vortex theory of planetary motion and arguments concerning the impossibility of vacua.		
	Pascal publishes work on vacua, confirming Toricelli's results.		
Desargues's theorem is published.	Pascal uses the barometer to show that the atmosphere has weight.		
	Drawing on the writings of Epicurus, Pierre Gassendi asserts that matter is made up of atoms.		
		Otto von Guericke constructs the first vacuum pump (published in 1672).	Archbishop James Ussher sets the date of creation at 4004 B.C. Harvard University is founded.
Pascal and Pierre de Fermat solve a basic problem of probability. Pascal publishes discoveries relating to binomial coefficients.	Guericke demonstrates atmospheric pressure using Magdeburg hemispheres.	The sealed thermometer is invented by Ferdinand II.	Thomas Hobbes publishes *Leviathan*, elaborating a materialistic doctrine of nature and human society.
Fermat claims to have proved a certain theorem but leaves no details of his proof (known as *Fermat's theorem*). Huygens publishes the first work on probability.		Huygens builds the first accurate pendulum clock, to determine longitude at sea.	Leopoldo de' Medici founds the Accademia del Cimento in Florence, the first scientific research institute since antiquity.
Christopher Wren finds the length of a cycloid.		Robert Hooke invents the spiral string for watches.	
	Hooke discovers Hooke's law: The extension of an elastic material is in proportion to the force exerted on it.		The Royal Society of London is founded.
The first book of statistics, showing the ages at which people (in London) are likely to die, is published.	Boyle's law (which he did not discover) is stated explicitly in the second edition of *Boyle's New Experiments Physico-Mechanical*.		

Date	Astronomy and Cosmology	Chemistry	Earth Sciences	Life Sciences
1663				
1664	Giovanni Borelli finds that the orbit of a comet is a parabola. Hooke discovers the Great Red Spot on Jupiter and Jupiter's rotation.			The view that animals are purely mechanical beings is advanced in a posthumous work of Descartes. Pancreatic juice is discovered by Regnier de Graaf.
1665				Hooke publishes *Micrographia*, a collection of diverse essays dealing with the microscopic structure of familiar substances, among which the cellular structure of cork is fully described and illustrated.
1666	Cassini observes the polar ice caps of Mars.			
1667	The observatory of the French Academy is founded in Paris. Jean Picard introduces the micrometer for telescope use and discovers anomalies in the positions of stars, which is later explained as being due to the aberration of light.			John Ray introduces a classification system for plants. Boyle shows that an animal can be kept alive by artificial respiration.
1668				Francesco Redi publishes *Observations on the Generation of Insects*, which shows that flies hatch from eggs that are deposited on meat by other flies.
1669		Johann Joachim Becher suggests that an "oily earth" causes fire, which later becomes the basis of the phlogiston theory.	Nicolaus Steno affirms the nature of fossils. Etna erupts, contributing to the development of geological ideas.	Jan Swammerdam describes the metamorphosis of insects, opposing spontaneous generation.
1671				
1672				Regnier de Graaf observes the ovarian follicles of mammals, wrongly presuming these to be the eggs of mammals.

Mathematics	Physics	Technology	Culture
		James Gregory gives the first description of a reflecting telescope.	Descartes's works are placed on the Roman Catholic Church's *Index of Prohibited Books.*
Isaac Newton invents his calculus.	Newton deduces the inverse-square gravitational force law from the falling of the Moon. Newton discovers that white light is made from a mixture of colors.		The Royal Society begins publication of *Philosophical Transactions*, the first journal of a strictly scientific nature. The *Journal des Savants* is founded in France. The Great Plague in London kills 75,000 people and closes Cambridge University for two years.
	Boyle argues that everything is built up of minute corpuscles.		The Académie Royale des Sciences is founded. It later becomes the Institut de France. London is ravaged by the Great Fire.
			(to 1672) Claude Perrault builds an observatory at Paris.
		Newton builds the first reflecting telescope; builds a second in 1671 for the Royal Society.	
	Erasmus Bartholin announces the discovery of double refraction in Iceland spar.		
	Using pendulum measurements, Jean Richer determines that gravity decreases with latitude.	Antoni van Leeuwenhoek starts constructing simple microscopes.	
	Guericke describes electric conduction and repulsion in his *Experimenta nova* (New Experiments). Newton presents his memoir "A New Theory about Light and Colours" to the Royal Society.	N. Cassegrain invents the reflecting telescope that is named for him.	

Date	Astronomy and Cosmology	Chemistry	Earth Sciences	Life Sciences
1673				Leeuwenhoek sends the first of his hundreds of letters to the Royal Society. Malpighi publishes *De formatione pulli in ovo* (On the Formation of the Chick in the Egg), a systematic study of the embryonic development of chicks using the microscope.
1674				Nicolas Malebranche elaborates the conception of *emboîtement* (encasement), which holds that each embryo is contained within the embryo of its parent.
1675	Construction of the Greenwich Royal Observatory, designed by Wren, begins. Cassini discovers that the rings of Saturn are not a single flat disk.			Malphigi's *Anatome plantarum* (Anatomy of Plants) is the first important treatise on plant anatomy.
1676	The Greenwich Royal Observatory is opened.			Leeuwenhoek observes "animalcules" (protozoa).
1677				Leeuwenhoek describes human spermatozoa, asserting that they are capable of developing into a child, with the egg providing nutrient.
1678				
1679	Edmund Halley publishes a catalog of stars observable from south of the equator.			
1680				
1682	Edmund Halley observes the comet that will be named after him, correctly predicting in 1705 that it will return in 1758.			John Ray describes thousands of plant species, classifying them on the basis of a new taxonomic system.

Mathematics	Physics	Technology	Culture
Gottfried Leibniz invents his calculus but does not elaborate his notations and findings fully until 1684.	Huygens gives the calculation of equivalent pendulum lengths and the laws of centripetal force.	Leibniz presents a calculating machine to the Royal Society.	
Leibniz is the first to use modern notation for an integral.	Newton publishes his second paper on light and colors, which advances a corpuscular theory of light. Boyle publishes the first book on electricity: *Experiments and Notes about the Mechanical Origine or Production of Electricity.* Ole Römer determines the speed of light to be finite, challenging one of the central principles of Cartesian science.		
Newton proves the binomial theorem. Leibniz discovers how to differentiate any integral or fractional power of x.	Hooke announces that the stretch of a spring varies directly with its tension (Hooke's law).		
Leibniz correctly determines the quotient rule for differentiation.			
	Huygens advances his wave theory of light.		
Leibniz introduces binary arithmetic, in which the symbols 0 and 1 are used to represent all numbers.	Hooke proposes an inverse-square law of gravity, thereby preempting Newton. Jean Richer describes the change in the period of the pendulum in different locations on Earth due to changes in gravity.		
		Clocks are equipped with hands to show minutes.	

Date	Astronomy and Cosmology	Chemistry	Earth Sciences	Life Sciences
1682				Nehemiah Grew describes the male and female parts of flowering plants.
1683				Leeuwenhoek observes bacteria.
1684				
1685			Thomas Burnet's *The Sacred Theory of the Earth* invokes geological speculation to support the scriptures (originally published in Latin in 1681).	
1686	Bernard Le Bouyer de Fontenelle publishes *Entretiens sur la pluralité des mondes* (Conversations on the Plurality of Worlds), insisting that life need not be confined to Earth.			Ray introduces the notion of species in *Historia plantarum*.
1687				
1689				
1690				
1692				
1693				Ray disputes Descartes's claim that animals are insentient and argues against the doctrine of spontaneous generation.
1694				Rudolph Jakob Camerarius publishes *De sexu plantarum epistola* (Letter on the Sex of Plants), which mounts a conclusive demonstration of the sexuality of plants.

Mathematics	Physics	Technology	Culture
Seki Kowa produces an explanation for the concept of determinants.			
Leibniz publishes the first account of differential calculus.	Hooke claims that he has found the laws governing the movement of the planets. Newton gives a brief summary of his theory in *De motu corporum* (On the Movement of Bodies), showing that planets moving under an inverse-square force law will obey Kepler's laws.		
John Wallis's *De algebra tractatus* (Treatise of Algebra) contains the first publication of Newton's binomial theorem.			
Leibniz describes the integral calculus in print for the first time.			
	Newton publishes *Philosophiae naturalis principia mathematica* (Mathematical Principles of Natural Philosophy), which presents his theories of motion, gravity, and mechanics and is the basis of much of modern physics.		
			John Locke publishes *Epistola de tolernatia* (Letter on Toleration), a masterful reflection of the enlightening role of science (letters on tolerance were written in English in 1690, 1692, and 1694).
Jakob Bernoulli uses the word *integral* for the first time to refer to the area under a curve.	Huygens proposes a theory of light as a longitudinal wave with vibration in the direction of its travel.		Locke lays down tenets for empiricist epistemology in his *Essay concerning Human Understanding*.
Leibniz introduces the terms *coordinate*, *abscissa*, and *ordinate*.			
Wallis publishes the first complete statement of Newton's version of the calculus.			

Date	Astronomy and Cosmology	Chemistry	Earth Sciences	Life Sciences
1694				The concept of genus is introduced by Joseph Pitton de Tournefort.
1696				
1697		Georg Ernst Stahl introduces the concept of phlogiston as the cause of combustion and rusting.		
1698				
1699			William Whiston publishes *A New Theory of the Earth*, arguing that the Irish megaceros deer (discovered in 1697) and a fossil elephant show that fossils are connected to Noah's flood.	
1700	Halley produces magnetic charts of the Atlantic and Pacific Oceans.			
1701				
1704				
1705	Halley predicts the periodicity of Halley's comet (confirmed in 1758).		Hooke's lectures on earthquakes are published posthumously, in which it is argued that earthquakes may have changed the surface of Earth since its creation.	
1706				
1707				
1712	John Flamsteed's star catalog is published at Newton's behest without his permission.			
1713				

Mathematics	Physics	Technology	Culture
The first textbook on differential calculus is published by Guillaume-François-Antoine de L'Hospital.			John Toland publishes *Christianity Not Mysterious*, the cornerstone of deism, claiming that even the Bible must be subject to rational scrutiny.
		Thomas Savery builds a steam-powered water pump for pumping water out of mines. Denis Papin builds a steam engine.	
			On the advice of Leibniz, Frederick I creates the Societé des Sciences in Berlin.
	Joseph Sauveur introduces the term *acoustics*.		Yale University is founded.
	Newton publishes *Opticks*, which asserts that light is particulate in nature.		
William Jones introduces the Greek letter π to represent the ratio of the circumference of a circle to its diameter.			
Newton publishes a precise formulation of Descartes's rule of signs.		Denis Papin modifies Thomas Savery's steam pump.	
Giovanni Ceva's *De re numeraria* (Concerning Money Matters) applies mathematics to economics for the first time.		Thomas Newcomen builds a practical piston-and-cylinder steam-powered water pump for pumping water out of mines.	
Nikolaus Bernoulli proposes the problem known as the Petersburg paradox. Jakob Bernoulli's book *Ars conjectandi* (The Art of Conjectures) is the first book to apply calculus to probability theory.	Second revised edition of Newton's *Principia*, incorporating the famous General Scholium and Newton's rules of reasoning, is published.		

Date	Astronomy and Cosmology	Chemistry	Earth Sciences	Life Sciences
1714	William Derham publishes *Astro-theology*, in which he uses astronomy as grounds for various theistic claims.			Cotton Mather discovers the remains of mastodons near Albany, New York.
1715				Thomas Fairchild reports the production of the first artificial hybrid plant.
1718	Halley measures stellar proper motions. Jacques Cassini publishes measurements which he claims support Descartes's prediction that Earth is elongated at the poles.	Georg Ernst Stahl publishes *Fundamenta chymiae dogmaticae et experimentalis*, advancing the phlogiston theory (first outlined in 1697). Etienne-François Geoffroy presents his *Tables of Affinities*, the first systematic record of the chemical reactivity of elements and compounds.		
1724				
1725			Luigi Marsili's *Histoire physique de la mer* (Physical History of the Sea) is the first treatise on oceanography.	
1727				Stephen Hales lays the foundation for plant physiology with *Vegetable Staticks* or *Statistical Essays.*
1728	James Bradley proposes the aberration of light to explain the observed shifts of stars' positions.			
1729				
1730				
1732	Pierre Louis Maupertuis predicts the shape of Earth using Newtonian principles.			
1733	Anders Celsius publishes his observations of the aurora borealis.			
1734				Réamur's *Mémoires pour servir a l'histoire des insectes* (History of Insects) is a foundational work in entomology.

Mathematics	Physics	Technology	Culture
		Daniel Gabriel Fahrenheit invents the mercury glass thermometer with a scale that is named after him.	The British Parliament passes a bill setting up a prize of £20,000 for the first person to develop an accurate way to find the longitude at sea.
Brook Taylor develops the calculus of finite differences, as well as the series named after him.		John Harrison builds an eight-day clock.	
Jakob Bernoulli's memoir contains the basic concepts of the calculus of variations.			
	Hermann Boerhaave argues that heat is a fluid.		The Academy of Sciences is founded at St. Petersburg, Russia.
			American scientists meet, at Benjamin Franklin's initiative, and form the Leathernapron Club, the nucleus for the American Philosophical Society (founded in 1743).
Leonhard Euler introduces *e* as the symbol for the base of the natural logarithms.	Stephen Gray identifies differences between insulators and conductors.	René-Antoine Ferchault de Réaumer builds an alcohol thermometer with a graduated scale from 0 to 80 to indicate freezing and boiling.	
			Matthew Tindal's *Christianity as Old as the Creation* advances an array of arguments in support of deism. Approximate date for the beginning of the period of enlightenment.
			First issue of *Poor Richard's Almanack* is published by Franklin.
Abraham De Moivre publishes his discovery of the distribution curve.	Charles-François de Cisternai Dufay advances the two-fluid theory of electricity.	Chester Moor Hall invents the achromatic lens refracting telescope.	
			Voltaire publishes a work that introduces Newtonian mechanics to a French audience.

Date	Astronomy and Cosmology	Chemistry	Earth Sciences	Life Sciences
1735	(to 1736) The Académie des Sciences organizes two great scientific expeditions to measure the meridian arc, which confirm Newton's theory that Earth is flattened at the poles.		Johann Georg Gmelin discovers permafrost on a voyage to Siberia.	Linnaeus publishes a first sketch of his *Systema naturae*, with its rigorous classification of animal, plant, and mineral kingdoms. Hales reports his investigations on blood flow and pressure in animals and the hydrostatics of sap in plants.
1736				
1737				
1738				
1739				
1742				
1743				La Mettrie publishes *Man Machine*, which claims that all physiological processes are mechanical in nature.
1744				Maupertuis revives the doctrine of pangenesis—that molecules from all parts of the body are gathered into the gonads—and advances his own version of epigeneticism.
1745			Georges-Louis Leclerc Buffon suggests that Earth was formed when a comet collided with the Sun.	
1746			Jean-Étienne Guettard produces the first geological map of France.	
1747	Alexis Clairaut elaborates the first approximate solution to the three-body problem.			Albrecht von Haller produces the first textbook in physiology.

Mathematics	Physics	Technology	Culture
Euler introduces the notation $f(x)$ for functions.			
Euler produces the first textbook on mechanics based on differential equations.		Harrison presents the first stable nautical chronometer, which allows for precise longitude determination at sea.	
Euler shows that the number e and its square are both irrational.			
	Maupertuis reports his measurements made in Lapland, supporting the Newtonian view that Earth is flattened at the poles.		
			The Royal Society of Edinburgh is founded. David Hume applies the experimental method to problems of psychology and human nature.
		Celsius proposes the centigrade thermometer scale.	
	Jean le Rond d'Alembert advances the principle that actions and reactions in a closed system of moving bodies are in equilibrium.		
	D'Alembert employs his principle to describe fluid motion. Maupertuis advances the principle of least action, which states that nature operates in such a way that action is at a minimum.		
	Petrus van Musschenbroek and Ewald von Kleist discover the principle of the Leyden jar, in which static electricity charges could be stored (announced by Musschenbroek in 1746).		
	D'Alembert develops the theory of complex numbers.	Euler works out the mathematics of the refraction of light.	Princeton University is founded. The first laboratory for experimental physics in the United States is founded.
D'Alembert presents the first general use of differential equations in mathematical physics.	(to 1754) Franklin elaborates the single-fluid theory of electricity. D'Alembert publishes his theory of vibrating strings.	Jean-Antoine Nollet constructs one of the first electrometers.	

Date	Astronomy and Cosmology	Chemistry	Earth Sciences	Life Sciences
1748	Bradley announces his discovery of the nutation of Earth's axis.			John Turberville Needham and Buffon conduct an experiment that supports the doctrine of spontaneous generation.
1749	D'Alembert provides a mathematical account of the regular changes in the orientation of Earth's axis.			Buffon begins publication of the 44-volume set, *Histoire naturelle*. He gives the modern definition of *species* as a group of organisms capable of breeding and producing offspring.
1750	Thomas Wright discusses galaxies and the shape of the Milky Way.			
1751				Linneaus proposes binary nomenclature in *Philosophia botanica*.
1752	Johann Tobias Mayer publishes tables of the motion of the Moon compared to the motion of the stars accurate enough for finding the longitude at sea.	Joseph Black conducts a series of experiments on alkaline and alkaline-earth carbonates, bringing pneumatic chemistry back into vogue.	Nicolas Desmarest proposes that England and France were once connected by a land bridge.	
1753				
1754		Black discovers "fixed air" (carbon dioxide).		Charles Bonnet discusses the nutritional value of plants.
1755	Immanuel Kant advances the nebular hypothesis of the origin of the solar system.			
1756	John Canton carries out the first observation of magnetic storms in Earth's magnetic field.		Black's *Experiments upon Magnesia, Quicklime, and Other Alkaline Substances* is the first work in quantitative chemistry.	
1758				
1759				Caspar Friedrich Wolff asserts the existence of a life force that is at the heart of living things.
1760				

Mathematics	Physics	Technology	Culture
Euler publishes *Analysis of Infinities*, an introduction to pure analytical mathematics.			
	Franklin installs a lightning rod on his home in Philadelphia.		David Hartley uses the term *psychology* to interpret the phenomena of mind by the theory of association.
Gabriel Cramer publishes his rule for solving systems of linear equations.			
	Franklin distinguishes between positive and negative electricity.		Hume composes *Dialogues concerning Natural Religion*, which challenges the belief in design according to divine plan. (1751 to 1772) Denis Diderot and D'Alembert coedit the *Grande Encyclopédie*, widely regarded as the most influential work of the eighteenth century. The first mental institution is opened in London.
Euler advances Euler's formula ($V - E + F = 2$), which was discovered earlier by Descartes. Franklin shows that lightning is electricity.	D'Alembert advances his principles of hydrodynamics.		
			The British Museum is founded around the library and collection of Hans Sloane.
			The University of Moscow is founded in Russia.
		John Dollond invents an achromatic lens system.	
	Frans Ulrich Theodosius Aepinus publishes *Tentamen theoriae electricitatis et magnetisimi.*	Harrison completes the marine chronometer that wins the British Board's prize for a practical way to find the longitude at sea.	
			The Kew Botanical Gardens are opened in London.

Date	Astronomy and Cosmology	Chemistry	Earth Sciences	Life Sciences
1761	Johann Heinrich Lambert advances a theory for the structure of the Milky Way in which stars are contained in giant spherical clusters that are confined to a region of small thickness.			
1762	Bradley finishes a catalog that lists the measured positions of 60,000 stars.			
1763			Jeremiah Dixon and Charles Mason begin their survey of the Pennsylvania-Maryland boundary that will result in the Mason-Dixon line. Antoine-Laurent Lavoisier and Guettard produce a mineralogical atlas of France.	
1764	Joseph-Louis Lagrange explains the librations of the Moon.			Bonnet advances his preformation theory.
1765				George Croghan discovers remains of mastodons in Kentucky; remains studied by Buffon.
1766		Henry Cavendish isolates hydrogen and defines it as pure phlogiston.		Haller shows that nerves stimulate muscles to contract and that all nerves lead to the spinal cord and brain.
1767				
1768				Spallanzani contends that spontaneous generation does not occur in tightly closed bottles that have been boiled for more than 30 minutes.
1769				Bonnet argues that the females of every species contain the germs of all future generations.
1770				
1771		Priestley discovers that plants convert carbon dioxide into oxygen.	Abraham G. Werner develops the theory of neptunism.	Rhinoceros remains are found in Siberia.
1772		Priestley reports that plants can restore air that has been made lifeless by animals or by fire. Priestley discovers how to make soda water from carbon dioxide and water. Lavoisier experiments on combustion.	James Cook proves that there is no large southern continent except Australia.	

Mathematics	Physics	Technology	Culture
	Black discovers latent heat.		
Thomas Bayes publishes *An Essay towards Solving a Problem is the Doctrine of Chances.* It includes Bayes's theorem, an important theorem in statistics. Gaspard Monge develops descriptive geometry.			
		James Watt develops his own improved steam engine, in which the condenser is separated from the cylinder.	
Lambert develops much of what is now called *non-Euclidean geometry.*	Euler advances a general treatment of the motion of rigid bodies, including the precession and nutation of Earth.	Horace-Bénédict de Saussure invents the electrometer.	Mathew Boulton founds the Lunar Society, which is devoted to promoting the arts and sciences.
	Priestly proposes that electrical force follows an inverse-square law.		
Euler publishes his discoveries on differential equations. Lambert proves that π is an irrational number.		Antoine Baumé invents the graduated hydrometer.	Cook sets sail for Tahiti with the naturalist Joseph Banks.
Euler lays the foundations for the calculation of optical systems.		Watt patents his first improved steam engine.	
Euler advances proof that Fermat's last theorem is true for $n = 3$.			
	Luigi Galvani discovers the action of electricity on the muscles of a dissected frog.		The first issue of *Transactions* is published by the American Philosophical Society.
Jean-Baptiste Louis Romé de l'Isle describes and identifies 110 crystal forms.			

Date	Astronomy and Cosmology	Chemistry	Earth Sciences	Life Sciences
1772		Nitrogen is isolated independently by Daniel Rutherford, Cavendish, Priestley, and Carl Wilhelm Scheele. Louis Bernard Guyton de Morveau reports that metals gain weight on calcinations.		
1774		Priestley discovers oxygen (dephlogisticated air) and visits Lavoisier in Paris (discovered independently by Scheele in 1772).	Werner advances method of classifying minerals by their physical characteristics.	Franz Anton Mesmer introduces hypnotism as a way of curing disease.
1775				Mesmer suggests that "animal magnetism" causes attraction between persons.
1776			James Keir claims that rock formations may have been caused by molten rock crystallizing as it cooled.	
1777		Scheele claims that air consists of oxygen and nitrogen.	Desmarest claims that basalt is made from lava emitted by volcanoes.	
1778		Scheele and Lavoisier discover that air is composed mostly of nitrogen and oxygen. Volta studies inflammable air from marshes and discovers methane gas.		
1779		Lavoisier proposes the name *oxygen* for the part of air that causes combustion.	Buffon claims that Earth is about 75,000 years old. Saussure coins the term *geology*.	Jan Ingen-Housz discovers that plants have two distinct respiratory cycles, a discovery that becomes clearer after the work of Lavoisier. Ingen-Housz reports that sunlight is essential for the production of oxygen by leaves. Spallanzani shows that for fertilization to take place, a sperm must be in physical contact with an egg.
1780		Scheele discovers lactic acid.		The skull of a prehistoric reptile is discovered in the Netherlands; it is identified as such by Cuvier in 1795.
1781	Herschel discovers Uranus. Charles Messier catalogs more than 100 star clusters and nebulas.	Lavoisier states the law of conservation of mass. Priestley creates water by igniting hydrogen and oxygen.		

Mathematics	Physics	Technology	Culture
		Alessandro Volta produces a device for storing a charge of static electricity. It leads to the modern electrical condenser. Watt obtains a patent for his version of the steam engine.	
	Pierre-Simon Laplace states that the future can be predicted completely if all the forces on all objects at any one time are known.		
Euler uses i to represent the square root of negative 1.		Charles Augustin Coulomb invents the torsion balance.	
	Lavoisier and Laplace publish a memoir on heat, which concludes that respiration is a form of combustion.		The American Society of Arts and Sciences is founded.
	Johan Carl Wilcke introduces the concept of specific heat.		Immanuel Kant publishes *The Critique of Pure Reason*, which asserts the impossibility of knowledge regarding the existence of atoms.

Date	Astronomy and Cosmology	Chemistry	Earth Sciences	Life Sciences
1782	John Goodricke explains the variation in the light from the star Angol, which is caused by an invisible companion star.			
1783	Herschel publishes his first list of double stars. Laplace advances a new method for solving the equations of motion of celestial bodies.	Lavoisier proposes new chemistry based on the notion of a chemical element. Lavoisier identifies a new gas as hydrogen.		
1784		Scheele discovers citric acid.		
1785	Herschel gives a description of the Milky Way.		James Hutton advances the principle of uniformitarianism.	
1786	Herschel publishes his *Catalogue of Nebulae*.	Kant declares that chemistry cannot be a science in the proper sense of the term.	Johann von Charpentier works out the details of a European ice age.	
1787		The first of four volumes of *Méthode de nomenclature chimique* is published by Lavoisier, Guyton, Claude Louis Berthollet, and Antoine François de Fourcroy. Jacques-Alexandre-César Charles discovers Charles's law, which states that the volume of a given mass of gas at constant pressure is directly proportional to its absolute temperature.		
1788			Hutton publishes his uniformitarian theory of geology.	
1789	Herschel completes his giant reflecting telescope.	Lavoisier publishes *Traité élémentaire de chimie*, which establishes the new chemical nomenclature.		Antoine-Laurent de Jussieu provides a new way of classifying plants into families that is still in use.
1790	Herschel discovers planetary nebulas.		Johann Friedrich Blumenbach describes 60 human crania.	
1791			William Smith notes the relationship between fossils and geologic strata.	
1793				Jean-Baptiste de Lamarck argues that fossils are the remains of organisms that were once alive.

Mathematics	Physics	Technology	Culture
Adrien-Marie Legendre describes polynomial solutions to differential equations. Legendre begins work on elliptic integrals and elliptic functions.	George Atwood determines the acceleration of a body in free fall.	Franklin invents bifocal lenses.	
Legendre advances the first statement of the law of quadratic reciprocity, but his proof is flawed.	Coulomb announces his inverse-square law of electrical attraction and repulsion.		
	Galvani discovers animal electricity and postulates that animal bodies are storehouses of electricity (published in 1789).	Abraham Bennet invents the gold-leaf electroscope.	
Lagrange publishes *Mécanique analytique* (Analytical Mechanics), which carries the study of mechanics to its highest level of generality.			The Linnean Society of London is founded.
	Galvani announces the discovery of animal electricity.	Bennet invents an electric induction machine.	The "Declaration of Rights of Man and of the Citizen" is passed in France.
			The journal *Annalen der Physik* is founded.
		The metric system of measurement is proposed in France.	
			The first chemical society is foundedin Philadelphia.

Date	Astronomy and Cosmology	Chemistry	Earth Sciences	Life Sciences
1794			Smith publishes the first large-scale map of England.	(to 1796) Erasmus Darwin publishes *Zoönomia*, which advances the notion that species are transformed by environmental factors.
1795			Hutton publishes *Theory of the Earth*. Cuvier identifies bones as belonging to a prehistoric reptile known as the Mosasaur.	James Braid renames *mermerism* as *hypnotism* and gives the practice medical respectability.
1796	Laplace presents his modified Kantian hypothesis on the origins of the solar system in his *Exposition du systéme du monde*.			Cuvier composes a memoir comparing fossil elephants to living species, showing that the two are different. Erasmus Darwin speculates that all living things derive from one filament in *Zoönomia*.
1797	Caroline Herschel discovers her eighth comet in seven years.	Joseph Louis Proust formulates law of definite proportions.	James Hall proves that igneous rocks cool to form crystalline rocks.	
1798	Bradley's catalog of stellar position is published. Laplace predicts the existence of black holes.			Thomas Malthus argues that an expanding population can only be controlled by famine, disease, and war. This idea influences Darwin and Wallace.
1799	Friedrich Wilhelm Humboldt observes the Great Leonid meteor shower. The first of five volumes of Laplace's *Traité de mécanique céleste* is published.		Humboldt identifies the Jurassic Period.	
1800		William Nicholson and Anthony Carlisle use electrolysis to separate water into hydrogen and oxygen. Humphry Davy discusses the effects of nitrous oxide. Nicholson discovers electrolysis.		Cuvier publishes *Leçons d'anatomie compare* (Lessons on Comparative Anatomy), the first complete work in the field of comparative anatomy. Karl Friedrich Burdach coins the term *biology* to denote the study of human morphology, physiology, and psychology.
1801	Johann Elert Bode publishes an atlas of 17,240 stars and nebulas. Giuseppe Piazzi discovers Ceres, the first known asteroid.	John Dalton formulates the law of partial pressures in gases: Each component of a gas mixture produces the same pressure as if it occupies the container alone.		Cuvier identifies 23 species of extinct animals. Lamarck elaborates an evolutionary theory based on law of increasing complexity, modified by the change of organs through continued use and loss through disuse.
1802		Dalton compiles his first table of atomic weights (published in 1805).	John Playfair publishes *Illustrations of the Huttonian Theory of the Earth*.	Gottfried Treviranus and Lamarck independently broaden the meaning of biology to include the study of all living things.

Mathematics	Physics	Technology	Culture
		Daniel Rutherford invents the first maximum-minimum thermometer	The École Polytechnique is opened in Paris.
			France adopts the metric system.
Gauss proves the law of quadratic reciprocity.			The Rumford Medals of the Royal Society and the American Association for the Advancement of Science are instituted.
Lagrande introduces the use of the infinite Taylor series derived from functions.			
Caspar Wessel introduces the vector representation of complex numbers.	Count Rumford proposes the idea that heat is a form of energy. Cavendish measures the gravitational constant.		
Gauss proves the fundamental theorem of algebra—that every polynomial equation has a solution.		Volta invents the electric battery, made of disks of zinc and silver (announced in 1800).	The Royal Institution is founded in London.
	Herschel discovers infrared radiation. Thomas Young advances a wave theory of light.		
Gauss proves that every natural number is the sum of at most three triangular numbers.	Johan Wilhelm Ritter discovers ultraviolet radiation. William Wollaston shows that frictional and galvanic electricity are the same. Young discovers interference.		
	Young uses the wave theory of light to explain optical interference. Wollaston observes dark lines in the solar spectrum.	Robert Hare invents the blowpipe.	

Date	Astronomy and Cosmology	Chemistry	Earth Sciences	Life Sciences
1802	Heinrich Olbers discovers Pallas, the second known asteroid.			
1803		Dalton advances atomic theory, contending that matter is composed of matter of different weights. Dalton states the law of multiple proportions: When two elements combine to form more than one compound, the weights of one element combine with a fixed weight to the other in a ratio of small whole numbers. Dalton devises a system of chemical symbols and arranges the relative weights of atoms in a table. Berthollet shows that reactions depend both on affinities and the amount of reacting substances. William Henry formulates the law that the mass of gas dissolved in a liquid is directly proportional to pressure.	Jean-Baptiste Biot concludes that meteorites do not originate on Earth.	
1804	Karl Ludwig Harding discovers the third asteroid, Juno. Olbers discovers the fourth asteroid, Vesta.		Biot and Joseph Louis Gay-Lussac ascend in a balloon to study terrestrial magnetism and Earth's atmosphere.	Saussure shows that plants require carbon dioxide from the air and nitrogen from the soil.
1805				The science of comparative anatomy is founded by Cuvier.
1806			Nicolas Louis Vauquelin and Jean-Bapsiste René Robinet isolate asparagines, the first known amino acid.	
1807		Davy discovers the element potassium by electrolysis. Jöns Jacob Berzelius classifies chemicals as either organic or inorganic. Jean Antoine Chaptal publishes the first book on industrial chemistry, *Chemistry Applied to the Arts*.		
1808	Siméon-Denis Poisson publishes his work on perturbations of planetary orbits.	Gay-Lussac publishes the law of combining gases in terms of rational and simple relations of volume. Davy isolates the alkaline-earth metals magnesium, calcium, strontium, and barium.		

Mathematics	Physics	Technology	Culture
Carl Gustav Jacobi works on elliptic functions.		Richard Trevithick builds a locomotive that pulls five loaded coaches along a track for 9 1/2 miles.	
	Laplace reports his theory of capillary forces.		
Legendre publishes the first printed description of the method of least squares.			
			Coal-gas lighting begins to illuminate the streets of London. The Geological Society, the world's first institution devoted solely to the study of geology, is founded.
	Étienne-Louis Malus discovers that reflected light is polarized.	Davy develops the first electric-powered lamp.	

Date	Astronomy and Cosmology	Chemistry	Earth Sciences	Life Sciences
1808		Dalton's atomic theory of chemical combinations is laid out in *A New System of Chemical Philosophy*.		
1809			The first geological survey of the United States is published by William Maclure.	Lamarck publishes *Zoological Philosophy*, detailing his theory of evolution by acquired characteristics.
1810		Davy shows that chlorine is an element.		
1811	Herschel advances his theory of development of stars from nebulas.	Amedeo Avogadro claims that equal volumes of gases should contain equal numbers of molecules. Gay-Lussac and Louis Jacques Thenard show that hydrocyanic acid does not contain oxygen, refuting Lavoisier's contention that oxygen is characteristic of acids. Berzelius introduces the modern system of chemical symbols.	Cuvier and Alexandre Brongniart publish a geological map of the Paris basin.	Johann Friedrich Meckel formulates the principle that ontogeny follows the stages of phylogeny.
1812		Berzelius suggests that atoms have electrical charges.		Cuvier discovers the fossil of a pterodactyl. Cuvier explains his theory of extinctions of animal groups in catastrophies.
1813				Augustin-Pyramus de Candolle's *Elementary Theory of Botany* introduces the term *taxonomy* to mean biological classification.
1814	Joseph Fraunhofer makes his first map of the solar spectrum.			
1815		William Prout claims that hydrogen is an atom and that all other atoms are built up from different numbers of the hydrogen atom (Prout's hypothesis).	Smith identifies rock strata on the basis of fossils.	
1817				Cuvier describes the entire animal kingdom, dividing it into four groups.
1818		Berzelius publishes his table of atomic weights.		
1819	Arago discovers that light from comet tails is polarized.		Eilhardt Mitscherlich formulates the law of isomorphism.	

Mathematics	Physics	Technology	Culture
Joseph Jean Baptiste Fourier presents the French Academy his work on representing functions by infinite trigonometric series.			The University of Berlin, the first university to place research ahead of education, is founded.
	Poisson advances a mathematical theory of heat based on the work of Fourier.		
	In his *Essai philosophique sur les probabilités*, Laplace draws out the deterministic ramifications of classical mechanics. David Brewster discovers the law named after him.	Dominique François Jean Arago invents the polarization filter. Wollaston invents the camera lucida.	
	Hans Christian Oersted suggests that electricity ought to be convertible to magnetism.		The French edition of Ptolemy's *Almagest*, prepared by Nicola Halma, is published, marking the beginning of the critical publication of classical works in the history of science.
			George Stephenson introduces his first steam locomotive.
	Augustin Fresnel demonstrates that transverse waves explain the diffraction of light.	Davy invents the safety lamp for coal miners.	
	Biot discovers biaxial crystals.		The *American Journal of Science and Arts* is founded.
	Oersted shows that an electric current is able to deflect a magnetic needle (published in 1820).		

Date	Astronomy and Cosmology	Chemistry	Earth Sciences	Life Sciences
1820				
1821			Ignatz Venetz proposes that glaciers once covered much of Earth.	
1822			William Daniel Conybeare and William Phillips identify the Carboniferous Period. Jean-Baptiste Julien Omalius d'Halloy identifies the Cretaceous Period. Friedrich Mohs introduces his system of classifying minerals.	Lamarck distinguishes between vertebrates and invertebrates.
1823	Herschel suggests that the Fraunhofer lines indicate the presence of metals in the Sun.		Daniell presents a study of the atmosphere and trade winds.	
1824				
1825	John Herschel describes a device for measuring the intensity of solar radiation.			Henry Walter Bates elaborates a theory of insect mimicry.
1826	Olbers formulates the Olbers paradox: If stars are evenly distributed through infinite space, why is the night sky dark? Samuel Heinrich Schwabe discovers the sunspot cycle.	Berzelius proposes a new system of chemical notation.	René-Joachim-Henri Dutrochet studies osmosis and the law relating to it.	
1827	Félix Savart shows that the binary star Zeta Ursae Majoris is governed by Newton's law of gravitation.	Pierre Berthelot synthesizes many organic compounds.		Karl Ernst von Baer reports that he has observed the ovicel in the ovary of a dog.

Mathematics	Physics	Technology	Culture
	Arago shows that a current can magnetize iron (the electromagnet). Ampère states the relationship between a magnetic field and the electric current that produces it (Ampère's law).	Johann Salomo Christoph Schweigger builds the first galvanometer.	The Royal Astronomical Society is founded in London.
	Michael Faraday shows that electrical forces can produce rotational motion of a needle (the principle of the electric motor). Thomas Seebeck converts heat into electricity (thermoelectricity).		
Karl Wilhelm Feuerbach publishes Feuerbach's theorem. Fourier lays the foundation for what is now known as *Fourier analysis*. Jean-Victor Poncelet lays the foundation for projective geometry.			The Congregation of the Holy Office abrogates the condemnation of Copernican theories that had been issued on March 5, 1616.
		William Sturgeon builds the first electromagnet. Charles Babbage begins construction of the Difference Engine, a machine for calculating logarithms and trigonometric functions.	The Royal Asiatic Society of Great Britain and Ireland is founded to study the civilization of the various regions of Asia.
Jakob Steiner develops inversive geometry. Gauss reports that he has discovered non-Euclidean geometry.	Sadi Carnot describes the Carnot cycle (the basis of the second law of thermodynamics).	Carnot publishes *Reflections on the Motive Power of Fire*, which advances the general principle of the steam engine. Fraunhofer builds a telescope that is mounted equatorially with a clock drive.	
Niels Henrik Abel discovers elliptic functions.	André Marie Ampère deduces law for force between current carrying conductors.	Richard August Carl Emil Erlenmeyer invents the flat-bottomed flask.	
Abel gives an example of an integral equation. Nikolai Lobachevsky reports his ideas on non-Euclidean geometry.	Georg Simon Ohm proposes what is now called *Ohm's law*, relating current, voltage, and resistance.		
Gauss introduces the subject of differential geometry, which describes features of surfaces by analyzing curves that lie on them. Homogeneous coordinates are invented by a number of mathematicians.	Robert Brown reports his observations on the phenomenon called *Brownian motion*. Ampère extends the inverse-square law to magnets. William Rowan Hamilton predicts conical refraction.		

Date	Astronomy and Cosmology	Chemistry	Earth Sciences	Life Sciences
1828			Paul Erman measures the magnetic field of Earth.	Von Baer advances his germ layer theory, according to which eggs develop to form four layers of tissues.
1829				
1830		Berzelius coins the term *isomerism* to describe chemicals of identical chemical composition but different structural properties.	Charles Lyell publishes the first volume of *Principles of Geology*, offering evidence for the uniformitarian theory of Earth's geological history.	
1831			The second volume of Lyell's *Principles of Geology* is published. James Clark Ross reaches the magnetic pole on June 1.	Charles Darwin begins his five-year voyage on the HMS *Beagle*. Robert Brown discovers the nucleus of the cell.
1832				
1833	A great meteor shower hits the United States.	Anselme Payen discovers the first enzyme, diatase, an extract of malt that accelerates the conversion of starch to sugar.	The third volume of Lyell's *Principles of Geology* is published.	Louis Agassiz publishes his research on fossilized fish.
1834		The British Association for the Advancement of Science recommends the adoption of the system of chemical symbols devised by Berzelius.	Friedrich August von Alberti identifies the Triassic Period in Earth's history.	Christian Jorgensen divides human history into a Stone Age, a Bronze Age, and an Iron Age.
1835			Roderick Impey Murchison identifies the Silurian Period. Adam Sedgwick identifies the Cambrian Period.	Darwin visits the Galápagos Islands. Quetelet develops the normal curve.
1836	Francis Baily describes Baily's beads.		Lyell proposes the divisions of the Eocene, Miocene, and Pliocene for the Tertiary Period.	

Mathematics	Physics	Technology	Culture
George Green introduces the theorem named after him, which reduces certain volume integrals to surface integrals.			
Lobachevsky develops hyperbolic geometry. Evariste Galois invents group theory. Julius Plücker establishes the principle of duality for projective geometry.	Gustave Gaspard de Coriolis coins the term *kinetic energy*.	Joseph Henry invents the first practical electric motor.	
	Henry discovers electromagnetic induction (the principle of the dynamo) but does not publish his discovery.	Louis Braille devises a form of communication by touch rather than sight. Joseph Jackson Lister develops an achromatic lens for the microscope.	Comte Montpellier establishes positivism as a philosophical school.
Adolphe Quetelet studies how the crime rate in France is affected by such factors as gender, education, climate, and season. Evariste Galois presents his third version of group theory, which is rejected by a prominent mathematician as unclear.	Faraday discovers electromagnetic induction and devises the first electrical generator. Faraday begins work on electrolysis.	Henry proposes an electromagnetic telegraph.	The British Association for the Advancement of Science is founded. The Royal Society starts publishing the Proceedings of the Royal Society.
János Bolyai publishes the second account of non-Euclidean geometry.		Samuel Morse invents the telegraph and a telegraphic alphabet. Babbage conceives of the first computer, the Analytical Engine. Charles Wheatstone invents the stereoscope.	
	Faraday announces the laws of electrolysis. Faraday and William Whewell introduce the terms *electrode*, *anode*, *ion*, *cathode*, *anion*, *cation*, *electrolyte*, and *electrolysis*.		At a meeting of the British Association for the Advancement of Science, Whewell proposes the term *scientist*.
Hamilton transforms the Lagrangian formulas to the form now called *Hamiltonians*.	Benoit-Pierre Chapeyron develops the second law of thermodynamics. Jean Charles Athanase Peltier discovers the Peltier effect.		
		Morse produces the first working model of the electric telegraph. Henry invents the electric relay.	
Theodor Schwann discovers the enzyme pepsin, the first known animal enzyme.			Daniell invents the Daniell cell.

Date	Astronomy and Cosmology	Chemistry	Earth Sciences	Life Sciences
1837	Friedrich Georg Struve publishes his catalog of double stars.			Darwin starts to assemble his argument for evolution by natural selection.
1838	Friedrich Bessel, Thomas Henderson, and Struve measure stellar parallax.			Jacob Matthias Schleiden proposes that all living plant tissue is made up of cells. Schwann shows that yeast is made of small living organisms.
1839				Schwann extends Schleiden's proposal to animals, laying the foundations of cell biology.
1840	John William Draper takes photographs of the Moon.			
1841	Bessel determines dimensions of Earth with geodetic degree measurements. Bessel identifies an unseen companion to Sirius that will come to be the first white dwarf.	Berzelius reports chemical allotrophy (two different forms of the same element).		
1842			Richard Owen coins the name *dinosaur*.	Darwin finishes a 35-page sketch of his theory but does not publish it. Darwin classifies coral reefs into three types.
1843	John Couch Adams uses irregularities in the orbit of Uranus to calculate the position of an unknown planet that we now know as Neptune.		Thomas Chrowder Chamberlin argues that there were many ice ages.	
1844				Robert Chambers publishes *Vestiges of the Natural History of Creation*, in which he argues for an evolutionary view. Darwin's "sketch" is now 230 pages.
1845	Urbain Jean Joseph Le Verrier postulates the existence of an eighth planet. William Parsons discovers the spiral form of galaxies. Le Verrier observes an excess precession of Mercury's orbit of 35 seconds of arc.			

Mathematics	Physics	Technology	Culture
Poisson establishes the rules of probability and describes the Poisson distribution.		Wheatstone and William Fothergill Cooke produce the first practical telegraph system.	
	Faraday discovers the Faraday dark space, a dark area before the cathode.		
Gauss lays the foundations of potential theory.		Charles Goodyear invents vulcanized rubber. Louis Daguerre reports his process for making photographs.	The Harvard College Observatory, the first official observatory in the United States, is founded.
	James P. Joule states that the amount of heat produced per second in any conductor is proportional to the product of the square of the current and the resistance of the conductor.	Giovanni Battista Amici invents the oil immersion microscope.	The first engineering professorship is established at the University of Glasgow.
			Quetelet establishes a statistical bureau in Belgium.
	Christian Doppler points out that the frequency of sound waves varies as the source moves closer or farther away from the observer (the Doppler shift). Julius Robert von Mayer states the law of conservation of energy.		
Hamilton invents quaternions, which make possible the application of arithmetic to three-dimensional objects. Arthur Cayley investigates spaces with more than three dimensions.	Joule determines a value for the mechanical equivalent of heat.	Morse builds the first long-distance electric telegraph line.	
Joseph Liouville finds the first transcendental numbers.		Morse patents his design for the telegraph.	
Augustine-Louis Cauchy proves the fundamental theorem of group theory, now known as *Cauchy's theorem.* Cayley publishes *Theory of Linear Transformations*, which lays the foundations for the school of pure mathematics.	Faraday discovers the effect of magnetic field on polarized light (the Faraday effect). Faraday discovers diamagnetism and paramagnetism.		The Royal College of Chemistry is founded in England.

Date	Astronomy and Cosmology	Chemistry	Earth Sciences	Life Sciences
1846	Johann Galle discovers the planet Neptune using predictions of its position by Leverrier and Adams.		William Thomson (later Lord Kelvin) calculates the age of Earth at 100 million years.	Hugo von Mohl identifies protoplasm as the principal living substance of a cell. The first experimental psychology laboratory is established in the United States.
1847				
1848	Henry shows that sunspots are cooler than neighboring surfaces.	Louis Pasteur founds the science of stereochemistry.		
1849				
1850		Thomas Graham founds the science of colloidal chemistry.	Johann von Lamont discovers periodic vibrations in the magnetic field of Earth. Matthew Fontaine Maury discovers clear indications of the Mid-Atlantic Ridge.	William Cranch Bond shows that plant and animal protoplasm are the same.
1851	Jean Bernard Léon Foucault demonstrates the rotation of Earth with a huge pendulum.			
1852		Edward Frankland publishes the first paper on organometallics and the first adumbration of the theory of valency.	Edward Sabine shows that changes in sunspots affect Earth's magnetic field.	
1853			James Henry Coffin describes three distinct wind zones that occur in the northern hemisphere.	

Mathematics	Physics	Technology	Culture
	In a short paper, "Thoughts on Ray Vibrations," Faraday discusses the possibility of doing away with the aether.		The Smithsonian Institution is founded in the United States.
George Boole publishes his views on symbolic logic. Ernst Eduard Kummer introduces the concept of an ideal into number theory. Kummer describes his invention of ideal complex numbers to show that Fermat's last theorem holds for the regular primes.	Hermann von Helmholtz proposes the law of the conservation of energy (independently of Mayer). Joule (independently of Helmholtz and Mayer) discovers the law of the conservation of energy.		
	Kelvin calculates the absolute zero point of temperature. Armand Fizeau confirms Doppler's claim that light should behave in the same way as sound.		The American Association for the Advancement of Science is extablished.
	Kelvin coins the term *thermodynamics*. Fizeau measures the velocity of light in air.		
	Rudolf Clausius states a form of the second law of thermodynamics.		
Joseph Liouville establishes the existence of transcendental numbers. Georg Bernard Riemann advances his notion now called *Riemann surfaces*.	Kelvin states that energy in a closed system tends to become unusable waste heat (the second law of thermodynamics). Kelvin proposes the concept of absolute zero. George Stokes describes mathematically how a small body falls through a fluid.		
	Faraday publishes "On the Physical Character of the Lines of Magnetic Force," charging that Newtonian science is an obstacle to the progress of science. Kelvin publishes *Physical Considerations Regarding the Possible Age of the Sun's Heat*, arguing that the universe will suffer "heat death."		
	William John Macquorn Rankine introduces the concept of potential energy.		

Date	Astronomy and Cosmology	Chemistry	Earth Sciences	Life Sciences
1854		David Alter predicts that each element can be identified from its spectrum. Alexander William Williamson explains how a catalyst works.		Alfred Russel Wallace sets off for the Malay Archipelago (Indonesia).
1855	Parsons observes the spiral structure of galaxies.	Charles-Adolphe Wurtz develops the method of synthesizing long-chain hydrocarbons.	Maury publishes first textbook of oceanography, *The Physical Geography of the Sea.*	Herbert Spencer publishes *The Principles of Psychology*, applying Lamarkian principles in a limited way. Claude Bernard advances his theories of homeostasis.
1856	Bond discovers that photographs of stars reveal their magnitudes.			The first Neanderthal skeleton is found in a cave near Dusseldorf, Germany. Pasteur reports that fermentation is caused by microorganisms (yeast). Bernard discovers glycogen.
1857	Bond photographs the double star Misar.			Spencer publishes *Progress, its Law and Cause*, elaborating a full-blown theory of social evolution. Johann Gregor Mendel begins his experiments on garden peas.
1858		August Kekulé von Stradonitz publishes his paper on bonding in carbon compounds. Stanislao Cannizzaro resurrects Avogadro's hypothesis. Archibald Scott Couper introduces the concept of bonds into chemistry. Kekulé shows that the four bonds of carbon can account for the various isomers of organic compounds.		Darwin and Wallace announce to the Linnean Society their theory of evolution by natural selection.
1859				Darwin's *On The Origin of Species* is published.
1860		Berthelot synthesizes new organic molecules (methane, ethyl alcohol, methyl alcohol) from the elements, adding another blow to vitalist theories. Cannizzaro revives Avogadro's hypothesis.		Thomas Henry Huxley defends Darwin's theory of evolution at an Oxford meeting of the British Association for the Advancement of Science. Agassiz attacks Darwin's theory of evolution.

Mathematics	Physics	Technology	Culture
Boole develops the form of symbolic logic known as *Boolean algebra*. Riemann generalizes the concept of non-Euclidean geometry, showing that many non-Euclidean geometries are possible.			
	Robert Bunsen and Gustave Robert Kirchhoff begin work on spectral analysis.	The Bunsen burner is invented by C. Desaga, a technician working with Bunsen. Johann Heinrich Wilhelm Geissler invents the mercury pump. Heinrich Daniel Ruhmkorff invents the version of the induction coil that bears his name.	
	Clausius establishes the kinetic theory of heat on a mathematical basis.	The first transatlantic cable is laid.	
Cayley founds the algebra of matrices.			
	With their newly constructed spectroscope, Kirchhoff and Bunsen discover that each element has its own distinct set of spectral lines.	Gaston Planté invents the storage battery.	
	James Clerk Maxwell and Ludwig Boltzmann develop the study of the statistical behavior of gases.		Kekulé organizes the First International Chemical Congress in Karlsruhe, Germany.

Date	Astronomy and Cosmology	Chemistry	Earth Sciences	Life Sciences
1860				Gustav Theodor Fechner develops psychology as an exact science. Mendel discovers his three law of heredity.
1861		Andrers Jonas Ångström discovers that hydrogen is found on the Sun.		
1862	William Huggins shows that the same elements exist in the stars as on Earth.		Francis Galton founds the modern method of mapping the weather.	Hermann von Meyer describes the *Archaeoptyrex* found in Jurassic limestone deposits, a fossil with a mixture of avian and reptilian features.
1863			Lyell argues for the existence of early humans.	
1864	Astronomers place the Sun's distance from Earth at 147 million kilometers. Huggins demonstrates that bright nebulas consist of gases.	John Alexander Reina Newlands proposes his law of octaves.		Pasteur demolishes the doctrine of spontaneous generation.
1865		Kekulé proposes a ring structure for benzene, reportedly after dreaming about six monkeys holding one another by the tail. Jean-Servais Stas produces the first modern table of atomic weights using oxygen as a standard.		Bernard attacks the existence of a life force and insists that the laboratory should be the place of medical education. Julius von Sachs shows that chlorophyll turns carbon dioxide and water into starch while releasing oxygen.
1866	The Great Leonid meteor shower occurs. Giovanni Virginio Schiaparelli establishes that there is a relationship between some comets and meteor showers.		The ninth edition of Lyell's *Principles of Geology* is published, with an endorsement of Darwin's evolutionary ideas.	Mendel publishes his laws of inheritance, which are unnoticed until 1900.
1867	Pietro Angelo Secchi classifies the spectra of stars into four classes.	Pierre Jules César Janssen discovers a line in the Sun's spectrum, confirmed by Norman Lockyer as a new element, helium.		Galton shows that the intellectual abilities of humans form a normal distribution.
1868				Edouard Lartet discovers the remains of Cro-Magnon man in a cave in France. Darwin elaborates the theory of pangenesis.
1869		Dmitri Ivanovich Mendeléev proposes his periodic table of the elements, predicting the existence of three new elements (gallium, scandium, and germanium).	Cleveland Abbe begins to send out weather bulletins from the Cincinnati Observatory.	

Mathematics	Physics	Technology	Culture
	Kirchhoff proposes the black-body problem, ascertaining a function for a physical system that absorbs all the rays that hit it.	The first great refracting telescopes are constructed.	
	Kelvin draws on thermodynamics to estimate the age of Earth (reported in *On the Age of the Sun's Heat* and *On the Secular Cooling of the Earth*).	Étienne Lenoir makes a gasoline-engine automobile.	The National Academy of Sciences is founded in the United States.
	Maxwell advances mathematical equations that describe the electromagnetic field and predict		
	Clausius coins the term *entropy* and states the first two laws of thermodynamics: The energy of the universe is constant, and the entropy of the universe tends to a maximum.		The London Mathematical Society is founded.
		Georges Leclanché invents the electrical battery (dry cell).	
			The scientific journal *Nature* is founded.

Date	Astronomy and Cosmology	Chemistry	Earth Sciences	Life Sciences
1870				
1871			Mendeleev announces that the gaps in his periodic table represent undiscovered elements, which are discovered in 1875, 1879, and 1885.	Darwin publishes *The Descent of Man*, elaborating his theory of sexual selection.
1872	Henry Draper photographs the spectrum of the star Vega.			Darwin characterizes emotion in evolutionary terms. Jean-Martin Charcot uses hypnosis as part of his treatment for therapy.
1873	Richard Anthony announces that the craters of the Moon were formed by impact of meteorites.			(to 1876) The British *Challenger* Deep Sea Expedition studies life in the depths of the oceans.
1874		Joseph Achille Le Bel and Jacobus Henricus van't Hoff independently propose that the molecule has a three-dimensional structure (stereochemistry).		
1875		In a zinc ore mined in the Pyrenees, Paul Émile Lecoq de Boisbaudran discovers gallium, in support of Mendeleev. Clemens Winkler discovers germanium, in support of Mendeleev.		
1876				
1877	Schiaparelli discovers "canals" on Mars. Asaph Hall discovers the two satellites of Mars.			Darwin publishes a diary of the development of his son, which is the first source of child psychology.
1878				
1879		Lars Fredrick Nilson discovers scandium, in support of Mendeleev.		Spencer applies the theory of evolution to social life. Cro-Magnon cave paintings are discovered in Altamira, Spain.
1880				Robert Koch uses solid cultures for growing microbes.

Mathematics	Physics	Technology	Culture
			Heinrich Schliemann discovers the ancient city of Troy, making the study of archeology a part of the popular consciousness.
	Eugen Goldstein submits that cathode radiation sustains a wave interpretation of matter. Cromwell Varley submits that cathode radiation sustains a particulate theory.		
Georg Cantor shows that a function can be represented by a trigonometric series. Charles Meray develops a method of describing limits without using irrational numbers.	Maxwell publishes his *Treatise on Electricity and Magnetism*, which fully develops the view that light is an electromagnetic phenomenon (first published in two memoirs of 1865 and 1868).		Cavendish Laboratory at Cambridge is completed after four years of construction.
Charles Hermite shows that *e* is a transcendental number.			The Zoological Station, a biological research facility, is founded in Naples by Anton Dohrn.
Carnot writes the first paper on set theory.	Boltzmann advances the basic principles of statistical mechanics, showing that the laws of mechanics and the theory of probability, when applied to the motions of atoms, can explain the second law of thermodynamics.		The Astrophysical Observatory of Potsdam is founded in Germany.
	William Crookes claims that cathode radiation points to a fourth state of matter that is particulate in nature.	Crookes invents the radiometer.	
Edouard Lucas devises a rule to show whether a given number is a prime.	Plücker describes the phenomenon of *cathode rays* and is the first to use the term.	Alexander Graham Bell patents the telephone.	Johns Hopkins University is founded as a graduate school.
Cantor proves that the number of points of a line segment is the same as the number of points in the interior of a square.		Thomas Alva Edison invents the phonograph.	
			The International Union of Geological Sciences is founded. Schliemann claims to have found the tomb of Agamemnon.
Richard Dedekind gives a precise definition for a field in mathematics. Gottlob Frege attempts to reduce mathematics to formal logic.	Josef Stefan discovers that the total radiation of a body is proportional to the fourth power of its absolute temperature.	Edison patents the incandescent light bulb.	The Geological Survey of the United States of America is established under the direction of Clarence Rivers King.
	Pierre Curie discovers the piezoelectric effect.		Konrad Friedrich Beilstein begins publication of the *Handbook of Chemistry*.

Date	Astronomy and Cosmology	Chemistry	Earth Sciences	Life Sciences
1880				Pasteur develops the germ theory of disease.
1881	Edward Emerson Barnard takes the first photograph of a comet.			Pasteur produces the first artificially prepared vaccine, against anthrax.
1882	David Gill gives birth to the idea of stellar cataloging by photography.		Balfour Stewart postulates the existence of the ionosphere to account for daily changes in Earth's magnetic field. Robert Koch discovers the bacterium that causes tuberculosis.	Walther Flemming reports his discovery of chromosomes and mitosis.
1883				Galton coins the term *eugenics*. George John Romanes publishes the first book on comparative psychology.
1884		Svante August Arrhenius proposes his disassociation theory.		
1885	A supernova appears that is visible to the naked eye.			Pasteur develops a vaccine against rabies.
1886				
1887				
1888				
1889	William Harkness measures the mass of the planets Mercury, Venus, and Earth. Hermann Vogel discovers spectroscopic binaries.			Galton introduces the formula for standard error. Paul Ehrlich establishes the field of immunology.
1890	Edward Pickering and Mina Fleming introduce the Harvard Classification.			William James describes psychology as a natural science.
1891	Maximilian Wolf makes the first discovery of an asteroid from photographs.	Alfred Werner develops his theory of secondary valence.		

Mathematics	Physics	Technology	Culture
		The interferometer is invented by Albert Michelson. The bolometer is invented by Samuel Pierpont Langley.	The first journal of psychology, the *Philosophische Studien* (Philosophical Studies), is founded.
Ferdinand Lindemann proves that π is a transcendental number.	Michelson determines the velocity of light to be 186,329 miles per second.		The Pearl Street Power Station brings electricity to New York City. Women are admitted to the Cavendish Laboratory staff on the same terms as men.
Cantor claims that every set can be made into a well-ordered set.	George F. FitzGerald points out that electromagnetic waves can be generated by varying an electric current.	Henry Ford builds his first car.	
		Parsons invents the multistage steam engine.	
	Johann Balmer discovers the formula for the hydrogen spectrum.		Gottlieb Daimler patents the first internal combustion engine.
	Crookes suggests that atomic weights measured by chemists are averages of the weights of different kinds of atoms of the same element.	Karl Benz develops the first working automobile with a gasoline-burning internal combustion engine.	
	Michelson and Edward W. Morley attempt to measure changes in the velocity of light produced by the motion of Earth through space. Their failure discredits the idea of the aether. Heinrich Hertz observes the photoelectric effect, in which a material gives off charged particles when it absorbs radiant energy.	Bunsen invents the vapor calorimeter.	The *Zeitschrift für Physikalische Chemie* is founded by Friedrich Ostwald, marking establishment of the discipline of physical chemistry.
Galton introduces the concept of correlation.	Hertz demonstrates the propagation of electromagnetic waves (radio waves).	George Eastman markets a hand-held box camera. Nikola Tesla patents the induction motor.	The Pasteur Institute is founded. The National Geographic Society is founded.
Giuseppe Peano translates the reformed Euclidean geometry into symbolic logic. Peano introduces much of the notation used for set theory and logic.			
		Edison, lifting ideas from others, patents the kinetescopic camera.	
	George Johnstone Stoney suggests the term *electron* for a unit of electricity.	The Tesla coil is invented by Tesla.	

Date	Astronomy and Cosmology	Chemistry	Earth Sciences	Life Sciences
1891				
1892	International mapping of stars begins. Barnard provides evidence that novas are exploding stars. Barnard discovers a fifth moon of Jupiter.			The cause of mosaic disease is identified as a virus. Galton discovers that fingerprints can be used for the purpose of identification. August Friedrich Leopold Weismann states that the germ plasm is unchanged from generation to generation and that changes in the body do not affect it.
1893	Greenwich refracting telescope is completed.			
1894	Percival Lowell founds an observatory at Flagstaff, Arizona.			Eugène Dubois announces the discovery of Java man, *Pithecanthropus erectus*, now known as *Homo erectus*.
1895		William Ramsay and Crookes identify the element helium on Earth.		Psychoanalysis is developed by Sigmund Freud.
1896				
1897	George Ellery Hale sets up the Yerkes Observatory, with the largest refracting telescope on Earth.			Eduard Buchner shows that vital processes can occur outside living cells, the beginnings of biochemistry. The first course in radiology is inaugurated. The Canadian Geological Survey discovers rich fossil beds containing Upper Cretaceous dinosaur fauna along the Red Deer River in Alberta.
1898				Trofim Lysenko, with the support of Joseph Stalin, organizes Soviet biological research around the idea that acquired characteristics can be inherited.

Mathematics	Physics	Technology	Culture
	Fitzgerald and Hendrik Lorentz formulate the Lorentz-Fitzgerald contraction.		
	(to 1904) Lorentz develops his electron theory.	Philipp Lenard invents a cathode-ray tube that permits rays to escape so that they can be studied in open air.	
	Ernst Mach states Mach's principle, the first constructive attack on the idea of Newtonian absolute space.		
	Guglielmo Marconi begins experimenting on communicating with radio waves.		
Henri Poincaré founds topology as a branch of mathematics.	Wilhelm Conrad Röntgen discovers x-rays, which are used immediately to visualize bodily structures.		
	Henri Becquerel discovers natural radioactivity in uranium salts. The Zeemen effect is discovered: A magnetic field exerts an influence on the character of the lines of the spectrum. Lorentz calculates the mass/charge ratio of the electron in an atom.	Marconi patents radio telegraphy. Charles Thomson Rees Wilson develops the first cloud chamber.	
David Hilbert introduces his theory of algebraic number fields.	J. J. Thomson shows that electrons are independent particles.		
	John Sealy Townsend measures the charge of an electron. Marie and Pierre Curie coin the term *radioactivity*. Marie and Pierre Curie isolate the radioactive elements radium and polonium.		

Date	Astronomy and Cosmology	Chemistry	Earth Sciences	Life Sciences
1899	William Wallace discovers that Polaris is a system of three stars.			The First International Congress of Genetics is held in London. Jacques Loeb demonstrates parthenogenesis.
1900	James Edward Keeler discovers that some nebulas have a spiral structure.			Hugo de Vries (the Netherlands) and Carl Franz Joseph Erich Correns (Germany) independently discover the laws of inheritance and relate them to Mendel's principles, marking the beginning of modern genetics.
1901				(to 1903) De Vries suggests that new species evolve by large-scale mutations.
1902		Léon Teisserenc de Bort discovers that Earth has two layers, the troposphere and the stratosphere.		Ivan Pavlov introduces his law of reinforcement or learning by conditioning. Walter Stanborough Sutton states that chromosomes may be the carriers of heredity.
1903				
1904			Vilhelm Frimann Koren Bjerknes publishes one of the first scientific studies of weather forecasting.	
1905	Percival Lowell predicts the existence of Pluto.		Daniel Barringer proposes that the large crater in Arizona was made by a meteor.	Wilhelm Ludvig Johannsen introduces the terms *genotype* and *phenotype*.
1906				

Mathematics	Physics	Technology	Culture
	Alpha and beta rays are shown by Ernest Rutherford to be distinct types of radiation. Rutherford notices thorium emanation. Kelvin reiterates his figures for the age of Earth, deduced from the probable cooling process, reducing his estimate to 24 million years. Paul Villard observes gamma rays.		
	Max Planck states that blackbodies radiate energy in quanta rather than continuously, thus beginning the science of quantum mechanics.	Emil Wiechert invents the inverted pendulum.	
		Marconi achieves the first transatlantic radio transmission.	Freud introduces the concept of the Freudian slip.
Beppo Levi makes the first clear statement of the axiom of choice. Bertrand Russell discovers the paradox that bears his name.	Rutherford and Frederick Soddy discover the law of radioactive decay.		Alfred Adler joins Freud to form the first psychoanalytic society.
	Rutherford and Soddy publish work on radioactivity as the transformation of atoms. Henri Poincaré introduces the fundamental idea of chaos.	The Wright brothers fly the first motor-driven airplane.	
	Thomson proposes his "plum pudding" model of the atom. Charles Glover Barkla discovers that x-rays are transverse waves.		
Leonard Eugene Dickson introduces cyclic algebras.	Albert Einstein completes his theory of special relativity. Einstein explains the photoelectric effect. Einstein explains Brownian motion as due to molecular collisions.		
	Soddy discovers that thorium has chemically indistinguishable variants which he later calls isotopes. Hermann Walther Nernst presents a formulation of the third law of thermodynamics, which states that matter tends toward random motion and that energy tends to dissipate at a temperature above absolute zero.		

Date	Astronomy and Cosmology	Chemistry	Earth Sciences	Life Sciences
1907			Bertram Borden Boltwood discovers how to use uranium to determine the age of rocks.	Thomas Hunt Morgan begins experiments with fruit flies.
1908	Hale discovers magnetic fields in sunspots.			Archibald Garrod introduces the idea of genetic diseases. William Bayliss describes the action of hormones.
1909			Fritz Pregl develops techniques for analyzing tiny amounts of organic chemicals.	Correns obtains the first proof for cytoplasmic inheritance. Johannsen introduces the term genes to describe Mendel's factors of inheritance.
1910				Morgan discovers that certain inherited characteristics are sex-linked.
1911			Augustus Edward Hough Love publishes *Some Problems of Geodynamics*, which presents his theory of Love waves (a type of surface seismic wave).	Morgan and colleagues produce the first chromosome maps of the fruit fly. Alfred Sturtevant produces the first chromosome map.
1912	Victor Franz Hess discovers the existence of cosmic rays during a balloon ascent. Vesto Slipher obtains the spectrum of the Andromeda galaxy.		Alfred Lothar Wegener proposes his theory of continental drift and the supercontinent of Pangaea.	Alfred Woodward and Charles Dawson find a skull of Piltdown man (a fraud, composed of a modern human cranium and an ape's jaw). Casimir Funk coins the term *vitamin*.
1913	Photoelectric photometry is introduced in astronomy. Henry Norris Russell presents his theory of stellar evolution.		Charles Fabry discovers the ozone layer in the upper atmosphere.	Hans Reck discovers rich deposits of early mammalian fossils, including Stone Age artifacts at Olduvai Gorge in East Africa.

Mathematics	Physics	Technology	Culture
Andrei Andreevich Markov develops theory of linked probabilities.			
Ernst Zermelo develops an axiomatic treatment of set theory.		Hans Wilhelm Geiger and Rutherford invent the Geiger counter, which counts individual alpha particles.	
	Rutherford and Thomas Royds demonstrate that alpha particles are doubly ionized helium atoms.		
Russell and Alfred North Whitehead begin work on *Principia mathematica*, an attempt to derive all of mathematics from pure logic.	William Henry Bragg discovers that x-rays and gamma rays cause rarified gases to conduct electricity. Thomson discovers that neon has two isotopes, verifying Soddy's prediction.		
	Heike Kanerlingh-Onnes discovers superconductivity in mercury cooled close to absolute zero. Rutherford proposes the nuclear model of the atom, in which the mass of the atom is concentrated in a positively charged nucleus balanced by orbiting electrons. John Archibald Wheeler coins the term *black hole*.		
	Bohr proposes that electrons orbit the nucleus in fixed orbits. Henry Gwyn Jeffreys Moseley shows that nuclear charge is the real basis for numbering the elements. Robert A. Millikan measures the fundamental unit of charge. William Henry and William Lawrence Bragg develop x-ray crystallography by showing that an orderly arrangement of atoms in crystals displays interference and diffraction patterns.		

Date	Astronomy and Cosmology	Chemistry	Earth Sciences	Life Sciences
1914	Walter Sydney Adams develops the method of spectroscopic parallax. Arthur Eddington proposes that spiral nebulas are galaxies. Slipher discovers evidence of interstellar dust clouds.		Beno Gutenberg discovers the boundary between Earth's mantle and the outer core.	John Broadus Watson proposes the use of animals as experimental subjects in psychology.
1915			Wegener introduces the idea of continental drift based on fossil and glacial evidence.	*The Mechanism of Mendelian Heredity* is published by Morgan, Sturtevant, Calvin Blackman Bridges, and Hermann Joseph Müller. Reginald Crundall Punnet produces a mathematical analysis of the effects of selection on populations that obey Mendelian laws of heredity. Frederick William Twort discovers viruses that feed on bacteria.
1916		Gilbert Newton Lewis develops his theory of shared electrons.		Félix d'Hérelle independently discovers viruses that feed on bacteria. He names them *bacteriophages*.
1917	Einstein advances model of a static universe and a cosmological constant that keeps it this way.			Joseph Grinnell introduces the notion of an ecological niche. Ferdinand Broili discovered the fossil remains of Seymouria, an organism displaying both amphibian and reptilian characteristics.
1919	Arthur Stanley Eddington leads a solar eclipse expedition which claims to detect gravitational deflection of light by the Sun. Frank Watson Dyson announces that the photographs of the sun taken by Eddington confirm Einstein's theory of gravitation.			Karl von Frisch discovers that bees communicate through body movement.
1920	Lockyer measures the diameter of the star Betelgeuse.			
1921	Edward Arthur Milne predicts the existence of a solar wind.			Hermann Rorschach introduces the inkblot test for the study of personality.
1922		Johannes Nicolaus Brønstead advances the modern concept of acids and bases.		

Mathematics	Physics	Technology	Culture
Felix Hausdorff introduces the idea of topological spaces.	Rutherford discovers the proton.		
	Einstein completes his theory of general relativity.	The mercury-vapor condensation pump is invented by Irving Langmuir.	
	Lewis and Langmuir formulate an electron shell model of chemical bonding. Arnold Sommerfeld proposes elliptical orbits for electrons in atoms. Bohr proposes the correspondence principle.		
			The National Research Council is established by the U.S. Academy of Sciences.
	Rutherford achieves the first artificial disintegration of a nucleus, the transmutation of the nitrogen nucleus by alpha-particle bombardment.		The International Astronomical Union is founded.
	Draper proposes the existence of the neutron.		The Copenhagen Institute of Theoretical Physics is established.
Emmy Noether develops the axiomatic approach to algebra.			
	Arthur H. Compton discovers that x-rays scattered by an atom have a shift in frequency (the Compton effect).	The Svedberg develops the ultracentrifuge.	

Date	Astronomy and Cosmology	Chemistry	Earth Sciences	Life Sciences
1923	Eddington advances the mass-luminosity law for stars. Edwin Powell Hubble shows that the galaxies are true independent systems.			
1924				
1925				Ronald Aymler Fisher publishes *Statistical Methods for Research Workers*, which becomes a standard text for research in genetics. Raymond Dart describes the Tuang skull.
1926	Eddington produces the first study of stellar structure. Hubble introduces a classification system for extragalactic nebulas.			Müller discovers that x-rays can induce mutations.
1927	Georges Lemaître introduces an early version of the Big Bang theory. Jan Oort demonstrates the spiral structure of the Milky Way.	Walter Heinrich Heitler provides a theoretical basis in quantum mechanics for covalent bonding.	Rudolf Geiger founds the study of microclimatology.	Davidson Black discovers evidence of Peking man.
1928				
1929	Hubble discovers the expansion of the universe.			
1930	Clyde W. Tombaugh discovers Pluto at the Lowell Observatory.			
1931	Subrahmanyan Chandrasekhar predicts the existence of white dwarf stars.		Earth is shown to be at least 2 billion years old.	Sewall Wright presents the first theoretical integration of Mendelism with Darwinism, illustrating the relations between selection pressure, mutation rates, inbreeding, and isolation.
1932			August Piccard launches the first balloon into the stratosphere. Edward G. Lewis finds the first *Ramapithecus*, the earliest known hominid fossil.	J. B. S. Haldane makes the first estimation of mutation rate in humans. Hans Adolf Krebs discovers the urea cycle. Morgan's *The Scientific Basis of Evolution* is published.

Mathematics	Physics	Technology	Culture
			The Bell Laboratories for research in physics are founded.
	Louis de Broglie suggests that material particles can also behave like waves.		(to 1936) A group of scientists and philosophers, which comes to be known as the Vienna Circle, develop the tenets of logical positivism.
	Patrick Maynard Stuart Blackett takes the first photographs of nuclear reactions. Wolfgang Pauli proposes the exclusion principle.	The first analog computer is developed.	The Scopes "monkey trial" is held in Dayton, Tennessee.
	Erwin Schrödinger states his non-relativistic quantum wave equation. Eugene Paul Wigner introduces group theory into quantum mechanics. Lewis introduces the name photon for the light quantum.	John Logie Baird transmits the first television signal.	
	Werner Heisenberg advances the uncertainty. Bohr formulates complementarity principle.		
John Von Neumann develops game theory.			
	Hermann Weyl formulates gauge invariance.	Ernest Orlando Lawrence and M. Stanley Livingston build a cyclotron.	
	Paul Dirac formulates a general mathematical theory which treats wave mechanics and matrix mechanics as special cases.		The Princeton Institute for Advanced Study is founded.
Kurt Gödel advances the undecidability theorem. Richard von Mises introduces the concept of sample space into probability theory.	Dirac proposes the positron.		
Von Neumann proves the ergodic theorem.	Sir James Chadwick discovers the neutron. John Douglas Cockcroft and Ernest Thomas Sinton Walton achieve the transmutation of elements by protons.	Ernst Ruska builds the first electron microscope.	

Date	Astronomy and Cosmology	Chemistry	Earth Sciences	Life Sciences
1932				
1933		Lewis obtains heavy water.		
1934	Walter Baade and Fritz Zwicky propose that neutron stars are formed during supernova eruptions. Walter Grotrian discovers zodiacal dust.			
1935				
1936				
1937				Theodosius Dobzhansky links evolution and genetic mutation in *Genetics and the Origin of the Species*. Krebs discovers the Krebs cycle.
1938	Hans Bethe claims that the energy produced by stars is caused by nuclear fusion. J. Robert Oppenheimer predicts the existence of pulsars.		G. S. Callendar shows that humans are causing an increase in the amount of carbon dioxide in Earth's atmosphere.	
1939	Oppenheimer identifies the conditions for the existence of black holes.			Latimeria, a living crossopterygian fish that was presumed to have become extinct in the Cretaceous Period, is caught off the coast of South Africa.
1940		Martin Kamen discovers carbon-14.		
1941				Gustaffson and co-workers produce agriculturally superior new strains of cereals by selection from mutants produced by x-rays. George Wells Beadle and Edward L. Tatum show that genes control chemical reactions in cells. Selman Waksman coins the term *antibiotic*.

Mathematics	Physics	Technology	Culture
	Carl David Anderson discovers positive electrons (positrons), the first form of antimatter to be discovered.		
	Enrico Fermi proposes the existence of the weak interaction.		
	Leo Szilard realizes that nuclear chain reactions may be possible. Fermi bombards uranium with neutrons and discovers the phenomenon of atomic fission, the basic principle of atomic bombs and nuclear power.	Lawrence and Livingston construct the cyclotron. The pH meter is invented.	
	Hideki Yukawa proposes the existence of the meson to explain nuclear forces.		
Alan Turing shows that there is no single way to prove or disprove all logical statements.	Anderson discovers the muon, an electronlike particle 200 times more massive than an electron.	Erwin Mueller develops the field-emission microscope.	
		The first radiotelescope is constructed.	
	Frédéric Joliot and Irene Curie-Joliot demonstrate the possibility of a chain reaction. Otto Hahn and Fritz Strassmann obtain the first atomic fission with uranium. Lise Meitner and Otto Frisch coin the term *fission*.		
Gödel shows that Cantor's continuum hypothesis is consistent with the axioms of set theory.			
			Franklin D. Roosevelt signs the order that leads to development of the nuclear fission bomb.

Date	Astronomy and Cosmology	Chemistry	Earth Sciences	Life Sciences
1942	Grote Reber makes the first radio maps of the universe.			
1943				
1944				Oswald T. Avery finds evidence indicating that DNA carries genetic properties.
1945				Karl Landsteiner's first study of chemical specificity in immunology is published.
1946				
1947			Willard Frank Libby develops carbon-14 dating.	Frisch shows that bees exploit the polarization of light for orientation. Tatum and Joshua Lederberg discover genetic recombination in the bacterium *Escherichia coli*.
1948	Hermann Bondi, Thomas Gold, and Fred Hoyle develop the steady-state theory of the cosmos. George Gamow and Robert Herman advance the Big Bang theory.			
1949	Fred L. Whipple suggests that comets are made of ice and rock dust.	Dorothy Crowfoot Hodgkin uses an electronic computer to ascertain the structure of an organic chemical.		
1950	Jan Hendrik Oort suggests that comets are caused by a cloud of material orbiting the Sun beyond the orbit of Pluto.		The first computer-assisted weather predictions are made by John von Neumann and a team of meteorologists.	Embryo transplants for cattle are carried out for the first time.
1951	A computer is used to calculate the planetary orbits of the five outer planets.			
1952	Martin Schwarzchild studies the evolution of stars.			Alfred Day Hershey and Martha Chase show, on the basis of their bacteriophage research, that DNA alone carries genetic information.

Mathematics	Physics	Technology	Culture
	Fermi makes the first controlled nuclear chain reaction. The Manhattan Project is launched.		
	Shin'ichirō Tomonaga develops quantum electrodynamics.	The first nuclear reactor is activated at Oak Ridge, Tennessee. The first electronic calculating device is constructed by Turing.	
		A quartz-crystal clock is installed at Greenwich Royal Observatory.	
		Arthur C. Clarke proposes the idea of communications satellites.	The first thermonuclear fission bomb is exploded at Alamogordo in the New Mexico desert. Hiroshima is bombed with a nuclear fission bomb on August 6. Nagasaki is bombed with a nuclear fission bomb on August 9.
		J. Presper Eckert and John W. Mauchley finish ENIAC, the first electronic computer.	
	Dennis Gabor elaborates the essential idea of holography.		The first supersonic flight takes place.
	Richard P. Feynman, Julian Seymour Schwinger, and Tomonaga develop quantum electrodynamics. The atomic clock is introduced. The transistor is invented by William B. Shockley, Walter H. Brattain, and John Bardeen.	Norbert Wiener elaborates his idea for a field of study to be called *cybernetics*.	
Claude Shannon advances his work on information theory.			
		The field ion microscope is developed.	
		Donald A. Glaser develops the bubble chamber to observe the behavior of subatomic particles, rendering the cloud chamber obsolete.	The first hydrogen bomb, Mike, is detonated. A polio epidemic affects many thousands of people in North America.

Date	Astronomy and Cosmology	Chemistry	Earth Sciences	Life Sciences
1952				The phenomenon of rapid eye movement is discovered.
1953		The first catalyst that combines monomers into a regular polymer is produced by Karl Ziegler.	Maurice Ewing discovers the great rift running down the middle of the Mid-Atlantic Ridge.	James Watson and Francis Crick construct a three-dimensional model of the DNA molecule. Rosalind Franklin produces x-ray studies of DNA that allow Watson and Crick to discern its structure. Harold C. Urey and Stanley Lloyd Miller find that several amino acids are formed when ammonia, methane, water vapor, and hydrogen are exposed to an electrical discharge for several days, pointing to a scenario for the origin of life in the primitive sea. Tars from tobacco smoke are shown to cause cancer in mice.
1954			Earth is estimated to be 5 to 6 billion years old.	Oxytocin becomes the first of the hormones to be synthesized.
1955				
1956		The human growth hormone is isolated.	The Mid-Oceanic Ridge is discovered.	
1957	*Sputnik I* is launched by the Soviet Union.			
1958			The Van Allen radiation belt that surrounds Earth is discovered.	
1959	The first photographs of the far side of the Moon are taken by the Soviet probe *Lunik III*.			Louis Seymour Leakey discovers the *Zinjanthropus boisei* fossil. C. E. Ford demonstrates that Turner's syndrome has a genetic basis.
1960				The atomic structure of hemoglobin is determined by Max Ferdinand Perutz.

Mathematics	Physics	Technology	Culture
			The transistor radio is introduced by Sony.
	Murray Gell-Mann introduces the concept of "strangeness" to explain the behavior of subatomic particles.		
	The Bevatron particle accelerator is completed at the Radiation Laboratory in Berkeley.		The Oppenheimer hearing is held. The European Center for Nuclear Research is founded.
Homological algebra, a blend of abstract algebra and algebraic topology, is developed by Henri Cartan and Samuel Eilenberg.	Antiprotons are produced for the first time. Neutrinos are observed for the first time.	The field ion microscope, which is used to create images of individual atoms, is developed.	
	The antineutron is discovered by a team of researchers. Clyde L. Cowan, Jr. and Frederick Reines detect the existence of the neutrino.	A team at IBM creates FORTRAN, the first computer programming language.	The United States explodes a hydrogen bomb at Bikini Atoll in the Pacific Ocean. The transatlantic telephone phone cable is made operative.
	Bardeen, Leon N. Cooper, and John Schrieffer advance the theory of superconductivity.		
		The first Xerox copier is introduced.	
		The first laser is developed by Theodore H. Maiman.	

Date	Astronomy and Cosmology	Chemistry	Earth Sciences	Life Sciences
1961	Yury Alekseyevich Gagarin of the Soviet Union is the first person to orbit Earth.		In a computer model of how the atmosphere behaves, Edward N. Lorenz discovers the mathematical first system with chaotic behavior.	The fossil remains of *Homo habilis* are located by Louis S. B. Leakey and Mary Douglas Leakey.
1962	*Mariner 2* becomes the first probe to travel to another planet when it reaches the locality of Venus.	Niel Bartlett shows that noble gases can be involved in the formation of compounds.		Rachel Carson's *Silent Spring* is published, sounding the alarm of chemical pollution.
1964				
1965	Astronomers discover that Mercury rotates on its axis with a period of 59 days. Arno Allan Penzias and Robert Woodrow Wilson discover the radio-wave remnants of the Big Bang.			
1966	A Soviet space probe lands successfully on the lunar surface.			
1967	Jocelyn Bell discovers the first pulsar.		S. Manabe and R. T. Wetherald issue a warning about the effect of greenhouse gases on global temperatures.	Biologically active DNA is synthesized for the first time.
1968				Lake Erie is declared to be biologically dead by a report from the U.S. House of Representatives.
1969	The first Moon landing is made by American astronauts. Neil Armstrong is the first person to walk on the lunar surface.			
1970		Human growth hormone is synthesized.		A gene (aniline-transfer RNA) is synthesized for the first time.
1971	*Mariner 9* returns spectacular pictures of Mars.			The use of DDT is restricted in the United States.
1972				Workers at Stanford University construct the first recombinant DNA molecules using restriction enzymes and DNA ligase.

Mathematics	Physics	Technology	Culture
	Gell-Mann and Yuval Ne'eman independently advance a classification scheme for subatomic particles that comes to be known as the *eightfold way*. Robert Hofstadter shows that protons and neutrons are made up of a positive central core surrounded by two shells of mesons.	Lasers are introduced in eye surgery.	The intrauterine device is introduced for birth control.
	L. B. Okun introduces the term *hadron* to refer to particles that are affected by the strong force.		
	Gell-Mann introduces the concept of quarks (particles that make up all baryons and mesons). Sheldon Glashow and James D. Bjorken introduce a fourth variety of quark, termed *charm*.		
			The Endangered Species Act is passed in the United States.
			The Fermi National Accelerator Laboratory (known as Fermilab) is founded near Chicago.
		Intel introduces the first microprocessor.	
	Gell-Mann introduces quantum chromodynamics.	The CAT (computerized axial tomography) scanner is introduced.	

Date	Astronomy and Cosmology	Chemistry	Earth Sciences	Life Sciences
1973				
1974				Donald Johanson and Maurice Taieb discover Lucy, a homonid more than 3 million years old.
1975	The Milky Way is shown to have a proper motion of 300 miles per second.			An international conference is held in Asilomar, California, urging strict guidelines regulating recombinant DNA research.
1976	*Viking 1* lands on the Martian surface.		Freon is shown to deplete Earth's ozone layer.	Genentech, the first genetic engineering company, is founded to use recombinant DNA methods to produce medically important drugs.
1980		The Nobel Prize in Chemistry is awarded for the creation of the first recombinant DNA molecules and the development of powerful DNA sequencing methods.	The discovery of a layer of clay at the Cretaceous-Tertiary boundary leads Walter Alvarez and others to conclude that the dinosaur extinctions were caused by a large object from space colliding with Earth.	Researchers transfer a gene from one mouse to another and it continues to function.
1981				
1983	The Milky Way is shown to move toward the "Great Attractor."			
1984	A planetary system is photographed around the star Beta Pictoris.			The technique of genetic fingerprinting is developed by Alec Jeffreys.
1985			The first deep-sea ocean vents are discovered in the Altantic Ocean on the Mid-Atlantic Ridge.	
1986	*Voyager 2* passes close to Uranus, sending back information about the planet and its satellites.			
1994				
1997				Ian Wilmut clones a lamb (Dolly) using DNA from an adult sheep.
2000	Research by Michael Malin and Kenneth Edgett reinforce theory that water once existed on Mars.			Global coalition of scientists completes the first "rough draft" of the human genome.

Mathematics	Physics	Technology	Culture
		The MRI (magnetic resonance imager) scanner is introduced.	
	Howard M. Georgi and Sheldon Lee Glashow develop the first of the grand unified theories which account for the strong, weak, and electromagnetic forces.		
The term *fractal* is introduced by Benoit Mandelbrot. The term *chaos* is introduced to refer to the sensitive dependence on initial conditions that frequently occurs in nonlinear systems.			
		The scanning tunneling microscope is invented.	
		The IBM personal computer is introduced.	Acquired immune deficiency syndrome (AIDS) is recognized by medical authorities.
	A team of physicists at CERN detect the sixth (top) quark.		
Princeton mathematician Andrew Wiles proves Fermat's last theorem.			
		NASA launches the first satellite in its Earth Observing System (EOS) program.	

Name Index

Note: Page numbers in *italics* indicate illustrations or photographs.

A

B

C

D

I

J

M

S

T

U

V

W

Subject Index

Note: Page numbers in **boldface** indicate main entries, those in *italics* indicate illustrations or photographs.

D

E

H

I

Q

R

S

T